Toxicity of Nanomaterials

Toxicity of Nanomaterials

Environmental and Healthcare Applications

Edited by
Prof. Suresh C. Pillai
Nanotechnology and Bio-Engineering Division, Faculty of Science,
Institute of Technology, Sligo, Ireland
Dr. Yvonne Lang
Nanotechnology and Bio-Engineering Division, Faculty of Science,
Institute of Technology, Sligo, Ireland

CRC Press
Taylor & Francis Group
Boca Raton London New York

CRC Press is an imprint of the
Taylor & Francis Group, an **informa** business

CRC Press
Taylor & Francis Group
6000 Broken Sound Parkway NW, Suite 300
Boca Raton, FL 33487-2742

Printed on acid-free paper

International Standard Book Number-13: 978-1-138-54430-7 (Hardback)

Visit the Taylor & Francis Web site at
http://www.taylorandfrancis.com

and the CRC Press Web site at
http://www.crcpress.com

Contents

Editors

Suresh C. Pillai obtained his PhD from Trinity College Dublin and then conducted his postdoctoral research at California Institute of Technology (Caltech), USA. Upon completion of this appointment he returned to Trinity College Dublin in April 2004 as a Research Fellow before joining CREST-DIT as a Senior Scientist. He joined IT Sligo in 2013 as a senior lecturer in nanotechnology and currently heads the Nanotechnology and Bio-Engineering Research Group. He is an elected fellow of the UK's Royal Microscopical Society (FRMS) and the Institute of Materials, Minerals and Mining (FIMMM). Prof. Pillai also completed an executive MBA from Dublin City University in 2009. He was responsible for acquiring more than €5 million direct R&D funding. He was awarded the 2019 'Boyle-Higgins' award by the Institute of Chemistry Ireland and received the 'Industrial Technologies Award 2011' for licensing functional coatings to Irish companies. He was also the recipient of the 'Hothouse Commercialisation Award 2009' from the Minister of Science, Technology and Innovation and the 'Enterprise Ireland Research Commercialization Award 2009'. He is an editor for the journal *Environmental Science and Pollution Research* (ESPR, Springer) and an Editororial Board Member for the *Chemical Engineering Journal and Applied Catalysis B* (Elsevier).

Yvonne Lang obtained her PhD from the National University of Ireland, Galway in 2014. Her doctoral research investigated the use of diatoms to fabricate polymeric structures with a defined geometry. Work emanating from this research has been published in peer-reviewed journals and has been presented at several key international conferences. She received the European Society for Biomaterials Julia Polak Educational Doctorate Award for her doctoral research. Dr. Lang's postdoctoral research focussed on varying aspects of nanoparticles including investigation of both therapeutic and environmental applications of nanoparticles, quantification of nanoparticles in biological samples and exploration of bioremediation approaches to detect and capture nanoparticles in environmental samples.

Dr. Lang is currently a lecturer in chemistry at the Institute of Technology, Sligo Ireland. She is a member of both the Nanotechnology and Bio-Engineering Research Division and the Centre for Environmental Research, Innovation and Sustainability at IT Sligo. She is a committee member of the Microscopy Society of Ireland and a past committee member of the Environmental Scientists Association of Ireland. She is actively involved in the co-ordination and delivery of outreach events to promote scientific activities to the public. Her research interests include: (1) investigation of diatom cultures and diatom products as decontamination agents for removal of organic pollutants from water; (2) interactions of microorganisms and microplastics and (3) impact of outreach activities on attitudes to habitat, health and well-being.

Contributors

Alexander Black
Division of Anatomy
School of Medicine
College of Medicine, Nursing and Health Sciences
National University of Ireland Galway
Galway, Ireland

Ailish Breen
Nanotechnology and Bio-engineering Research Group
Department of Environmental Science
School of Science
Institute of Technology Sligo
and
Centre for Precision Engineering, Materials and Manufacturing Research (PEM)
Institute of Technology Sligo
Sligo, Ireland

Saoirse Dervin
Nanotechnology and Bio-Engineering Research Group
Department of Environmental Science
Institute of Technology Sligo
and
Centre for Precision Engineering, Materials and Manufacturing Research (PEM)
Institute of Technology Sligo
Sligo, Ireland

Peter Dockery
Division of Anatomy
School of Medicine
College of Medicine, Nursing and Health Sciences
National University of Ireland Galway
and
Centre for Microscopy and Imaging
National University of Ireland Galway
Galway, Ireland

Andy Fogarty
Department of Life and Physical Sciences
Athlone Institute of Technology
Athlone, Ireland

Priyanka Ganguly
Nanotechnology and Bio-engineering Research Group
Department of Environmental Science
School of Science
Institute of Technology Sligo
and
Centre for Precision Engineering, Materials and Manufacturing Research (PEM)
Institute of Technology Sligo
Sligo, Ireland

Honey John
Department of Polymer Science and Rubber Technology
Cochin University of Science and Technology
and
Inter University Centre for Nanomaterials and Devices
Cochin University of Science and Technology
Kochi, India

M. K. Kavitha
Inter University Centre for Nanomaterials and Devices
Cochin University of Science and Technology
Kochi, India

Vignesh Kumaravel
Department of Environmental Science
School of Science
Institute of Technology Sligo
Sligo, Ireland

Yvonne Lang
Lecturer in Chemistry
Institute of Technology Sligo
Sligo, Ireland

Chen-zhong Li
Nanobioengineering & Bioelectronics Lab
Department of Biomedical Engineering
Florida International University
Miami, Florida

Emma McDermott
Division of Anatomy
School of Medicine
College of Medicine, Nursing and Health Sciences
National University of Ireland Galway
and
Centre for Microscopy and Imaging
National University of Ireland Galway
Galway, Ireland

Iain Murray
Department of Life and Physical Sciences
Athlone Institute of Technology
Athlone, Ireland

A. Joseph Nathanael
Department of Chemical Engineering
Yeungnam University
Gyeongsan, Republic of Korea

Amirali Nilchian
Nanobioengineering & Bioelectronics Lab
Department of Biomedical Engineering
Florida International University
Miami, Florida

T. P. Nisha
Department of Polymer Science and Rubber Technology
Cochin University of Science and Technology
Kochi, India

Tae Hwan Oh
Department of Chemical Engineering
Yeungnam University
Gyeongsan, Republic of Korea

Suresh C. Pillai
Nanotechnology and Bio-engineering Research Group
Department of Environmental Science
School of Science
Institute of Technology Sligo
and
Centre for Precision Engineering, Materials and Manufacturing Research (PEM)
Institute of Technology Sligo
Sligo, Ireland

Sami Rtimi
Swiss Federal Institute of Technology (EPFL)
Lausanne, Switzerland

Mehenur Sarwar
Nanobioengineering & Bioelectronics Lab
Department of Biomedical Engineering
Florida International University
Miami, Florida

Meera Sathyan
Department of Polymer Science and Rubber Technology
Cochin University of Science and Technology
Kochi, India

Alanna Stanley
Division of Anatomy
School of Medicine
College of Medicine, Nursing and Health Sciences
National University of Ireland Galway
Galway, Ireland

Kerry Thompson
Division of Anatomy
School of Medicine
College of Medicine, Nursing and Health Sciences
National University of Ireland Galway
and
Centre for Microscopy and Imaging
National University of Ireland Galway
Galway, Ireland

Aine M. Whelan
School of Chemical and Pharmaceutical Sciences
Dublin Institute of Technology
Dublin 2, Ireland

1

Cytotoxicology Studies of 2-D Nanomaterials

Priyanka Ganguly,
Ailish Breen, and
Suresh C. Pillai

1.1 Introduction

Engineered nanomaterials (NMs) are of significant interest in manufactured products. The application of these nano-dimension materials in electronics, pharmacy, and nanomedicines has an associated significant concern over their environmental impact (Erol et al., 2017; Lalwani et al., 2013). The fate and transformation of these NMs inside the human body as well as in the environment depends upon their physicochemical properties. The change or alteration in even a single set of factors such as the size, shape, or surface properties of the NM can result in a new toxicity pattern (Farré et al., 2009; Gottschalk et al., 2009; Mahmoudi et al., 2013; Podila and Brown, 2013; Powers et al., 2006). Two-dimensional nanomaterials (2D NMs) are an interesting class of NMs as the absence of the third dimension aids in exhibiting interesting attributes (Li et al., 2017; Naguib and Gogotsi, 2015). Remarkable optical and electronic characteristics have been displayed by graphene and other sister NMs such as ternary metal dichalcogenides (TMD) (Castro Neto and Novoselov, 2011; Novoselov et al., 2005). TMDs are made up of a hexagonal layer of metal atoms (M) sandwiched in between two layers of chalcogen atoms (X) (An and Yu, 2011; Coleman et al., 2011). This layered structure has strong covalent bonding within each

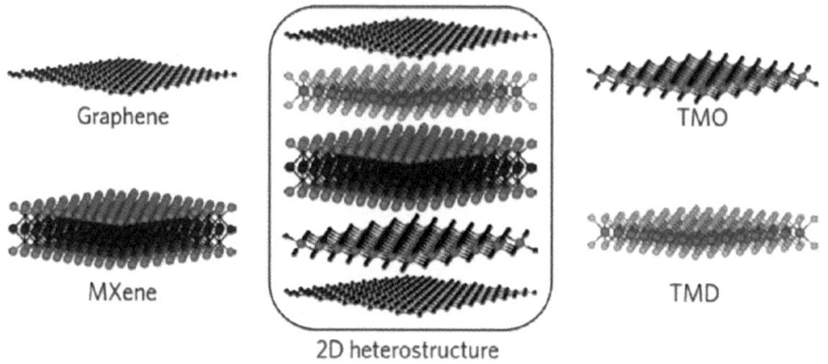

FIGURE 1.1 Schematic illustration of different kinds of 2D NMs such as graphene, transition metal carbides, nitrides (MXenes), transition metal oxides (TMO), and transition metal dichalcogenides (TMD). (From Pomerantseva, E. and Gogotsi, Y., *Nature Energy*, 2, 17089, 2017.)

layer (X-M-X), while a weak Van der Waals force holds the TMD sheets (MX_2) (Butler et al., 2013). The exfoliation of these TMD into thin sheets provides several catalytically active surface sites for various functional applications. In the case of 2D NMs, the mobility of the charge carriers is restricted along the thickness while being permitted to transfer along the plane. The large planar area makes these materials extremely sensitive to external stimuli. Figure 1.1 exhibits the most commonly reported 2D NMs (Khan et al., 2017).

The evolution of entirely new 2-dimensional materials with excellent features also increases the aspect of uncertainty and the necessity of understanding the toxicity profiles of these NMs. There exist multiple reports and reviews detailing various types of assays and studies on the toxicity of NMs, yet there still exists a definite deficiency in bringing them under a common domain (Chng and Pumera, 2015; Ganguly et al., 2018). Moreover, reviews summarising the cytotoxicity studies of 2D NMs are not yet reported. Standard parameters and guidelines to underscore the toxicity profile of NMs in one common domain are absent. Toxicity assaying is not a new topic but the use of NMs at an ever-increasing scale has led to questions being raised on whether the present research capacity or standard could cope in finding an appropriate answer to the toxicity profiles of such new materials. Keeping up with the rapidly changing manufacturing industries and their products has certainly raised the bar and posed a critical challenge for toxicology researchers. Safety standards and guidelines to handle such materials are critical; improper data on toxicity assessment can be fatal.

The present chapter details the different kinds of commonly observed types of assays for toxicity. The chapter also details the various physiological outcomes observed by NM interaction. Moreover, the parameters influencing the toxic profiles of 2D structures are described. Different cellular internalisation pathways are summarised. Additionally, a brief discussion of recent studies of cytotoxicity reports on 2D NMs is presented. Finally, a section detailing the safety guidelines is summarised.

1.2 Cell Death

Understanding toxicity and its impact is essential for commercialising NMs. Therefore, it is vital to understand the mechanism behind cell death. The fatal mechanism of the cell is basically chartered into three types: apoptosis, autophagy, and necrosis (Figure 1.2) (Kroemer et al., 2009; Nunes et al., 2014).

1.2.1 Apoptosis

Apoptosis is commonly referred to as the type-1 cell death and relies on several cell surviving signals. The distinct property of this cell death is the complete disintegration of the cellular morphology,

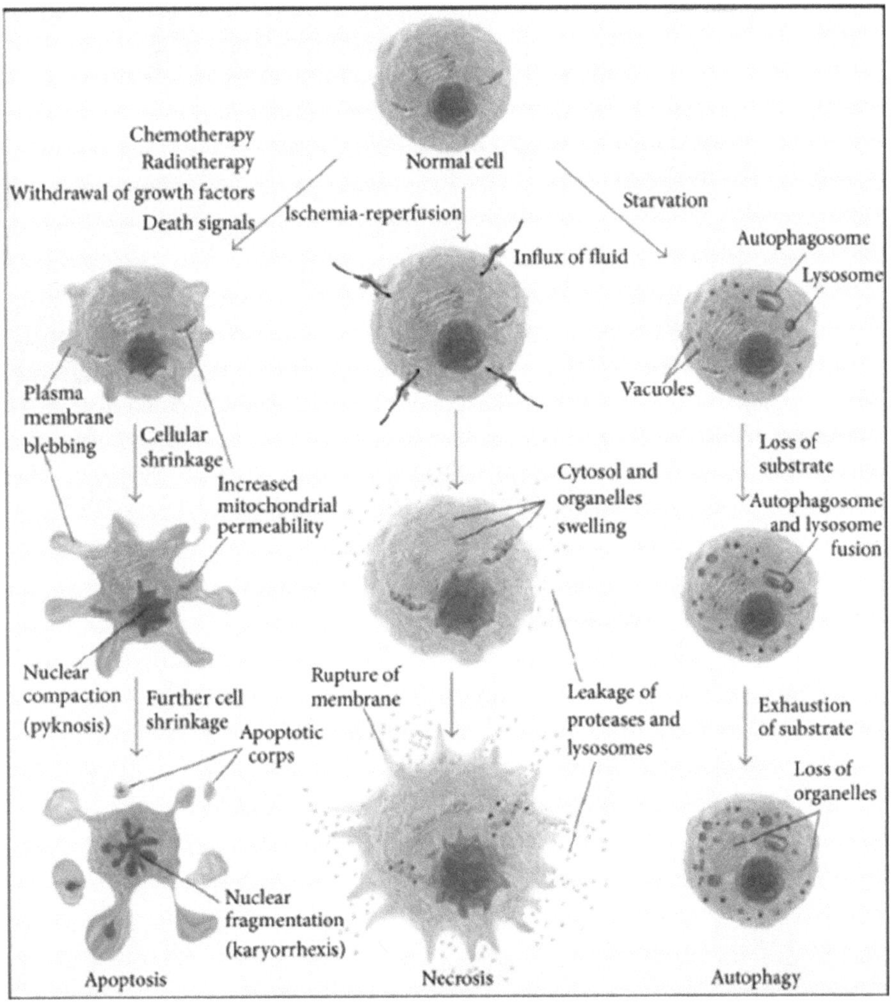

FIGURE 1.2 Schematic illustration defining various types of cell death. The apoptosis pathway is represented with the characteristic cellular shrinkage and formation of the apoptotic bodies without leakage of contents. In the middle, the necrotic pathway shows the cytosol and organelle swelling and rupture of the plasma membrane with subsequent leakage of cellular contents. On the right, autophagy is illustrated with the appearance of vacuoles, the autophagosome, and its fusion with the lysosome, which ends in organelle digestion. (From Nunes, T. et al., *BioMed Res. Int.*, 2014.)

development of apoptotic bodies, chromatic condensation, etc. The sudden stop of these cell surviving signals triggers the apoptosis process. Intracellular proteases known as caspases are initiated once the cell surviving signals cease to exist. There exist 14 types of caspases inside a mammalian body. These caspases together initiate the cell death process (Enari et al., 1998; Galluzzi et al., 2008; Garrido and Kroemer, 2004; Kerr et al., 1972; Roach and Clarke, 2000).

1.2.2 Autophagy

Autophagy is called the type-2 cell death. The formation of a double-layered membrane containing vacuoles called autophagosomes initiates the type-2 cell death mechanism. The double-layered membrane engulfs the cytoplasmic components and later on combines with lysosomes to form autophagolysosomes to cause degradation (Mizushima, 2005; Tasdemir et al., 2008). This process is an intracellular degradation activity which has several pathophysiological important functions, such as tumour separation, starvation adaptation, organelle clearance, and antigen presentation. Various forms of secondary stress arise inside the cell during cell differentiation, such as nutrient starvation, which initiates the process of autophagy (Baehrecke, 2002; González-Polo et al., 2005; Levine and Kroemer, 2008; Mizushima, 2007).

1.2.3 Necrosis

Necrosis is the type-3 cell death process known as uncontrolled cell death process. It arises due to physicochemical stress initiated by an alteration in the nucleus—for example, karyolysis, pyknosis, etc.—and in the cytoplasm (condensation, rupture of cytosol, and further disintegration). Necrosis is the process that arises 12–20 hours after the cell death; hence, it cannot be necessarily defined as cell death. However, the other two types of cell death initiate the process of necrosis. Necrosis results in cytoplasmic swelling, dilation of organelles, and rupture of the plasma membrane, which results in the outflow of cytosolic content and damage of the outside cellular environment (Festjens et al., 2006; Golstein and Kroemer, 2007; Majno and Joris, 1995).

1.3 Different Types of Assay

1.3.1 *In Vitro* Assay

One of the major toxicity assessment techniques, *in vitro* assay is used to evaluate the toxicity of materials (Figure 1.3) (Beloica et al., 2015). Different cell lines are exposed against several xenobiotic agents and incubated for definite time intervals. This potentially helps in calculating the dosage of the exposure.

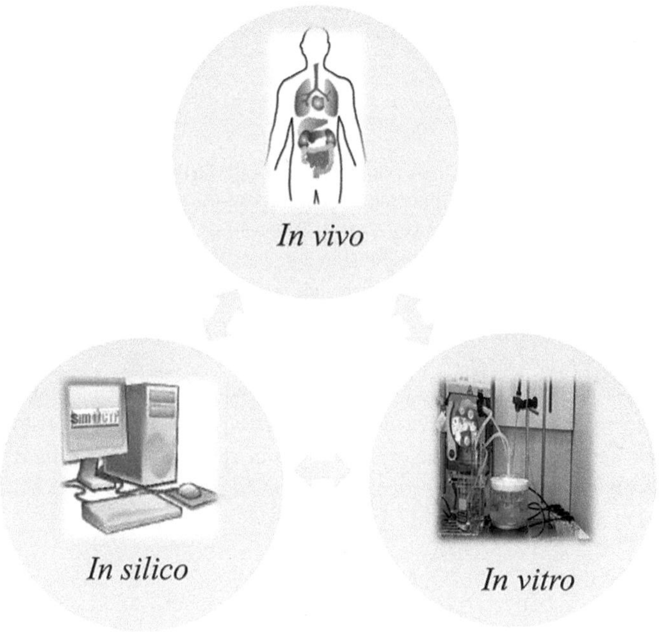

FIGURE 1.3 Illustration defining different types of toxicity assays. (From Beloica, S. et al., *Eur. J. Pharm. Sci.*, 75, 151–159, 2015.)

Different assays are used to measure the proliferation and cellular metabolism such as 3-(4,5-Dimethyl-2-thiazolyl)-2,5-diphenyl-2H tetrazolium bromide (MTT) and 2-(2-methoxy-4-nitrophenyl)-3-(4-nitrophenyl)-5-(2,4-disulfophenyl)-2H-tetrazolium, or mono sodium (WST-8). These techniques are rapid and cost-effective and do not require the use of animals, which reduces any form of ethical conflicts. However, these processes are supposed to mimic the cellular environment, which does not always correlate to the physiological outcomes (Ciappellano et al., 2016; Lewinski et al., 2008; Roggen, 2011; Sharifi et al., 2012; Stockert et al., 2012; Wataha et al., 1991).

1.3.2 *In Vivo* Assay

This toxicity assessment technique is considered the most reliable method to date (Figure 1.3). A small amount/dose of the xenobiotic component is administered in model animals such as mice. The component's transportation, metabolism, distribution, and finally removal are the key aspects for the concern. This method requires monetary as well as time investments. However, the results obtained are highly efficient and reliable. Concerns of ethical conflicts persist with this kind of preclinical trial (Aillon et al., 2009; Fielden and Kolaja, 2008; Filip et al., 2015; Lee et al., 2016; Wen et al., 2017).

1.3.3 *In Silico* Assay

Utilising several theoretical models to judge the physicochemical properties of some xenobiotic materials is one of the novel routes of assaying (Figure 1.3) (Hindman and Ma, 2018). Formation of quantitative structures to collectively develop toxicity assaying models is one of the key elements. These quantitative structures are gathered from existing literature and mathematical modelling studies. This method of toxicity assaying is cost effective, less time consuming, and free from any possible ethical conflicts. A wide gap exists in predicting the behavioural pattern using theoretical modelling based upon the properties of this ever-changing material. Additional experimental validation using one of the other two assaying techniques is essential (Aires et al., 2017; Bell et al., 2018; Gao et al., 2018; George et al., 2011; Gleeson et al., 2012; Laomettachit et al., 2017; Sizochenko et al., 2018).

1.4 Physiological Impacts Due to Nanomaterials Interaction

1.4.1 Oxidative Stress and ROS Generation

The ROS levels in the human body are critical for different metabolic functions (Jornot et al., 1998; Sharifi et al., 2012). Variations in the ROS levels can cause alteration of key cellular events such as signal transduction and protein redox potential. ROS are generated when the cellular redox potential is comparable to the band edge potentials of the interacting nanoparticles (Choi, 2016; Nel et al., 2006; Stahl et al., 1998). The neutralising mechanism is initiated inside the cell, which is defined as oxidative stress. Redox signalling pathways like mitogen-activated protein kinase (MAPK) cascades are enabled in this condition (Jornot et al., 1998). Several cellular processes such as cell proliferation and differentiation are controlled by these protein cascades (Apel and Hirt, 2004). Proinflammatory cytokines and chemokines are expressed with the help of these protein cascades. At tier 1, lower levels of oxidation stress aid in the expression of genetic antioxidant response, which results in the formation of antioxidant enzymes. These enzymes are responsible for the activation of the transcription factor Nrf2 (Li et al., 2004; Sun et al., 2017). These transcription factors are responsible for tier 2 stress mitigation by expressing many anti-inflammatory, cytoprotective, and antioxidant enzymes for lungs. These antioxidants are protein elements used to reduce the oxidative stress experienced in the cell (Li et al., 2002). The antioxidants are of two types: the primary defence, which involves enzymes like superoxide dismutase, GSH, catalase, and thioredoxin reductase; and the secondary defence, which involves reduced glutathione (Fridovich, 1995). Less-reactive hydrogen peroxide is formed on reaction with the primary defence antioxidants with highly reactive superoxide (Sharifi et al., 2012). On the other hand, GSH, a non-protein thiol, exists in two forms: oxidised form glutathione disulphide (GSSG) and reduced GSH. These thiols aid in balancing the cellular redox levels. GSH reductase helps in maintaining the levels of both the oxidative and reduced versions of GSH. Several enzymes corresponding to redox system such as GCLC, GSH peroxidase, or MnSOS start to flow as a response to the inflammatory mediators generated against the oxidative stress. The variation in the levels of GSH and GSSG results in such oxidative stress. In addition, the decreased levels of GSH results in enhanced membrane permeability and activation of NF-κB (Biswas and Rahman, 2009; Chakravarthi et al., 2006; Park et al., 1998; Rahman et al., 2001). Tier 3 levels of oxidative stress are activated when the first two tiers are crossed. This results in disruption of mitochondrial permeability, which subsequently affects electron transport. This finally results in cell death such as apoptosis or necrosis (Kumagai et al., 1997; Meng et al., 2009). Figure 1.4 is a schematic illustration providing a summarized overview of oxidative stress (Biswas and Rahman, 2009). Even though oxidative stress is considered fatal, researchers have found ways of utilising this feature as the potential armour of immunotherapy for cancer and drug delivery. Li and co-workers recently demonstrated the use of a lipid as an envelope to carry small interference ribonucleic acid (siRNA). Growth of tumour and cancerous cells are promoted by the formation of new lymph vessels, known as lymphangion genesis, and new blood vessels, known as angiogenesis. Therefore, controlling the signals for the growth factors responsible for these processes by siRNA could provide a promising strategy to control cancerous cell

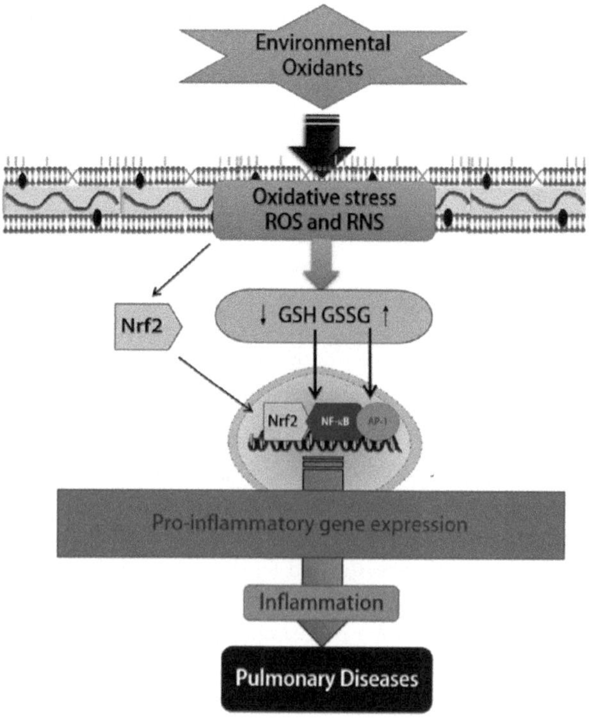

FIGURE 1.4 The oxidative stress results in the variation of the redox levels and leads to several pulmonary diseases. The modulation of GSH/GSSS levels results in the activation of proinflammatory genes like NF-κB. Various thiol compounds such as N-acetyl-L-cysteine (NAC) and N-acystelyn (NAL) supply cysteine for biosynthesis. (Adapted from Biswas, S.K. and Rahman, I., *Mol. Aspects Med.*, 30, 60–76, 2009.)

growth. Nanocarriers potentially responsive to ROS in the atmosphere can be utilised for timely release and delivery of such siRNA (Li et al., 2018).

1.4.2 Inflammation

Inflammation is another defensive action observed in response to toxicological attack by a xenobiotic component. The type of external stimuli determines the effect of inflammation on the affected tissue. Figure 1.5 provides a summarised glance of the inflammatory pathway. Any xenobiotic material is a potential inducer, which is detected by inflammatory cells. Mediators like chemokines, cytokines, eicosanoids, and products of proteolytic cascades are generated on the uptake of this xenobiotic element. These soluble factors or mediators result in the flow of plasma protein and leukocytes (neutrophils) along the venules at the site of infected cells. Potential ROS, reactive nitrogen species, cathepsin, elastase, and proteinase 3 are initiated as a cleaning function as the neutrophils interact with the infected cells. However, these ROS on the cleansing process are not capable of distinguishing between the infected and healthy cells which lead to collateral damage. However, the repair phase is initiated soon after the inflammatory response (Bhattacharya et al., 2013; Biswas and Rahman, 2009; Medzhitov, 2008).

1.4.3 Non-oxidant Routes

Cellular injury can be investigated using several nonoxidant methods. Nanoparticle dissolution is one of the major forms of toxicity in cells. The particle dissolution is certainly a thermodynamic property and requires negative surface energy. The surface morphology, surface energy states, and surface area of the nanoparticles are the key concerns that influence its solubility (Brunner et al., 2006; Franklin et al., 2007). Limbach et al. explained the concept of cytotoxicity by particle dissolution using a mechanism known as the Trojan horse mechanism (Limbach et al., 2007). The oxidative stress generated on exposure to the metal oxides is 25 times higher than exposure to metal solutions alone. It was explained that, since the selective permeability prevents the entrance of metal ions, a controlled oxidative stress was generated. On the other hand, the metal oxide nanoparticles at a suitable pH required for dissolution can enter the cytosol. Further leaching of the metal ion results in inducing potential cytotoxicity (Latiff et al., 2015).

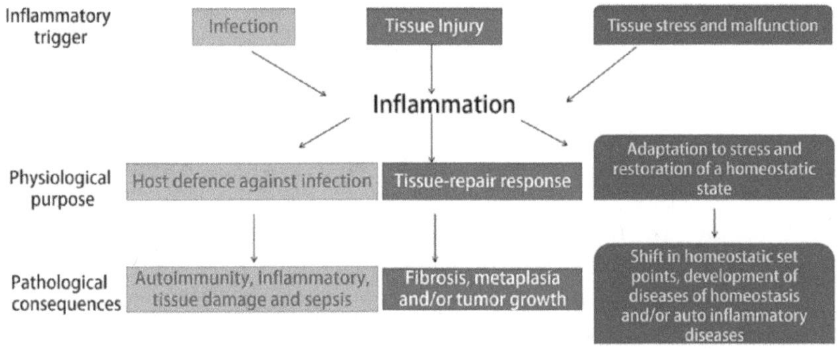

FIGURE 1.5 Schematic illustration of the various inflammatory pathways. (Adapted from Medzhitov, R., *Nature*, 454, 428–435. 2008.)

1.5 Physicochemical Properties of Nanomaterials Affecting the Cytotoxicity

The physicochemical properties of the NMs such as the morphology, aspect ratio, surface charge, and functional groups are some of the key factors responsible for inducing toxicity (Jakus et al., 2015; Wang et al., 2014). Non-toxic materials in bulk could be toxic in lower dimension. There exist other factors that contribute to the toxicity of any materials; the current section is limited the discussion of a few major ones.

1.5.1 Layer Thickness and Exfoliation

The layer thickness plays a significant role in inducing toxicity. The exfoliation state of the layered structures determines the biological interactions. Chng et al. studied the influence of exfoliation of bulk MoS_2 using three different lithium sources (Methyllithium [Me-Li], n-butyllithium [n-BuLi], and tert-butyllithium [t-Bu-Li]). It was observed that the exfoliation and toxicity of the materials are dependent. Hence, the n-BuLi and t-Bu-Li with more layered sheets had a higher rate of toxicity compared to less exfoliated Me-Li. The enhanced surface area and presence of surface-active sites are some of the reasons for the change in the toxicity profile of the samples (Chng et al., 2014). However, the reports postulate the reasons without elaborating on the influence of the other physical parameters. In another study, the authors dealt with three different types of MoS_2. The layered structures were obtained by chemical exfoliation or ultrasonically assisted exfoliation with Pluronic F87 as a surfactant. An aggregated MoS_2 was also studied. The cytotoxicity of the samples was assessed on THP-1 and BEAS-2B cell lines and illustrated no change in the viability. However, pro-inflammatory and pro-fibrogenic responses were generated *in vitro* by the aggregated NMs. Similarly, on a study on mice, the same materials induced acute lung inflammation, while the other two forms remained unaffected. Thus, stable dispersions of exfoliated NMs are less or non-toxic compared to the aggregated versions of the samples. This ensures the biocompatibility of the structures and their successful applications for biomedical and sensing (Wang et al., 2015b). Shah et al. also confirmed the biocompatibility nature of the 2D MoS_2 structures. They used the sulforhodamine B (SRB) assay and electrical impedance spectroscopy (EIS) on rat pheochromocytoma cells (PC12) and rat adrenal medulla endothelial cells (RAMEC). The EIS results displayed no change in the resistance value, on increase in concentration, while the same was validated by the toxicity assay (Shah et al., 2015). In another study, the authors investigated the toxicity of MoS_2 and WS_2 nanosheets prepared through different techniques such as mechanical exfoliation and grown by chemical vapour deposition (CVD). The toxicity was determined using live-dead cell assays, ROS generation assays, and direct assessment of cell morphology of NM-exposed human epithelial kidney cells (HEK293F). The live/dead assay and the ROS assay showed no significant change to the viability and levels of the ROS generated in the cytosolic environment, respectively. However, scanning electron micrograph (SEM) images showed strong adhesion of the nanosheets with the cell structure. Hence, this proved the biocompatibility of the exfoliated nanosheets irrespective of the two different synthesis techniques in this case (Appel et al., 2016). In a different study, the authors studied the cytotoxicity of MoS_2 and boron nitride nanosheets. The study evaluated human hepatoma HepG2 cells to determine the different toxic endpoints. The results illustrated the decrease in viability at a NM concentration of less than 30 μg/mL. The toxicity induces serious effects on intracellular ROS generation (≥2 μg/mL), mitochondrial depolarization (≥4 μg/mL), and membrane integrity (≥8 μg/mL for MoS_2 and ≥2 μg/mL for boron nitride). Additionally, the study also established that at lower NM concentrations (0.2–2 μg/mL) of MoS_2 and boron nitride, NMs could enhance the plasma membrane fluidity and stop the transmembrane ATP binding cassette efflux transporter activity. This transforms the NMs as a chemosensitizer, leading to damage to the plasma membrane. The release of Mo or B species may be considered as the possible reason that both NMs obstruct the efflux pump activities

(Liu et al., 2017a). Moore et al. investigated the cytotoxicity profile of MoS_2 nanoflakes of a different dimension in different cell lines under various exposure routes (inhalation, A549; ingestion, AGS; monocyte, THP-1). The use of sodium cholate as a surfactant to isolate flakes of different sizes (50 nm, 117 nm, and 177 nm) was studied and the toxicity remained unaffected for all three sizes at a concentration of 1 $\mu g\ mL^{-1}$. Transmission electron micrograph (TEM) results suggest the intake of nanosheets inside the cellular environment but inflammatory responses were observed. It was observed that the cytokine production (IL-6, IL-10, IL-13, TNF-α, and IL-1β) is size-dependent and increases the levels in THP1 cells. The smallest dimension flake induces the highest cytokine production. However, the synthesis precursors of MoS_2 samples led to the formation of endotoxin contamination. Thus, it is challenging to understand the cause of the enhanced toxicity, as it is promoted by the size of the flakes or the enhanced surface area, which provides sites for endotoxin adsorption. The increased surface area of the smaller nanoflakes provides the space for endotoxins, which further enhances the production of inflammatory cytokines (Moore et al., 2017). Fu et al. synthesized black phosphorus nanosheets using a liquid exfoliation technique and studied the effect of lateral sizes of sheets isolated at varying centrifuge speeds. It was observed that the sheets are cytocompatible irrespective of size. However, the photothermal ablation ability of the materials is certainly size dependent. The higher photothermal conversion was attained for sheets with larger lateral dimension (Fu et al., 2017b). Thus, the importance of the lateral dimension, concentration, and synthesis mode of the 2D NMs is well established in the previous discussion. The NMs in the 2D framework have enhanced surface area, but these surfaces may promote the adsorption or attraction of unwanted contaminants which contribute to increased toxicity.

1.5.2 Surface Functionalisation and Structural Form

Apart from lateral dimension and the mode of synthesis, there exist a few more important parameters of concern for 2D NMs. As explained in the previous section, the enhanced surface areas of the 2D materials are the active sites for adsorption. More often, various synthesis techniques result in surface functionalisation. This attribute could be utilised to use the materials as potential substrates for various structural forms. Suhito et al. studied the toxicity of graphene oxide (GO), MoS_2, WS_2, and boron nitride. These 2D NMs were coated on cell-culture substrates (human mesenchymal stem cells) by a drop-casting method and acute toxicity was not found for any of the different 2D materials at a lower concentration range (<5 μg/mL). However, the 2D material-modified substrates demonstrated an enhanced cell adhesion, spreading, and proliferation when compared to a non-treated substrate. It was observed that the important differentiation lineages of stem cells, known as the adipogenesis, are highly enhanced by using WS_2–, MoS_2–, or BN-coated substrates when compared with both non-treated and GO-coated substrates. In case of osteogenesis, GO displayed the best result among the four different 2D materials (Suhito et al., 2017). In another study, the authors prepared MoS_2 thin films on substrates to explore their effects on neural stem cells and neural differentiation. The biocompatibility of thin films was examined by immunostaining and real-time polymerase chain reaction. The results exhibit that the thin films possessed flexibility, cell seeding ability, and electrical properties, which promoted the growth of living tissue engineering scaffolds in nerve regeneration, even in the absence of growth factors (Wang et al., 2017a). McManus et al. studied the biocompatibility of thin films created from printed 2D NM inks. The cell viability of A549 and skin (human keratinocytes: HaCaT) cells was evaluated using the lactate dehydrogenase (LDH) assay and propidium iodide (PI)/annexin V staining was utilized to differentiate cell viability by flow cytometry. No difference in the viability measures was observed for either of the assaying techniques (McManus et al., 2017). In a more recent study, Wang et al. synthesized a 3D printed scaffold with MoS_2 nanosheets for a bifunctional application of tumour therapy and tissue generation (Wang et al., 2017b). Bioceramic scaffolds were utilized to grow MoS_2 nanosheets through a hydrothermal process. The MoS_2 loading in the scaffold determined the effect it generated for the tumour therapy. In addition, the scaffold promoted osteogenesis and angiogenesis (Wang et al., 2017b).

The influence of specific ligands or functional groups does influence biocompatibility and their applications. The medical application of black phosphorus nanosheets is discussed in Section 1.7. In one such study, the authors demonstrated the influence of titanium sulfonate as a potential ligand. The presence of this ligand makes the sheets effective in uptake by macrophages inside the cytosol and further reduces cytotoxicity and proinflammation (Qu et al., 2017). In another report, the authors studied the biodegradability and the elimination pathway of bare MoS_2 and functionalized MoS_2 sheets (Kurapati et al., 2017).

1.6 Modes for Cellular Uptake of 2D Nanomaterials

NM uptake is governed by several parameters such as the NM's adsorption, distribution, metabolism, and excretion (Buzea et al., 2007; Oberdörster et al., 2005). The major challenge for the 2D NMs is to sense and target the intracellular genetic products. The limited delivery of these NMs into the cytosol is another hurdle to cross for these small packages, as the route of uptake is perceived to play a minor role in toxicity. Therefore, it is of paramount interest to understand the different pathways of ingestion and their fate inside the cytosolic environment (Figure 1.6) (Oh and Park, 2014) (Buzea et al., 2007).

1.6.1 Endocytosis

Endocytosis is an ingestion process utilised by the cells to ingest NMs and other extracellular components. In this process, an inward folding of the plasma membrane encloses the NMs present outside the cytosolic environment. The folding detaches from the plasma membrane to form intracellular vesicles (Decuzzi and Ferrari, 2007). The ingested foreign object/material is delivered to lysosomes for further

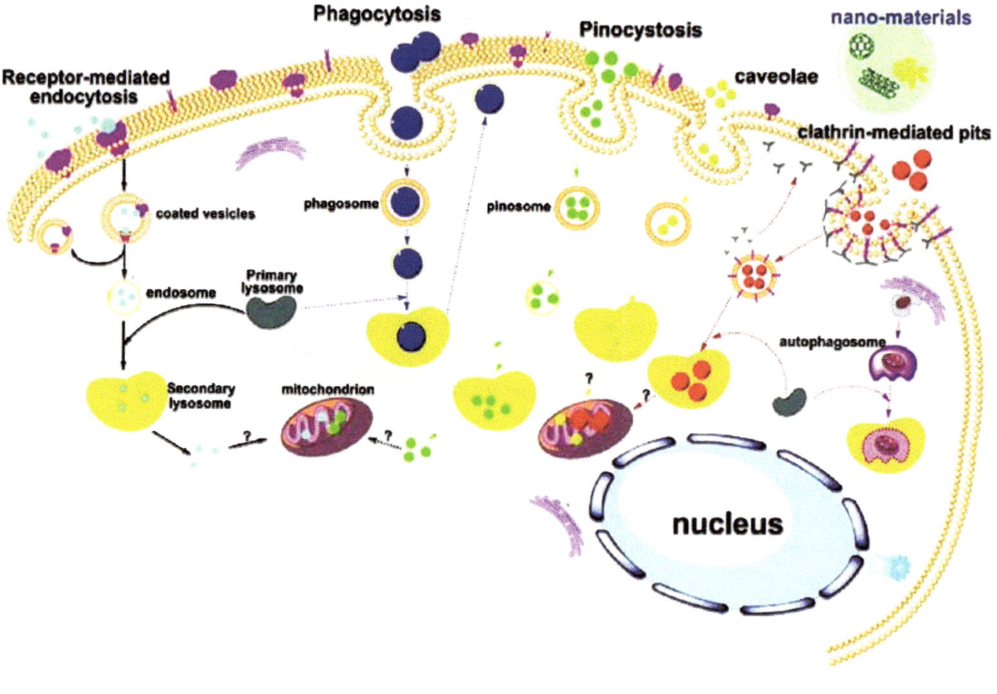

FIGURE 1.6 Illustration of various cell uptake pathways. (From Zhao, F. et al., *Small*, 7, 1322–1337, 2011.)

degradation (Goldstein et al., 1979). There exist several pathways utilised by the cells to ingest NMs: phagocytosis, pinocytosis, clathrin-mediated endocytosis, and caveolae-mediated endocytosis (Cao et al., 2017; Rabinovitch, 1995; Zhao et al., 2011).

1.6.2 Clathrin-Mediated Endocytosis

Clathrin-coated vesicles are utilised to internalise NMs of size usually <100 nm. It is a receptor-mediated endocytosis pathway, where the plasma membrane undergoes inward budding and forms vesicles. The vesicles are layered with various protein receptors permitted to internalise the specific molecule (Sorkin and Puthenveedu, 2013). In this energy-dependent process the clathrin does not interact with the membrane or the ingested particles. It completely depends on the protein receptors and the accessory proteins present on the walls of the vesicles. The accessory proteins are the cytoplasmic proteins which are later subjected to reuse for another endocytosis cycle. The internalised NMs experience organization in the endosomes and are later sent to the surface or delivered to other mature endosomes like lysosomes (McMahon and Boucrot, 2011). The uptake of nutrients, activation of signalling pathways, regulation of surface expression of proteins, and retrieval of proteins deposited after vesicle fusion are some of the functions associated with clathrin-mediated endocytosis (Chen et al., 1998; Liu et al., 2001; McMahon and Boucrot, 2011; Motley et al., 2003; Sikora et al., 2017).

1.6.3 Caveolae/Raft-Dependent Endocytosis

NMs of size <200 nm are taken up *via* the caveolae-dependent endocytosis. This endocytosis method is a clathrin-independent intake method which is a mixture of pinocytosis and endocytosis facilitated by caveolae and glycolipid rafts. The caveolae are cholesterol and sphingolipid-rich invaginations of the plasma membrane and the glycolipid rafts are cholesterol-rich membrane fractions. The presence of the integral membrane protein caveolin helps to distinguish the invaginated domains of the plasma membrane. This technique of intake is able to avert the digestion in lysosomes of the ingested NMs. Thus, this receptor-independent endocytosis could be utilised effectually for drug or DNA delivery applications (Nabi and Le, 2003; Pelkmans et al., 2005; Richard et al., 2005; Singh et al., 2003). Rejman et al. reported the significance of the size of the particles in different endocytosis routes (Rejman et al., 2004). This study revealed that the entry portal is determined by the size of the particles. Therefore, an appropriate evaluation of other kinetic factors of internalisation could expand the drug delivery efficacy.

1.6.4 Phagocytosis

NMs of size <200 nm are taken up *via* phagocytosis. It is utilised by several mammalian cells such as the mononuclear phagocytes, macrophages, and neutrophils to eliminate infectious particles or cellular debris. These specialised cells have advanced their operational ability and contribute in the intake of nutrients, development and remodelling of tissues, and immune response and inflammation. The ligands on the particles interact with the receptors of the cell to initiate the internalisation process (Sbarra and Karnovsky, 1959; Stossel, 1974). The ingestion process of the NMs by phagocytosis process is speeded up by the presence of specialised molecules such as antibodies, labelled on the surface of the ingested particle; this technique of labelling is termed opsonization. Additionally, this results in polymerization of actin and the ingestion of the particles via an actin-based mechanism. The lysosomes (contains digesting enzymes) combine with the phagocytic cells to form phagolysosomes. The nature of the surface interaction between the particle and the phagosome membrane determines the time for this fusion (Aderem and Underhill, 1999; Allen and Aderem, 1996). Various proteases and nicotinamide adenine dinucleotide phosphate (NADPH) oxidases are used for the complete disintegration of the ingested particles inside the compartment. The remaining cell debris after the disintegration process is excreted via exocytosis (Buzea et al., 2007). The existence of different types of receptors and the alteration of the individual fate of the ingested particles make this process of phagocytosis extremely complex in nature. The key defining task of these cells is to distinguish between potential pathogens and itself. Nevertheless, this task is completed by numerous phagocytic receptors which have the ability to differentiate between them (Champion and Mitragotri, 2006; Indik et al., 1995).

1.6.5 Pinocytosis/Macropinocytosis

Pinocytosis is an endocytic internalisation process occurring in all different cell types, which uptake particles from a few to several 100 nm. This route results in the development of membrane-based vesicles from the cell surface that internalise the fluid and the solute from the external atmosphere. These internalised pinocytic vesicles combine along with the lysosome (Saha et al., 2013; Steinman et al., 1976).

1.7 Cytotoxicity Reports of Two-Dimensional Materials

Tuning the dimension has been an exciting domain to improve the efficacy of NMs. Graphene and other sister 2D NMs have displayed favourable results against various applications (Ding et al., 2017; Singh et al., 2011; Tu et al., 2017). The high surface area and the presence of surface-active sites make these materials an attractive option for drug delivery. The effective functionalisation and the large surface provide room for enhanced drug loading ability. The near-infrared region (NIR) (700–1300 nm) is an important zone of the electromagnetic spectrum. Biological tissues are inactive in this region, and hence various *in vivo* and *in vitro* applications are utilised, employing materials that are optically active in this domain. Photothermal ablation utilises NIR absorbing material to convert energy into light or heat. This heat is further utilised for tissue ablation for advanced photothermal therapy. However, these applications remain futile if the toxicity of the materials is dominant and affects the viability of healthy cells. The present section reports on 2D NMs and their composites discussing their toxicity profile and cell death mechanism. Ju et al. reported the synthesis of Cu-doped $g\text{-}C_3N_4$ for photodynamic therapy (Ju et al., 2016). The C_3N_4 sheets demonstrated the overexpression of GSH which largely diminished the ROS generated, which reduced the therapeutic efficiency of the sheets (Figure 1.7). The authors incorporated redox-active Cu metal ions with C_3N_4, which resulted in the reduction of molecular oxygen to

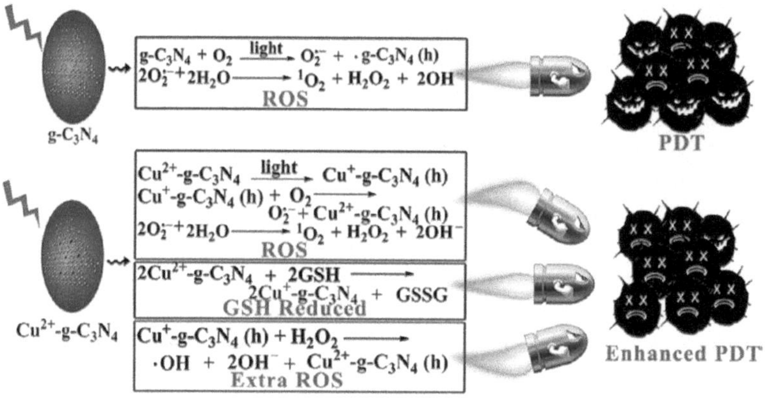

FIGURE 1.7 Schematic illustration of $g\text{-}C_3N_4$ and $Cu^{2+}\text{–}g\,C_3N_4$ nanosheets for photodynamic therapy by enhanced ROS generation and GSH depletion. (From Ju, E. et al., *Angew. Chem.*, 128, 11639–11643, 2016.)

potential ROS species. The effective combination of the two components resulted in controlled levels of GSH which further aided the cancer therapy (Ju et al., 2016).

Chou et al. reported the application of chemically modified MoS_2 nanosheets for an efficient NIR photothermal transducer. The photothermal ablation property of the nanosheets was demonstrated using an *in vitro* study of the viability of HeLa cells following exposure to a concentration of 40 ppm for up to 24 hours. Assaying with MTT revealed that the viability remained approximately at 80% without being irradiated to any NIR source. On the other hand, the viability was zero when the cell lines with the same dosage were subjected to NIR irradiation for 20 minutes (Chou et al., 2013). The cytotoxicity profile of exfoliated MoS_2, WS_2, and WSe_2 using human lung carcinoma cells (A549) was studied by Teo et al. The toxicity order follows as $WSe_2 > MoS_2 > WS_2$ and Tungsten selenide (WSe_2) was found to be the most toxic amongst the TMDs evaluated. The presence of chalcogenides in the outer layers of the TMDs is attributed as the plausible reason for the toxicity, as they are the elements interacting with the cellular environment. The study of toxicity on H_2S and H_2Se has revealed H_2Se to be more toxic and the authors have reported the same reason for the enhanced toxicity by selenium. Therefore, the presence of Se results in the TMD (WSe_2) to be more toxic than the rest (Teo et al., 2014). The same authors also demonstrated the toxicity of layered GaS and GaSe. GaS is found to be more toxic than the other known transition metal chalcogenides. The high solubility of the gallium ions in the cytosolic environmental pH is ascribed to the enhanced cytotoxicity (Latiff et al., 2015). The application of NMs for drug delivery necessitates cytotoxicity assessment (Zhang et al., 2017; Dönmez Güngüneş et al., 2017; Fu et al., 2017a). Dhenadhayalan and co-workers studied the use of 2D nanosheets of MoO_3, MoS_2, and $MoSe_2$ as biomarkers for prostate cancer. Human embryonic kidney 293T (HEK) cells were utilized for the viability measurements. The cells exposed to a range of dosages of nanosheets (up to 100 μg/mL) for 48 hours illustrated a viability of 80% (Dhenadhayalan et al., 2017). In a similar attempt, drug delivery applications were studied using lipid-functionalized WS_2 and MoS_2 nanosheets. The hydrogen bonding aided drug loading and the Van der Waal forces supported the functionalization of liposomes. HeLa cells were exposed with different dosage of nanosheets up to 50 μg/mL concentration. The exposed cell lines illustrated non-toxic features on being assayed by MTT, which exhibits the biocompatibility of the nanosheets (Liu and Liu, 2017). In another report, Wang and co-workers reported the synthesis of 2D MoS_2 nanosheets composite with Bi_2S_3 functionalized with polyethylene glycol (PEG) (Figure 1.8). The incorporation of PEG enhances the solubility and biocompatibility nature of the composite. The introduction of Bi atoms results in the coupling with the S atoms, which further prevents the release of toxic H_2S gas. Moreover, the Bi atoms also aid in computed tomography (CT) and the MoS_2 nanosheets promote the photoacoustic (PA) imaging and the photothermal transformation. The reported composite is utilized for both *in vitro* and *in vivo* CT and PA imaging-guided combined tumour photothermal therapy and sensitized radiotherapy. The cytotoxicity of the nanosheet composites was investigated using the cell counting kit-8 assay on L929 cells. Incubation of the cells up to a concentration of 0.15 mg/mL for 24 hours illustrated a nontoxic nature and demonstrated the excellent cytocompatibility of the sheets. The cell morphology was observed using phase-contrast micrograph and remained unaffected (Wang et al., 2015a).

Feng et al. synthesized multifunctional Bi_2WO_6 nanosheets for drug delivery applications. The bismuth tungstate nanosheets were PEGylated and loaded with anticancer drug Doxorubicin. HeLa cells were exposed to various concentrations up to 800 μg/mL for 24 hours and demonstrated a viability of 97%. Furthermore, the cells were incubated with functionalized NMs loaded with drug at a concentration of 200 μg/mL to understand the cellular uptake using confocal laser scanning microscopy. The phagocytic internalization of the NMs results in an increase in Bi content within 6 hours of incubation, while the content declined on reaching the 12 hours time point (Feng et al., 2018). PEGylated WS_2 nanosheets have been reported for *in vivo* enhanced X-ray CT and PA tomography. The WS_2 nanosheets exhibited an enhanced NIR absorption which is utilized for the 100% removal of the tumour via *in vivo* photothermal ablation. The toxicity profile of the nanosheets remained

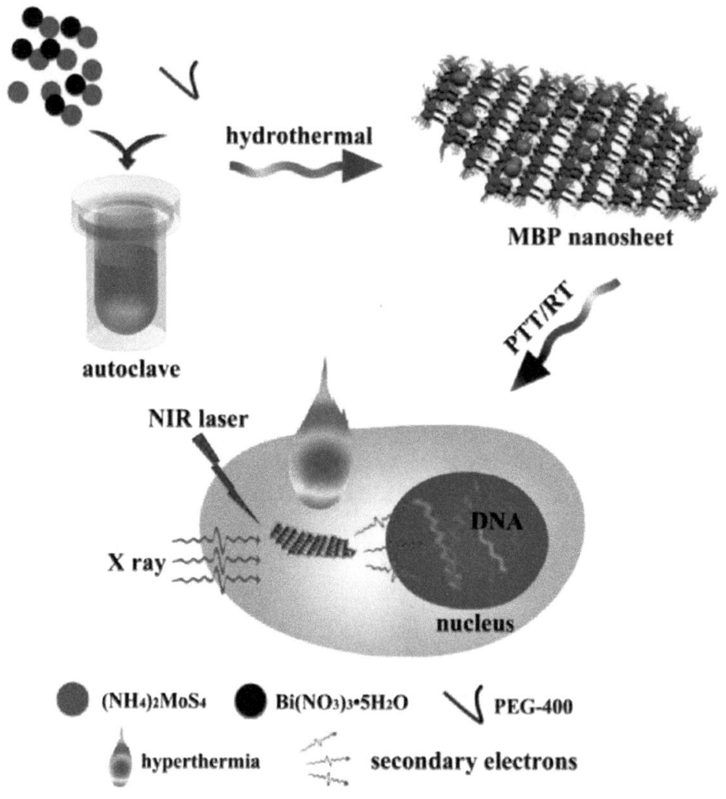

FIGURE 1.8 Schematic illustration of the solvothermal synthesis of PEGylated MoS_2/Bi_2S_3 composite and tumour photothermal therapy and radiotherapy. (From Wang, S. et al, *Advanced Materials*, 27, 2775–2782, 2015a.)

unaltered after being exposed for 24 hours up to a concentration of 0.1 mg/mL against the cell lines of 4T1 (murine breast cancer cells), HeLa (human epithelial carcinoma cells), and 293T (human embryo kidney cells). The LDH assay of the cell lines with the nanosheets did not show any difference compared to the untreated cells, which provides evidence of the biocompatible nature of the sample (Cheng et al., 2014). In another study, PEGylated WS_2 nanoflakes were doped with gadolinium metal ions (Gd^{3+}). The strong NIR absorption makes the nanosheets excellent for PA imaging and CT. The addition of the Gd^{3+} metal ions with a paramagnetic property aided in magnetic resonance imaging. The relative viabilities of 4T1 murine breast cancer cells were used to access the *in vitro* photothermal property. Exposure of the cell lines to a high concentration of nanoflakes up to 0.1 mg/mL for up to 48 hours resulted in no change in the viability, while irradiating the same cell lines with a NIR source for 5 minutes destroyed the complete viable cells (Yin et al., 2014). In a recent study, nanosheets of black phosphorous were utilized for photodynamic/photothermal/chemotherapy of cancer. These metal-free nanosheets had a higher surface to volume ratio compared to graphene and MoS_2 because of its puckered lattice arrangement, which subsequently enhances the drug loading ability to 950% (by weight) compared to other reported 2D architectures. The efficient electronic structure and the strong absorption in the NIR makes them a suitable candidate for photothermal treatment. The biocompatibility of the nanosheets was evaluated by exposing various cell lines 4T1, HeLa, L929, and A549 cells to the NMs at a concentration of 200 μg/mL. The MTT assay illustrated no cytotoxicities and thus indicated biocompatibility on the NMs (Chen et al., 2017). Song et al. assessed the toxicity profile of layered black phosphorus against L-929 fibroblasts. The authors found the toxicity profile is concentration and time-dependent, which is led by the oxidative stress-mediated enzyme activity reduction and membrane disruption. Interestingly, these NMs did not display substantial cytotoxicity at concentrations below 4 μg/mL (Song et al., 2018a). Therefore, it is suggested that layered BPs can be effectively utilized as therapeutic delivery carriers and imaging agents. Novel drug delivery systems utilising various external stimuli such as pH, heat, light, or magnetic energy have been reported widely but the problem of drug resistance in the cytosolic environment is significant (Tian et al., 2013). Recently, Dong et al. dealt with the major issue of multidrug resistance. The authors synthesized hyaluronic acid (HA) and polyethyleneimine (PEI) functionalized MoS_2 nanosheets to deal with drug-resistant breast cancer (MCF-7-ADR) cell lines. The over expression of P-glycoprotein results to the multi-drug resistance. The high NIR to photothermal efficiency, selective fluorescence, and high surface area makes MoS_2 an interesting choice amongst the NMs. While hyaluronic acid is a biodegradable polysaccharide which is extremely compatible to the cytosolic environment and moreover, the tumour environment already contains hyaluronidase, which can further help in degradation and selective targeting for the therapeutic application. These polysaccharides also target CD44 receptors, responsible for the overexpression of drug-resistant cancerous cells. The A549, MCF-7, and MCF-7-ADR cell lines show no toxicity on exposure to nanosheets for 24 hours when evaluated by the live/dead stain and cell counting kit-8 (Dong et al., 2018). In another study, the PEGylated black phosphorus nanosheets were studied for their efficient theranostic delivery. The nanosheets were in-taken by endocytosis and were determined by measuring the final concentration in the lysosomes. The photothermal efficiency of the nanosheets was determined by the cytotoxicity evaluation of the HeLa cells by MTT assay. Incubating these carcinomic cell lines for an hour with a concentration as low as 25 μg/mL resulted in less than 10% cellular viability (Tao et al., 2017). Liu et al. reported the synthesis of tantalum sulphide nanosheets as theranostic nanoplatforms for CT imaging-guided combinatorial chemo-photothermal therapy. The *in vitro* analysis of HeLa and PC3 cells for 48 hours with a maximum dosage up to 0.5 mg/mL resulted in no change in the viability. Similarly, the LDH measurement remained consistent when compared with the values obtained from the untreated cell lines. The NMs were internalized through caveolae- and clathrin-mediated endocytosis pathways inside the cancerous cells. On irradiation with NIR source the viability gradually decreased (Liu et al., 2017b). In another study, the authors illustrated the importance of targeted drug delivery and the challenges behind the internalization process. Perfluorinated functionalized WSe_2 nanosheets were synthesized, as these compounds are hydrophobic and lipophobic. This feature assists the phase separation in both polar and non-polar cytosolic environment. The functionalized nanosheets have been taken up by the cellular endosomal pathway and they exhibited low toxicity. Assaying with MTT on exposed cell lines for hours illustrated no major toxic influence. These NMs helped in delivery of DNA probes labelling the miRNA. The miRNA are the noncoding RNAs responsible for genetic expression to target messenger RNAs, which blocks the ribosomes translating and further results in termination. The overexpression of miRNA results in different pathogenic processes such as the tumour progression and metastasis (Song et al., 2018b).

1.8 Safety Guidelines for Handling Nanomaterials

The stability of 2D NMs has distinct consequences to their toxicity and endpoint. Therefore, understanding and evaluating techniques to handle NMs is extremely important. The use of NMs in general in consumer goods has risen exponentially. This warrants the need for a defined set of instructions to use and mitigate the problem of toxicity at the same time. A recent book chapter highlighted the global actions taken in regards to the health and safety concerns (Table 1.1) (Viswanath and Kim, 2016).

TABLE 1.1 Global Strategies to Address Human Health and/or Environmental Safety Aspects of Nanomaterials.

Organization	Objectives
The Organization for Economic Co-operation and Development (OECD)	• Evaluation of risk assessment approaches for manufactured NMs through information exchange and the identification of opportunities to strengthen and enhance risk assessment capacity
National Institute for Occupational Safety and Health (NIOSH)	• Development and implementation of commercial nanotechnology through conducting strategic planning and research to provide national and world leadership for incorporating research findings of the implications and applications of nanotechnology into good occupational safety and health practice
EU NanoSafetyCluster	• Works on the projects addressing all aspects of nanosafety including toxicology, ecotoxicology, exposure assessment, mechanisms of interaction, risk assessment, and standardization • Conducts workshops and seminars to educate people, particularly all nanotechnology workers
Federal Institute for Materials Research and Testing (BAM)	• BAM is involved in EU-funded (FP7) projects: NanoDefine (FP7): The aim of NanoDefine is to support the governance challenges associated with the implementation of the NM legislation by addressing the issues on the availability of suitable measuring techniques, reference material, validated methods, acceptable for all stakeholders, and delivering an integrated and interdisciplinary approach NanoValid (FP7): The main objective of NanoValid is the development of new reference methods and certified reference materials, including methods for characterization, detection/ quantification, dispersion control, and labelling, as well as hazard identification, exposure, and risk assessment of ENs
Federal Ministry of Education and Research (BMBF)	• NanoCare—Safe Handling of Manufactured Nanomaterials • Studying the effects of ENM on humans and the environment
Federal Environment Agency (Umweltbundesamt, UBA)	• NanoCare—Safe Handling of Manufactured Nanomaterials • Investigating Impacts on Health and the Environment
Federal Institute for Risk Assessment (Bundesinstitut für Risikobewertung, BfR)	• To demonstrate and establish new principles and ideas based on data from value chain implementation studies • To establish Safe-by-Design as a fundamental pillar in the validation of a novel manufactured NM • Establishing NM grouping/ classification strategies according to toxicity and biological effects for supporting risk assessment • Grouping of nanostructured materials for protection of workers, consumers, the environment, and risk minimisation
Federal Institute of Occupational Safety and Health (BAuA)	• Grouping of nanostructured materials for protection of workers, consumers, the environment, and risk minimization
Federal Research Institute of Nutrition and Food (Max Rubner-Institut, MRI)	• Detection and characterization of NMs in complex matrices such as food, etc. • Research on nano-sized carrier systems for bioactive compounds • Interaction of NMs with compounds of the food matrix
Modena Cost	• The specific objectives are to study the synthesis of engineered NMs (ENM) with controlled composition, size, area, and nano-texture • To develop strategies to immobilise ENM in matrices and on substrates with minimum effect on the desired properties and surface reactivity, and to identify the relevant datasets for Quantitative Nanostructure-Toxicity Relationships (QNTR) modelling

Source: Viswanath, B. and Kim, S., *Rev. Environ Contam Toxicol.*, 242, 61–104, 2016.

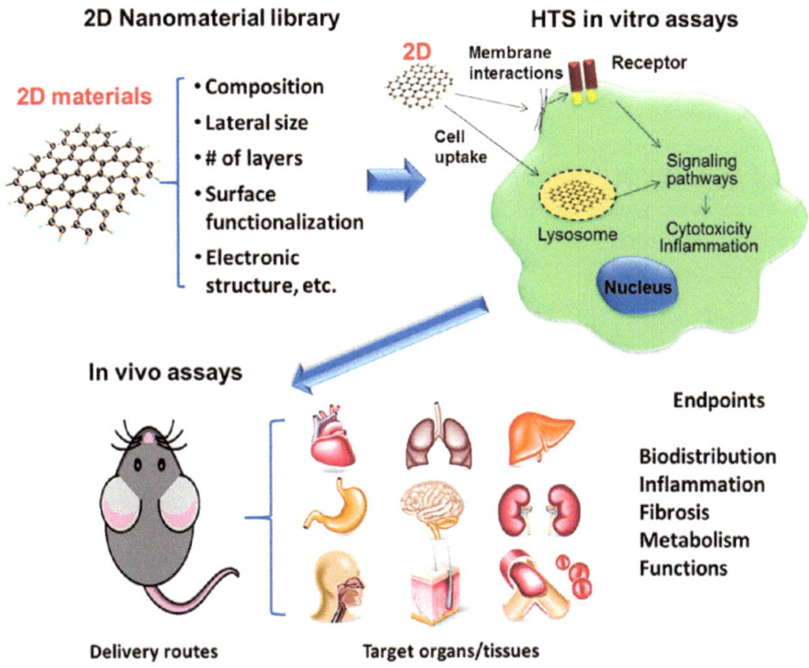

FIGURE 1.9 A schematic illustration for a proposed approach for the hazard assessment of 2D materials utilising a potential library, *in vitro* high throughput screening (HTS) assays, and *in vivo* assays. (From Guiney, L.M. et al., *ACS Nano*, 12, 6360–6377, 2018.)

The existence of NM libraries detailing the physical parameters and the biological environment can help to enhance material selection and commercial applications (Figure 1.9) (Guiney et al., 2018).

1.9 Conclusion and Outlook

The rise in the use of 2D NMs in various applications demands the necessity to understand their health and environmental effects. The physicochemical properties of these materials govern the nature of their cytosolic interaction. In this chapter, the role of the surface and morphological parameters is discussed in detail by highlighting several recent studies. The mode of intake is also a domain of particular interest as it dictates the targeted delivery and nanotoxicity. It is of paramount interest to understand that the cytotoxicity of the material cannot be governed or categorized by any set of precise factors; as explained earlier, a minute change in the biological or physiological parameter of the material or the environment can totally change the toxicity profile. The 2D NMs are still a very new class of materials, which still require several combinations of characterization to evaluate the prime cause of toxicity. Careful *in vivo* assays to substantiate the results obtained from the *in vitro* analysis are yet to be developed for 2D NMs. The need for interdisciplinary collaboration for the formation of libraries detailing the ever-changing profiles of the NMs and also a possible theoretical model enabled to predict the toxicity impact is highly desirable. These steps can ensure the structuring of the mitigation process to define a safer world for NM use.

Acknowledgements

Author PG would like to thank the IT Sligo President's bursary for providing financial support.

References

Aderem, A. and D. M. Underhill. 1999. Mechanisms of phagocytosis in macrophages, *Annual Review of Immunology*, 17(1): 593–623.

Aillon, K. L., Y. Xie, N. El-Gendy, C. J. Berkland, and M. L. Forrest. 2009. Effects of nanomaterial physicochemical properties on in vivo toxicity, *Advanced Drug Delivery Reviews*, 61(6): 457–466.

Aires, A., J. F. Cadenas, R. Guantes, and A. L. Cortajarena. 2017. An experimental and computational framework for engineering multifunctional nanoparticles: Designing selective anticancer therapies, *Nanoscale*, 9(36): 13760–13771.

Allen, L.-A. H. and A. Aderem. 1996. Mechanisms of phagocytosis, *Current Opinion in Immunology*, 8(1): 36–40.

An, X. and J. Yu. 2011. Graphene-based photocatalytic composites, *RSC Advances*, 1(8): 1426–1434.

Apel, K. and H. Hirt. 2004. Reactive oxygen species: Metabolism, oxidative stress, and signal transduction, *Annual Review of Plant Biology*, 55: 373–399.

Appel, J. H., D. O. Li, J. D. Podlevsky, A. Debnath, A. A. Green, Q. H. Wang, and J. Chae. 2016. Low cytotoxicity and genotoxicity of two-dimensional MoS_2 and WS_2, *ACS Biomaterials Science & Engineering*, 2(3): 361–367.

Baehrecke, E. H. 2002. How death shapes life during development, *Nature Reviews Molecular Cell Biology*, 3(10): 779–787.

Bell, D. R., S.-G. Kang, T. Huynh, and R. Zhou. 2018. Concentration-dependent binding of CdSe quantum dots on the SH3 domain, *Nanoscale*, 10(1): 351–358.

Beloica, S., S. Cvijić, M. Bogataj, and J. Parojčić. 2015. In vitro-in vivo-in silico approach in biopharmaceutical characterization of ibuprofen IR and SR tablets, *European Journal of Pharmaceutical Sciences*, 75: 151–159.

Bhattacharya, K., F. T. Andón, R. El-Sayed, and B. Fadeel. 2013. Mechanisms of carbon nanotube-induced toxicity: Focus on pulmonary inflammation, *Advanced Drug Delivery Reviews*, 65(15): 2087–2097.

Biswas, S. K. and I. Rahman. 2009. Environmental toxicity, redox signaling and lung inflammation: The role of glutathione, *Molecular Aspects of Medicine*, 30(1): 60–76.

Brunner, T. J., P. Wick, P. Manser, P. Spohn, R. N. Grass, L. K. Limbach, A. Bruinink, and W. J. Stark. 2006. In vitro cytotoxicity of oxide nanoparticles: Comparison to asbestos, silica, and the effect of particle solubility, *Environmental Science & Technology*, 40(14): 4374–4381.

Butler, S. Z., S. M. Hollen, L. Cao, Y. Cui, J. A. Gupta, H. R. Gutiérrez, T. F. Heinz, S. S. Hong, J. Huang, and A. F. Ismach. 2013. Progress, challenges, and opportunities in two-dimensional materials beyond graphene, *ACS Nano*, 7(4): 2898–2926.

Buzea, C., I. I. Pacheco, and K. Robbie. 2007. Nanomaterials and nanoparticles: Sources and toxicity, *Biointerphases*, 2(4): MR17–MR71.

Cao, Y., Y. Gong, L. Liu, Y. Zhou, X. Fang, C. Zhang, Y. Li, and J. Li. 2017. The use of human umbilical vein endothelial cells (HUVECs) as an in vitro model to assess the toxicity of nanoparticles to endothelium: A review, *Journal of Applied Toxicology*, 37(12): 1359–1369.

Castro Neto, A. and K. Novoselov. 2011. Two-dimensional crystals: Beyond graphene, *Materials Express*, 1(1): 10–17.

Chakravarthi, S., C. E. Jessop, and N. J. Bulleid. 2006. The role of glutathione in disulphide bond formation and endoplasmic-reticulum-generated oxidative stress, *EMBO Reports*, 7(3): 271–275.

Champion, J. A. and S. Mitragotri. 2006. Role of target geometry in phagocytosis, *Proceedings of the National Academy of Sciences of the United States of America*, 103(13): 4930–4934.

Chen, H., S. Fre, V. I. Slepnev, M. R. Capua, K. Takei, M. H. Butler, P. P. Di Fiore, and P. De Camilli. 1998. Epsin is an EH-domain-binding protein implicated in clathrin-mediated endocytosis, *Nature*, 394(6695): 793–797.

Chen, W., J. Ouyang, H. Liu, M. Chen, K. Zeng, J. Sheng, Z. Liu, Y. Han, L. Wang, and J. Li. 2017. Black phosphorus nanosheet-based drug delivery system for synergistic photodynamic/photothermal/chemotherapy of cancer, *Advanced Materials*, 29(5): 1603864.
Cheng, L., J. Liu, X. Gu, H. Gong, X. Shi, T. Liu, C. Wang, X. Wang, G. Liu, and H. Xing. 2014. PEGylated WS_2 nanosheets as a multifunctional theranostic agent for in vivo dual-modal CT/photoacoustic imaging guided photothermal therapy, *Advanced Materials*, 26(12). 1886–1893.
Chng, E. L. K. and M. Pumera. 2015. Toxicity of graphene related materials and transition metal dichalcogenides, *RSC Advances*, 5(4): 3074–3080.
Chng, E. L. K., Z. Sofer, and M. Pumera. 2014. MoS_2 exhibits stronger toxicity with increased exfoliation, *Nanoscale*, 6(23): 14412–14418.
Choi, S. K. 2016. Mechanistic basis of light induced cytotoxicity of photoactive nanomaterials, *NanoImpact*, 3: 81–89.
Chou, S. S., B. Kaehr, J. Kim, B. M. Foley, M. De, P. E. Hopkins, J. Huang, C. J. Brinker, and V. P. Dravid. 2013. Chemically exfoliated MoS_2 as near-infrared photothermal agents, *Angewandte Chemie*, 125(15): 4254–4258.
Ciappellano, S. G., E. Tedesco, M. Venturini, and F. Benetti. 2016. In vitro toxicity assessment of oral nanocarriers, *Advanced Drug Delivery Reviews*, 106: 381–401.
Coleman, J. N., M. Lotya, A. O'Neill, S. D. Bergin, P. J. King, U. Khan, K. Young, A. Gaucher, S. De, and R. J. Smith. 2011. Two-dimensional nanosheets produced by liquid exfoliation of layered materials, *Science*, 331(6017): 568–571.
Decuzzi, P. and M. Ferrari. 2007. The role of specific and non-specific interactions in receptor-mediated endocytosis of nanoparticles, *Biomaterials*, 28(18): 2915–2922.
Dhenadhayalan, N., K. Yadav, M. I. Sriram, H.-L. Lee, and K.-C. Lin. 2017. Ultra-sensitive DNA sensing of a prostate-specific antigen based on 2D nanosheets in live cells, *Nanoscale*, 9(33): 12087–12095.
Ding, D., Y. Xu, Y. Zou, L. Chen, Z. Chen, and W. Tan. 2017. Graphitic nanocapsules: Design, synthesis and bioanalytical applications, *Nanoscale*, 9(30): 10529–10543.
Dong, X., W. Yin, X. Zhang, S. Zhu, X. He, J. Yu, J. Xie, Z. Guo, L. Yan, and X. Liu. 2018. Intelligent MoS_2 nanotheranostic for targeted and enzyme-/pH-/NIR-responsive drug delivery to overcome cancer chemotherapy resistance guided by PET imaging, *ACS Applied Materials & Interfaces*, 10(4): 4271–4284.
Dönmez Güngüneş, Ç., Ş. Şeker, A. E. Elçin, and Y. M. Elçin. 2017. A comparative study on the in vitro cytotoxic responses of two mammalian cell types to fullerenes, carbon nanotubes and iron oxide nanoparticles, *Drug and Chemical Toxicology*, 40(2): 215–227.
Enari, M., H. Sakahira, H. Yokoyama, K. Okawa, A. Iwamatsu, and S. Nagata. 1998. A caspase-activated DNase that degrades DNA during apoptosis, and its inhibitor ICAD, *Nature*, 391(6662): 43–50.
Erol, O., I. Uyan, M. Hatip, C. Yilmaz, A. B. Tekinay, and M. O. Guler. 2017. Recent advances in bioactive 1D and 2D carbon nanomaterials for biomedical applications, *Nanomedicine: Nanotechnology, Biology and Medicine*, 14(7): 2433–2454.
Farré, M., K. Gajda-Schrantz, L. Kantiani, and D. Barceló. 2009. Ecotoxicity and analysis of nanomaterials in the aquatic environment, *Analytical and Bioanalytical Chemistry*, 393(1): 81–95.
Feng, L., D. Yang, S. Gai, F. He, G. Yang, P. Yang, and J. Lin. 2018. Single bismuth tungstate nanosheets for simultaneous chemo-, photothermal, and photodynamic therapies mediated by near-infrared light, *Chemical Engineering Journal*, 351: 1147–1158.
Festjens, N., T. Vanden Berghe, and P. Vandenabeele. 2006. Necrosis, a well-orchestrated form of cell demise: Signalling cascades, important mediators and concomitant immune response, *Biochimica et Biophysica Acta (BBA)—Bioenergetics*, 1757(9–10): 1371–1387.
Fielden, M. R. and K. L. Kolaja. 2008. The role of early in vivo toxicity testing in drug discovery toxicology, *Expert Opinion on Drug Safety*, 7(2): 107–110.

Filip, A., M. Potara, A. Florea, I. Baldea, D. Olteanu, P. Bolfa, S. Clichici, L. David, B. Moldovan, and L. Olenic. 2015. Comparative evaluation by scanning confocal Raman spectroscopy and transmission electron microscopy of therapeutic effects of noble metal nanoparticles in experimental acute inflammation, *RSC Advances*, 5(83): 67435–67448.

Franklin, N. M., N. J. Rogers, S. C. Apte, G. E. Batley, G. E. Gadd, and P. S. Casey. 2007. Comparative toxicity of nanoparticulate ZnO, bulk ZnO, and $ZnCl_2$ to a freshwater microalga (Pseudokirchneriella subcapitata): The importance of particle solubility, *Environmental Science & Technology*, 41(24): 8484–8490.

Fridovich, I. 1995. Superoxide radical and superoxide dismutases, *Annual Review of Biochemistry*, 64(1): 97–112.

Fu, C., F. He, L. Tan, X. Ren, W. Zhang, T. Liu, J. Wang, J. Ren, X. Chen, and X. Meng. 2017a. MoS_2 nanosheets encapsulated in sodium alginate microcapsules as microwave embolization agents for large orthotopic transplantation tumor therapy, *Nanoscale*, 9(39): 14846–14853.

Fu, H., Z. Li, H. Xie, Z. Sun, B. Wang, H. Huang, G. Han, H. Wang, P. K. Chu, and X.-F. Yu. 2017b. Different-sized black phosphorus nanosheets with good cytocompatibility and high photothermal performance, *RSC Advances*, 7(24): 14618–14624.

Galluzzi, L., N. Joza, E. Tasdemir, M. Maiuri, M. Hengartner, J. Abrams, N. Tavernarakis, J. Penninger, F. Madeo, and G. Kroemer. 2008. No death without life: Vital functions of apoptotic effectors, *Cell Death & Differentiation*, 15(7): 1113–1123.

Ganguly, P., A. Breen, and S. C. Pillai. 2018. Toxicity of nanomaterials: Exposure, pathways, assessment, and recent advances, *ACS Biomaterials Science & Engineering*, 4(7): 2237–2275.

Gao, Y., J. Feng, L. Kang, X. Xu, and L. Zhu. 2018. Concentration addition and independent action model: Which is better in predicting the toxicity for metal mixtures on zebrafish larvae, *Science of the Total Environment*, 610: 442–450.

Garrido, C. and G. Kroemer. 2004. Life's smile, death's grin: Vital functions of apoptosis-executing proteins, *Current Opinion in Cell Biology*, 16(6): 639–646.

George, S., T. Xia, R. Rallo, Y. Zhao, Z. Ji, S. Lin, X. Wang, H. Zhang, B. France, and D. Schoenfeld. 2011. Use of a high-throughput screening approach coupled with in vivo zebrafish embryo screening to develop hazard ranking for engineered nanomaterials, *ACS Nano*, 5(3): 1805–1817.

Gleeson, M. P., S. Modi, A. Bender, R. L Marchese Robinson, J. Kirchmair, M. Promkatkaew, S. Hannongbua, and R. C Glen. 2012. The challenges involved in modeling toxicity data in silico: A review, *Current Pharmaceutical Design*, 18(9): 1266–1291.

Goldstein, J. L., R. G. Anderson, and M. S. Brown. 1979. Coated pits, coated vesicles, and receptor-mediated endocytosis, *Nature*, 279(5715): 679–685.

Golstein, P. and G. Kroemer. 2007. Cell death by necrosis: Towards a molecular definition, *Trends in Biochemical Sciences*, 32(1): 37–43.

González-Polo, R.-A., P. Boya, A.-L. Pauleau, A. Jalil, N. Larochette, S. Souquère, E.-L. Eskelinen, G. Pierron, P. Saftig, and G. Kroemer. 2005. The apoptosis/autophagy paradox: Autophagic vacuolization before apoptotic death, *Journal of Cell Science*, 118(14): 3091–3102.

Gottschalk, F., T. Sonderer, R. W. Scholz, and B. Nowack. 2009. Modeled environmental concentrations of engineered nanomaterials (TiO_2, ZnO, Ag, CNT, fullerenes) for different regions, *Environmental Science & Technology*, 43(24): 9216–9222.

Guiney, L. M., X. Wang, T. Xia, A. E. Nel, and M. C. Hersam. 2018. Assessing and mitigating the hazard potential of two-dimensional materials, *ACS Nano*, 12(7): 6360–6377.

Hindman, B. and Q. Ma. 2018. Carbon nanotubes and crystalline silica induce matrix remodeling and contraction by stimulating myofibroblast transformation in a three-dimensional culture of human pulmonary fibroblasts: Role of dimension and rigidity, *Archives of Toxicology*, 1–15.

Indik, Z. K., J.-G. Park, S. Hunter, and A. Schreiber. 1995. The molecular dissection of Fc gamma receptor mediated phagocytosis, *Blood*, 86(12): 4389–4399.

Jakus, A. E., S. L. Taylor, N. R. Geisendorfer, D. C. Dunand, and R. N. Shah. 2015. Metallic architectures from 3D-printed powder-based liquid inks, *Advanced Functional Materials*, 25(45): 6985–6995.

Jornot, L., H. Petersen, and A. F. Junod. 1998. Hydrogen peroxide-induced DNA damage is independent of nuclear calcium but dependent on redox-active ions, *Biochemical Journal*, 335(1): 85–94.

Ju, E., K. Dong, Z. Chen, Z. Liu, C. Liu, Y. Huang, Z. Wang, F. Pu, J. Ren, and X. Qu. 2016. Copper (II)-Graphitic carbon nitride triggered synergy: Improved ROS generation and reduced glutathione levels for enhanced photodynamic therapy, *Angewandte Chemie*, 128(38): 11639–11643.

Kerr, J. F., A. H. Wyllie, and A. R. Currie. 1972. Apoptosis: A basic biological phenomenon with wide-ranging implications in tissue kinetics, *British Journal of Cancer*, 26(4): 239.

Khan, A. H., S. Ghosh, B. Pradhan, A. Dalui, L. K. Shrestha, S. Acharya, and K. Ariga. 2017. Two-dimensional (2D) nanomaterials towards electrochemical nanoarchitectonics in energy-related applications, *Bulletin of the Chemical Society of Japan*, 90(6): 627–648.

Kroemer, G., L. Galluzzi, P. Vandenabeele, J. Abrams, E. Alnemri, E. Baehrecke, M. Blagosklonny, W. El-Deiry, P. Golstein, and D. Green. 2009. Classification of cell death: Recommendations of the Nomenclature Committee on Cell Death 2009, *Cell Death & Differentiation*, 16(1): 3–11.

Kumagai, Y., T. Arimoto, M. Shinyashiki, N. Shimojo, Y. Nakai, T. Yoshikawa, and M. Sagai. 1997. Generation of reactive oxygen species during interaction of diesel exhaust particle components with NADPH-cytochrome P450 reductase and involvement of the bioactivation in the DNA damage, *Free Radical Biology and Medicine*, 22(3): 479–487.

Kurapati, R., L. Muzi, A. P. R. De Garibay, J. Russier, D. Voiry, I. A. Vacchi, M. Chhowalla, and A. Bianco. 2017. Enzymatic biodegradability of pristine and functionalized transition metal dichalcogenide MoS_2 nanosheets, *Advanced Functional Materials*, 27(7): 1605176.

Lalwani, G., A. M. Henslee, B. Farshid, L. Lin, F. K. Kasper, Y.-X. Qin, A. G. Mikos, and B. Sitharaman. 2013. Two-dimensional nanostructure-reinforced biodegradable polymeric nanocomposites for bone tissue engineering, *Biomacromolecules*, 14(3): 900–909.

Laomettachit, T., I. Puri, and M. Liangruksa. 2017. A two-step model of TiO_2 nanoparticle toxicity in human liver tissue, *Toxicology and Applied Pharmacology*, 334: 47–54.

Latiff, N., W. Z. Teo, Z. Sofer, Š. Huber, A. C. Fisher, and M. Pumera. 2015. Toxicity of layered semiconductor chalcogenides: Beware of interferences, *RSC Advances*, 5(83): 67485–67492.

Lee, U., C. J. Yoo, Y. J. Kim, and Y. M. Yoo. 2016. Cytotoxicity of gold nanoparticles in human neural precursor cells and rat cerebral cortex, *Journal of Bioscience and Bioengineering*, 121(3), 341–344.

Levine, B. and G. Kroemer. 2008. Autophagy in the pathogenesis of disease, *Cell*, 132(1): 27–42.

Lewinski, N., V. Colvin, and R. Drezek. 2008. Cytotoxicity of nanoparticles, *Small*, 4(1): 26–49.

Li, B. L., M. I. Setyawati, L. Chen, J. Xie, K. Ariga, C.-T. Lim, S. Garaj, and D. T. Leong. 2017. Directing assembly and disassembly of 2D MoS_2 nanosheets with DNA for drug delivery, *ACS Applied Materials & Interfaces*, 9(18): 15286–15296.

Li, N., J. Alam, M. I. Venkatesan, A. Eiguren-Fernandez, D. Schmitz, E. Di Stefano, N. Slaughter, E. Killeen, X. Wang, and A. Huang. 2004. Nrf2 is a key transcription factor that regulates antioxidant defense in macrophages and epithelial cells: protecting against the proinflammatory and oxidizing effects of diesel exhaust chemicals, *The Journal of Immunology*, 173(5): 3467–3481.

Li, N., M. Wang, T. D. Oberley, J. M. Sempf, and A. E. Nel. 2002. Comparison of the pro-oxidative and proinflammatory effects of organic diesel exhaust particle chemicals in bronchial epithelial cells and macrophages, *The Journal of Immunology*, 169(8): 4531–4541.

Li, Y., H. Bai, H. Wang, Y. Shen, G. Tang, and Y. Ping. 2018. Reactive oxygen species (ROS)-responsive nanomedicine for RNAi-based cancer therapy, *Nanoscale*, 10(1): 203–214.

Limbach, L. K., P. Wick, P. Manser, R. N. Grass, A. Bruinink, and W. J. Stark. 2007. Exposure of engineered nanoparticles to human lung epithelial cells: Influence of chemical composition and catalytic activity on oxidative stress, *Environmental Science & Technology*, 41(11): 4158–4163.

Liu, J., Y. Sun, D. G. Drubin, and G. F. Oster. 2001. Clathrin-mediated endocytosis. In *Endocytosis*, M. Marsh (Ed.), Oxford, UK: Oxford University Press.
Liu, S., Z. Shen, B. Wu, Y. Yu, H. Hou, X.-X. Zhang, and H.-Q. Ren. 2017a. Cytotoxicity and efflux pump inhibition induced by molybdenum disulfide and boron nitride nanomaterials with sheet-like structure, *Environmental Science & Technology*, 51(18): 10834–10842.
Liu, Y., X. Ji, J. Liu, W. W. Tong, D. Askhatova, and J. Shi. 2017b. Tantalum sulfide nanosheets as a theranostic nanoplatform for computed tomography imaging-guided combinatorial chemo-photothermal therapy, *Advanced Functional Materials*, 27(39): 1703261.
Liu, Y. and J. Liu. 2017. Hybrid nanomaterials of WS_2 or MoS_2 nanosheets with liposomes: Biointerfaces and multiplexed drug delivery, *Nanoscale*, 9(35): 13187–13194.
Mahmoudi, M., A. M. Abdelmonem, S. Behzadi, J. H. Clement, S. Dutz, M. R. Ejtehadi, R. Hartmann, K. Kantner, U. Linne, and P. Maffre. 2013. Temperature: The 'ignored' factor at the nanobio interface, *ACS Nano*, 7(8): 6555–6562.
Majno, G. and I. Joris. 1995. Apoptosis, oncosis, and necrosis. An overview of cell death, *The American Journal of Pathology*, 146(1): 3–15.
McMahon, H. T. and E. Boucrot. 2011. Molecular mechanism and physiological functions of clathrin-mediated endocytosis, *Nature Reviews Molecular Cell Biology*, 12(8): 517–533.
McManus, D., S. Vranic, F. Withers, V. Sanchez-Romaguera, M. Macucci, H. Yang, R. Sorrentino, K. Parvez, S.-K. Son, and G. Iannaccone. 2017. Water-based and biocompatible 2D crystal inks for all-inkjet-printed heterostructures, *Nature Nanotechnology*, 12(4): 343.
Medzhitov, R. 2008. Origin and physiological roles of inflammation, *Nature*, 454(7203): 428–435.
Meng, H., T. Xia, S. George, and A. E. Nel. 2009. A predictive toxicological paradigm for the safety assessment of nanomaterials, *ACS Nano*, 3(7): 1620–1627.
Mizushima, N. 2005. The pleiotropic role of autophagy: From protein metabolism to bactericide, *Cell Death and Differentiation*, 12(S2): 1535–1541.
Mizushima, N. 2007. Autophagy: Process and function, *Genes Development*, 21(22): 2861–2873.
Moore, C., D. Movia, R. J. Smith, D. Hanlon, F. Lebre, E. C. Lavelle, H. J. Byrne, J. N. Coleman, Y. Volkov, and J. McIntyre. 2017. Industrial grade 2D molybdenum disulphide (MoS_2): An in vitro exploration of the impact on cellular uptake, cytotoxicity, and inflammation, *2D Materials*, 4(2): 025065.
Motley, A., N. A. Bright, M. N. Seaman, and M. S. Robinson. 2003. Clathrin-mediated endocytosis in AP-2–depleted cells, *The Journal of Cell Biology*, 162(5): 909–918.
Nabi, I. R. and P. U. Le. 2003. Caveolae/raft-dependent endocytosis, *The Journal of Cell Biology*, 161(4): 673–677.
Naguib, M. and Y. Gogotsi. 2015. Synthesis of two-dimensional materials by selective extraction, *Accounts of Chemical Research*, 48(1): 128–135.
Nel, A., T. Xia, L. Mädler, and N. Li. 2006. Toxic potential of materials at the nanolevel, *Science*, 311(5761): 622–627.
Novoselov, K., D. Jiang, F. Schedin, T. Booth, V. Khotkevich, S. Morozov, and A. Geim. 2005. Two-dimensional atomic crystals, *Proceedings of the National Academy of Sciences of the United States of America*, 102(30): 10451–10453.
Nunes, T., C. Bernardazzi, and H. S. de Souza. 2014. Cell death and inflammatory bowel diseases: Apoptosis, necrosis, and autophagy in the intestinal epithelium, *BioMed Research International*, 2014.
Oberdörster, G., A. Maynard, K. Donaldson, V. Castranova, J. Fitzpatrick, K. Ausman, J. Carter, B. Karn, W. Kreyling, and D. Lai. 2005. Principles for characterizing the potential human health effects from exposure to nanomaterials: Elements of a screening strategy, *Particle and Fibre Toxicology*, 2(1): 8.
Oh, N. and J.-H. Park. 2014. Endocytosis and exocytosis of nanoparticles in mammalian cells, *International Journal of Nanomedicine*, 9(Suppl 1): 51–63.

Park, E.-M., Y.-M. Park, and Y.-S. Gwak. 1998. Oxidative damage in tissues of rats exposed to cigarette smoke, *Free Radical Biology and Medicine*, 25(1): 79–86.

Pelkmans, L., E. Fava, H. Grabner, M. Hannus, B. Habermann, E. Krausz, and M. Zerial. 2005. Genome-wide analysis of human kinases in clathrin-and caveolae/raft-mediated endocytosis, *Nature*, 436(7047): 78–86.

Podila, R. and J. M. Brown. 2013. Toxicity of engineered nanomaterials: A physicochemical perspective, *Journal of Biochemical and Molecular Toxicology*, 27(1): 50–55.

Pomerantseva, E. and Y. Gogotsi. 2017. Two-dimensional heterostructures for energy storage, *Nature Energy*, 2(7): 17089.

Powers, K. W., S. C. Brown, V. B. Krishna, S. C. Wasdo, B. M. Moudgil, and S. M. Roberts. 2006. Research strategies for safety evaluation of nanomaterials. Part VI. Characterization of nanoscale particles for toxicological evaluation, *Toxicological Sciences*, 90(2): 296–303.

Qu, G., W. Liu, Y. Zhao, J. Gao, T. Xia, J. Shi, L. Hu, W. Zhou, J. Gao, and H. Wang. 2017. Improved biocompatibility of black phosphorus nanosheets by chemical modification, *Angewandte Chemie*, 129(46): 14680–14685.

Rabinovitch, M. 1995. Professional and non-professional phagocytes: An introduction, *Trends in Cell Biology*, 5(3): 85–87.

Rahman, I., B. Mulier, P. S. Gilmour, T. Watchorn, K. Donaldson, P. K. Jeffery, and W. MacNee. 2001. Oxidant-mediated lung epithelial cell tolerance: The role of intracellular glutathione and nuclear factor-kappaB, *Biochemical Pharmacology*, 62(6): 787–794.

Rejman, J., V. Oberle, I. S. Zuhorn, and D. Hoekstra. 2004. Size-dependent internalization of particles via the pathways of clathrin-and caveolae-mediated endocytosis, *Biochemical Journal*, 377(1): 159–169.

Richard, J. P., K. Melikov, H. Brooks, P. Prevot, B. Lebleu, and L. V. Chernomordik. 2005. Cellular uptake of unconjugated TAT peptide involves clathrin-dependent endocytosis and heparan sulfate receptors, *Journal of Biological Chemistry*, 280(15): 15300–15306.

Roach, H. and N. Clarke. 2000. Physiological cell death of chondrocytes in vivo is not confined to apoptosis, *Bone & Joint Journal*, 82(4): 601–613.

Roggen, E. L. 2011. In vitro toxicity testing in the twenty-first century, *Frontiers in Pharmacology*, 2: 3.

Saha, K., S. T. Kim, B. Yan, O. R. Miranda, F. S. Alfonso, D. Shlosman, and V. M. Rotello. 2013. Surface functionality of nanoparticles determines cellular uptake mechanisms in mammalian cells, *Small*, 9(2): 300–305.

Sbarra, A. J. and M. L. Karnovsky. 1959. The biochemical basis of phagocytosis I. Metabolic changes during the ingestion of particles by polymorphonuclear leukocytes, *Journal of Biological Chemistry*, 234(6): 1355–1362.

Shah, P., T. N. Narayanan, C.-Z. Li, and S. Alwarappan. 2015. Probing the biocompatibility of MoS_2 nanosheets by cytotoxicity assay and electrical impedance spectroscopy, *Nanotechnology*, 26(31): 315102.

Sharifi, S., S. Behzadi, S. Laurent, M. L. Forrest, P. Stroeve, and M. Mahmoudi. 2012. Toxicity of nanomaterials, *Chemical Society Reviews*, 41(6): 2323–2343.

Sikora, B., P. Kowalik, J. Mikulski, K. Fronc, I. Kaminska, M. Szewczyk, A. Konopka et al. 2017. Mammalian cell defence mechanisms against the cytotoxicity of $NaYF_4$:(Er, Yb, Gd) nanoparticles, *Nanoscale*, 9(37): 14259–14271.

Singh, R. D., V. Puri, J. T. Valiyaveettil, D. L. Marks, R. Bittman, and R. E. Pagano. 2003. Selective caveolin-1–dependent endocytosis of glycosphingolipids, *Molecular Biology of the Cell*, 14(8): 3254–3265.

Singh, V., D. Joung, L. Zhai, S. Das, S. I. Khondaker, and S. Seal. 2011. Graphene based materials: Past, present and future, *Progress in Materials Science*, 56(8): 1178–1271.

Sizochenko, N., A. Mikolajczyk, K. Jagiello, T. Puzyn, J. Leszczynski, and B. Rasulev. 2018. How the toxicity of nanomaterials towards different species could be simultaneously evaluated: A novel multi-nano-read-across approach, *Nanoscale*, 10(2); 582–591.

Song, S.-J., Y. Shin, H. Lee, B. Kim, D.-W. Han, and D. Lim. 2018a. Dose-and time-dependent cytotoxicity of layered black phosphorus in fibroblastic cells, *Nanomaterials*, 8(6): 408.

Song, Y., X. Yan, G. Ostermeyer, S. Li, L. Qu, D. Du, Z. Li, and Y. Lin. 2018b. Direct cytosolic microRNA detection using single-layer perfluorinated tungsten diselenide nanoplatform, *Analytical Chemistry*, 90(17): 10369–10376.

Sorkin, A. and M. A. Puthenveedu. 2013. Clathrin-mediated endocytosis, in *Vesicle Trafficking in Cancer*, 1–31: Springer.

Stahl, W., A. Junghans, B. de Boer, E. S. Driomina, K. Briviba, and H. Sies. 1998. Carotenoid mixtures protect multilamellar liposomes against oxidative damage: Synergistic effects of lycopene and lutein, *FEBS Letters*, 427(2): 305–308.

Steinman, R. M., S. E. Brodie, and Z. A. Cohn. 1976. Membrane flow during pinocytosis, *The Journal of Cell Biology*, 68: 665–687.

Stockert, J. C., A. Blázquez-Castro, M. Cañete, R. W. Horobin, and Á. Villanueva. 2012. MTT assay for cell viability: Intracellular localization of the formazan product is in lipid droplets, *Acta Histochemica*, 114(8): 785–796.

Stossel, T. P. 1974. Phagocytosis, *New England Journal of Medicine*, 290(13): 717–723.

Suhito, I. R., Y. Han, D.-S. Kim, H. Son, and T.-H. Kim. 2017. Effects of two-dimensional materials on human mesenchymal stem cell behaviors, *Biochemical and Biophysical Research Communications*, 493(1): 578–584.

Sun, X., Y. Yang, J. Shi, C. Wang, Z. Yu, and H. Zhang. 2017. NOX4-and Nrf2-mediated oxidative stress induced by silver nanoparticles in vascular endothelial cells, *Journal of Applied Toxicology*, 37(12): 1428–1437.

Tao, W., X. Zhu, X. Yu, X. Zeng, Q. Xiao, X. Zhang, X. Ji, X. Wang, J. Shi, and H. Zhang. 2017. Black phosphorus nanosheets as a robust delivery platform for cancer theranostics, *Advanced Materials*, 29(1): 1603276.

Tasdemir, E., L. Galluzzi, M. C. Maiuri, A. Criollo, I. Vitale, E. Hangen, N. Modjtahedi, and G. Kroemer. 2008. Methods for assessing autophagy and autophagic cell death, *Autophagosome and Phagosome*, 29–76.

Teo, W. Z., E. L. K. Chng, Z. Sofer, and M. Pumera. 2014. Cytotoxicity of exfoliated transition-metal dichalcogenides (MoS_2, WS_2, and WSe_2) is lower than that of graphene and its analogues, *Chemistry-A European Journal*, 20(31): 9627–9632.

Tian, J., Y. Sang, G. Yu, H. Jiang, X. Mu, and H. Liu. 2013. A Bi_2WO_6-based hybrid photocatalyst with broad spectrum photocatalytic properties under UV, visible, and near-infrared irradiation, *Advanced Materials*, 25(36): 5075–5080.

Tu, Z., V. Wycisk, C. Cheng, W. Chen, M. Adeli, and R. Haag. 2017. Functionalized graphene sheets for intracellular controlled release of therapeutic agents, *Nanoscale*, 9(47): 18931–18939.

Viswanath, B. and S. Kim. 2016. Influence of nanotoxicity on human health and environment: The alternative strategies, *Reviews of Environmental Contamination and Toxicology Volume* 242, 61–104.

Wang, N., F. Wei, Y. Qi, H. Li, X. Lu, G. Zhao, and Q. Xu. 2014. Synthesis of strongly fluorescent molybdenum disulfide nanosheets for cell-targeted labeling, *ACS Applied Materials & Interfaces*, 6(22): 19888–19894.

Wang, S., X. Li, Y. Chen, X. Cai, H. Yao, W. Gao, Y. Zheng, X. An, J. Shi, and H. Chen. 2015a. A facile one-pot synthesis of a two-dimensional MoS_2/Bi_2S_3 composite theranostic nanosystem for multi-modality tumor imaging and therapy, *Advanced Materials*, 27(17): 2775–2782.

Wang, S., J. Qiu, W. Guo, X. Yu, J. Nie, J. Zhang, X. Zhang, Z. Liu, X. Mou, and L. Li. 2017a. A nanostructured molybdenum disulfide film for promoting neural stem cell neuronal differentiation: Toward a nerve tissue-engineered 3D scaffold, *Advanced Biosystems*, 1(5): 1600042.

Wang, X., T. Li, H. Ma, D. Zhai, C. Jiang, J. Chang, J. Wang, and C. Wu. 2017b. A 3D-printed scaffold with MoS_2 nanosheets for tumor therapy and tissue regeneration, *NPG Asia Materials*, 9(4): e376.

Wang, X., N. D. Mansukhani, L. M. Guiney, Z. Ji, C. H. Chang, M. Wang, Y. P. Liao, T. B. Song, B. Sun, and R. Li. 2015b. Differences in the toxicological potential of 2D versus aggregated molybdenum disulfide in the lung, *Small*, 11(38): 5079–5087.

Wataha, J. C., C. Hanks, and R. G. Craig. 1991. The in vitro effects of metal cations on eukaryotic cell metabolism, *Journal of Biomedical Materials Research*, 25(9): 1133–1149.

Wen, Y., W. Zhang, N. Gong, Y.-F. Wang, H.-B. Guo, W. Guo, P. C. Wang, and X.-J. Liang. 2017. Carrier-free, self-assembled pure drug nanorods composed of 10-hydroxycamptothecin and chlorin e6 for combinatorial chemo-photodynamic antitumor therapy in vivo, *Nanoscale*, 9(38): 14347–14356.

Yin, W., L. Yan, J. Yu, G. Tian, L. Zhou, X. Zheng, X. Zhang, Y. Yong, J. Li, and Z. Gu. 2014. High-throughput synthesis of single-layer MoS_2 nanosheets as a near-infrared photothermal-triggered drug delivery for effective cancer therapy, *ACS Nano*, 8(7): 6922–6933.

Zhang, Y., W. Xiu, Y. Sun, D. Zhu, Q. Zhang, L. Yuwen, L. Weng, Z. Teng, and L. Wang. 2017. RGD-QD-MoS_2 nanosheets for targeted fluorescent imaging and photothermal therapy of cancer, *Nanoscale*, 9(41): 15835–15845.

Zhao, F., Y. Zhao, Y. Liu, X. Chang, C. Chen, and Y. Zhao. 2011. Cellular uptake, intracellular trafficking, and cytotoxicity of nanomaterials, *Small*, 7(10): 1322–1337.

2

Effects of Nanocrystal Morphologies on Cytotoxicity

Aine M. Whelan

2.1 Introduction to Nanomaterials and Morphologies

Nanoparticle (NP) shape has a profound effect on many properties of nanomaterials, including their optical, electrical, and magnetic properties. Thus, control of shape is of paramount importance when utilizing NPs for applications such as biological imaging or drug delivery. While the medical applications of shaped NPs will be discussed in greater detail later in this chapter, it is necessary first to consider how NPs of controlled morphology may be prepared.

A number of methods have been reported for synthesis of particles of controlled shape. The polyol reduction method has been used to prepare Ag and Au NPs of defined morphologies and optical properties (Kim et al. 2004, Wiley et al. 2005). This process involves the thermal reduction of the metal salt in poly(ethylene glycol) solvent, in the presence of a polymeric stabiliser, usually poly(vinyl pyrrolidone). The adsorption of the stabiliser molecules to particular crystal facets in the growing particles has been identified as the critical factor in determining the final particle morphology (Enrique et al. 2007). The drawback of the polyol method is that the particles are prepared in organic solvent, rather than in the aqueous phase. Generally, for biological applications of nanomaterials, it is desirable to have NPs in aqueous suspension.

Seed-mediated methods have been used to prepare gold nanorods in aqueous solution (Jana et al. 2001b, Pérez-Juste et al. 2005). The crystalline nature of the seed particle was found to influence the yield of nanorods obtained, with citrate-capped multiply twinned gold seeds resulting in lower yields than when cetyl trimethyl ammonium bromide (CTAB)-capped single crystal gold seeds were used.

This seeding method has been further adapted for the synthesis of silver nanowires and nanorods (Jana et al. 2001a) and triangular silver nanoplates (Ledwith et al. 2007).

The synthesis of CdSe NPs of controlled morphology was achieved by injection of a tri-alkyl phosphine solution of dimethyl cadmium and selenium powder into hot trioctyl phosphine oxide (TOPO) with addition of hexyl phosphonic acid (Peng et al. 2000). Manipulation of the growth kinetics allowed for the synthesis of CdSe nanorods – wurtzite CdSe possesses an intrinsic anisotropy due to its unique c-axis. Therefore, when an extremely high monomer concentration is present, growth is generally faster along this axis, promoting the growth of rod-shaped particles.

A general method for the preparation of CdS, CdSe, CdTe quantum rods, and quantum rod heterostructures was devised by Shieh et al. (Shieh et al. 2005). A Cd-n-tetradecylphosphonic acid (TDPA) complex was synthesized and then mixed with the desired chalcogen-trioctylphosphine (TOP) precursor. The particle shape was varied from spherical to rodlike by periodic injection of the chalcogen TOP precursor, with CdTe and CdSe nanorods of highest aspect ratio obtained after five injections.

Generally, semiconductor nanocrystals of defined morphology can be prepared by the use of shaping ligands, which can exert their effects by either (i) preferentially coordinating with surface atoms on specific crystal planes, thus modifying the surface energy and chemical reactivity of these planes, or by (ii) preferentially binding to certain crystal planes and sterically inhibiting the growth of these planes (Sandeep and Thomas 2006).

Template-assisted synthetic methods have been used to prepare a variety of anisotropic magnetic particles. For example, nanoporous aluminium oxide and polycarbonate membranes were used to direct the growth of magnetic metallic and metal oxide NPs of high aspect ratio (Ji et al. 2003, Sui et al. 2004, Byrne et al. 2009). Alternatively, magnetically directed assembly can be used to prepare magnetic nanochains and nanosheets, as illustrated in Figure 2.1. Once the chains are assembled, they can be coated with silica or polymer in order to improve the mechanical strength of the chain.

The morphology of noble metal NPs has a critical effect on their optical properties, with differently shaped silver NPs exhibiting different UV-visible absorption spectra as illustrated in Figure 2.2.

Similarly to silver nanorods, gold nanorods also display multiple peaks in their absorption spectra with two separate surface plasmon resonance bands, as well as transverse (TSPR) and longitudinal plasmon bands (LSPR) (Zeng et al. 2011). The transverse band and the longitudinal band are observed in the visible region and in the near-infrared (NIR) region, respectively (Fei et al. 2014). The fact that these materials absorb in the NIR means that the maximum amount of light can penetrate tissues, thus making such materials suitable for *in vivo* applications. Additionally, gold nanorods can convert light energy into heat by local SPR -this allows the application of these materials in photothermal therapy (Fei et al. 2014). The use of NPs for photothermal therapy will be discussed in greater detail later in this chapter.

The shape of NPs has also been shown to affect the rates of endocytosis of these materials. Molecular modelling was used to simulate the lipid membrane translocation process for gold NPs of

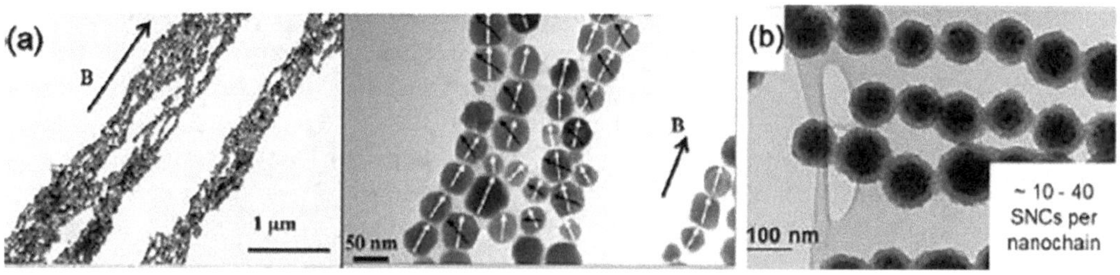

FIGURE 2.1 (a) Nanochains synthesised by alignment of magnetic cobalt NPs in a magnetic field. The black arrows indicate the crystallographic [1 1 1] direction and the small white arrows indicate the orientation of the particles' magnetic moment. (b) Chains prepared by alignment of silica-coated iron oxide spheres. (Reproduced from Lisjak, D. and Mertelj, A., *Prog. Mater. Sci.*, 95, 286–328, 2018. With permission.)

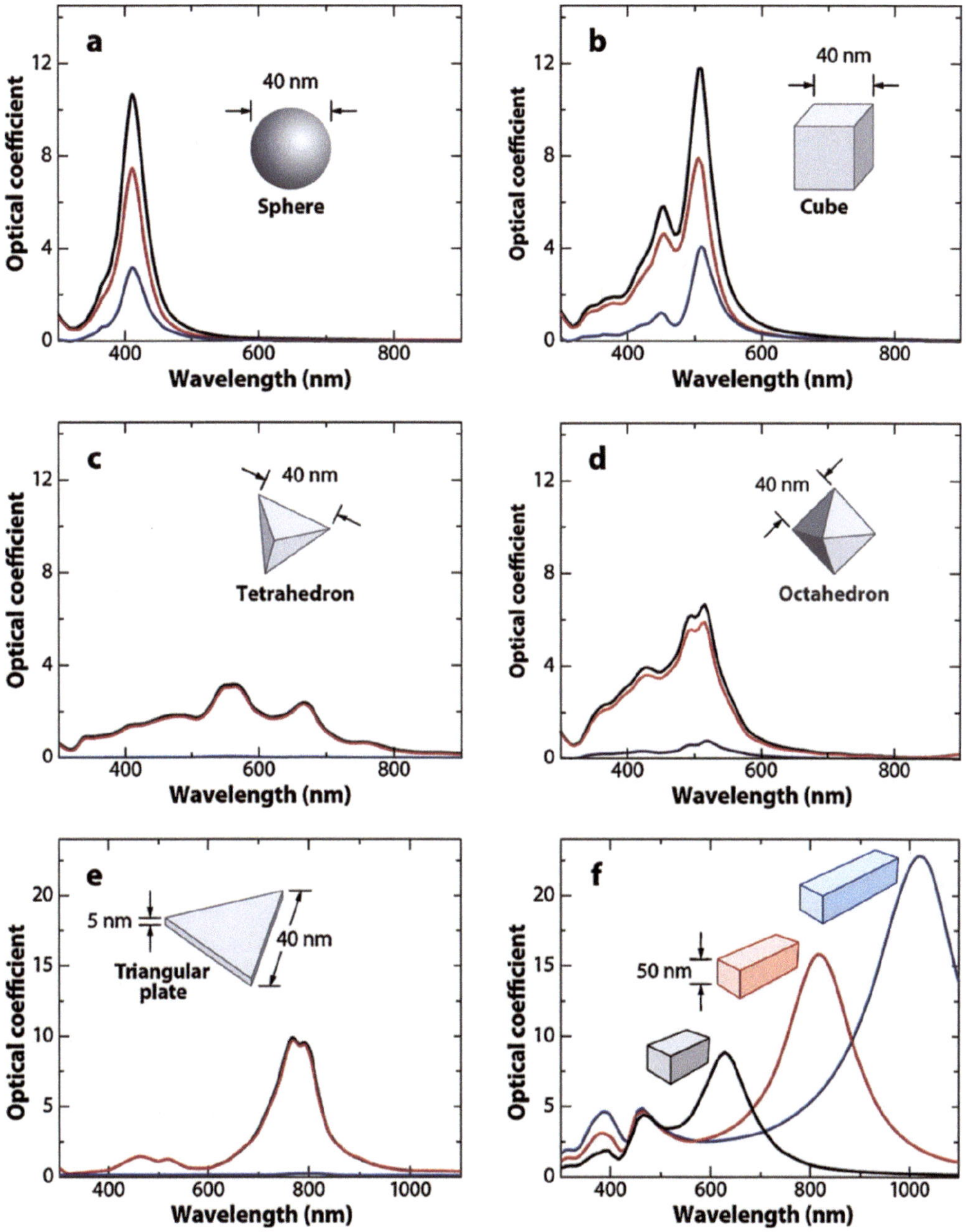

FIGURE 2.2 Extinction (black), absorption (red), and scattering (blue) spectra calculated for silver NPs of various morphologies: (a) sphere, (b) cube, (c) tetrahedron, (d) octahedron, and (e) triangular plate. (f) Extinction spectra of silver rods with aspect ratios of 2 (black), 3 (red), and 4 (blue). (Reproduced from Petryayeva, E. and Krull, E.U.J., *Anal. Chim. Acta*, 706, 8–24, 2011. With permission.)

various morphologies (cone, cube, rod, rice, pyramidal, and sphere shaped, as shown in Figure 2.3) (Nangia and Sureshkumar 2012). The rate constants and half lives for internalization of the different shapes were calculated and it was found that the rice shape was most easily internalized, followed by the spherical particles ($t_{1/2}$ = 4.3 ms), while the cube-shaped particles took the longest time to cross the cell membrane ($t_{1/2}$ = 8.7 s). It was suggested that the presence of facets on the particles

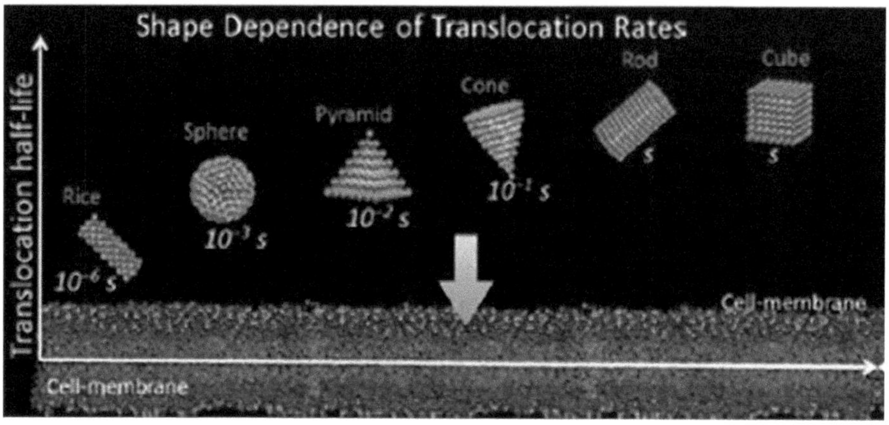

FIGURE 2.3 The six NP morphologies used in the molecular modelling of lipid membrane translocation process. (Reproduced from Nangia, S. and Sureshkumar, R., *Langmuir*, 28, 17666–17671, 2012. With permission.)

promotes stronger adhesion of these particles to the cell membrane by providing interacting flat index planes for particle-membrane interactions.

Particle geometry also has implications for the properties of semiconductor NPs (quantum dots). A study of the optical properties of spherical and elongated nanocrystals showed that the nanorods had enhanced absorption cross sections relative to the spheres and, therefore, improved photostability (Htoon et al. 2003). Particle shape was also found to affect the bleach recovery rates of CdSe nanospheres vs nanorods, due to the larger number of excitation states in the rods (Mohamed et al. 2001). Furthermore, while spherical quantum dots exhibit plane polarized emission, nanorods have polarized emission along the particle long axis (Hu et al. 2001). Finally, it has been shown that the emission of CdSe nanorods can be switched on and off via application of an external electric field (Rothenberg et al. 2005).

Particle shape has also been shown to affect the magnetic properties of NPs. Elongated magnetic NPs display an anisotropic response to a magnetic field (Pečko et al. 2015). When the magnetic field is applied parallel to the particle's long axis, a square-like hysteresis loop is observed with large coercivity (H_c) and the saturation magnetization (M_s) equals the remanent magnetization (M_r). However, when the field is applied perpendicular to the long axis of the particle, hysteresis is not observed and $M_r << M_s$.

2.2 Medical Applications of Nanoparticles of Varying Morphologies

There has been a wide range of uses of shaped NPs in medicine, including imaging, photothermal therapy, hyperthermal therapy, drug delivery, and diagnostics.

2.2.1 Imaging Applications

NPs offer many advantages for improved optical imaging of cells – semiconductor NPs (quantum dots) are brighter and more photostable than organic fluorescent dyes and the use of metal NPs can also allow the SERS effect to be exploited. Suarasan and coworkers explored the use of gold NPs of different shapes (spheres, nanorods, and triangles) for imaging of NIH:OVCAR-3 ovary cancer cells (Suarasan et al. 2018). Gold NPs of different shapes were prepared by seed mediated synthesis, then coated in gelatine to improve their biocompatibility. The particles were incubated with the cells for 24 hours, then two-photon excitation fluorescence lifetime imaging microscopy (TPE FLIM) was used to examine the cells. The advantage of this technique is that NIR laser irradiation can be used for the visualization of the cells. This avoids the cell autofluorescence or photodamage that can occur when cells are irradiated with other wavelengths. Interestingly, it was observed that the strongest TPE photoluminescence was obtained from the gold nanotubes while the spherical NPs showed the weakest intensity photoluminescence. This enhancement of luminescence intensity was attributed to the 'lightning rod effect' (Imura et al. 2006).

Optical coherence tomography is a technique which uses NIR light for real-time imaging in tissues up to a few millimetres in depth. It relies on variation in optical scattering and absorption. While organic dyes can be used as spectroscopic optical coherence tomography (SCOT) contrast agents, they are not ideal due to issues with chemical and optical stability. The use of gold nanorods as an alternative SCOT contrast agent for imaging of excised breast carcinoma was investigated (Oldenburg et al. 2009). The nanorods absorption could be tuned to NIR frequencies (755 nm) by variation of the aspect ratio. The nanorods were of average length 50 nm, which allowed permeation into the tumour tissue.

Biocompatible surface-silanized rod-shaped CdSe/CdS/ZnS core/shell NPs were developed as fluorescent biological labels for the imaging of live MDA-MB-231 human breast cancer cells (Fu et al. 2007). They were much brighter than the corresponding spherical NPs and thus demonstrated the importance of particle morphology when designing NP-based biological probes.

2.2.2 Photothermal Therapy

The photothermal technique has been used to destroy cancerous tumours. Upon the excitation of a surface plasmon, nonradiative relaxation can occur via a number of mechanisms including electron–phonon and phonon–phonon coupling. This results in the generation of heat that can be transmitted to the area surrounding the NP (Jain et al. 2008). This thermal energy can be used to induce tumour tissue ablation (Neal et al. 2004). The optimum size of gold nanorods to be used in phototherapy has been investigated (Mackey et al. 2014). Three different sizes of gold nanorods (38 × 11, 28 × 8, and 17 × 5 nm) were compared and it was calculated that the 28 × 8 nm gold nanorods would be the most effective for generation of heat via plasmonic means. *In vitro* experiments confirmed that the 28 × 8 nm gold nanorods were the most effective photothermal contrast agent for plasmonic photothermal therapy (PPTT) of human oral squamous cell carcinoma. After 2 minutes irradiation with a NIR cw laser, cell viability was reduced to below 20%. The effectiveness of these nanorods was attributed to this size of particle offering the most efficient heat conversion for the amount of light absorbed.

Chitosan-capped star-shaped gold/silver NPs were synthesized and their potential as photothermal therapeutic agents was investigated (Cheng et al. 2012). Oral cancer cells were incubated with the particles for 24 hours then irradiated with a 150 mW NIR femtosecond pulse laser with a wavelength of 800 nm. These cells underwent necrotic cell death. The authors commented that, compared with gold nanorods, these star-shaped particles provided a greater amount of dissipated heat for tumour ablation.

A nanorod-based agent for targeted cancer cell destruction was created by coating Au-Ag nanorods with aptamers (Huang et al. 2008). In contrast to gold nanoshells or nanorods which required high power laser irradiation (1×10^5–1×10^{10} W/m^2) to induce cell death via photothermal means, the novel aptamer Au-Ag nanorods required lower laser power (8.5×10^4 W/m^2) to induce 93% cell death of target cells. These conjugates were highly selective in a mixed-cell suspension; 87% of control cells remained intact after laser exposure, as indicated in Figure 2.4.

2.2.3 Hyperthermal Therapy

One of the first applications of anisotropic NPs for hyperthermal therapy was demonstrated when magnetic nickel nanowires were exposed to human embryonic kidney cells (HEK-293) *in vitro*. Application

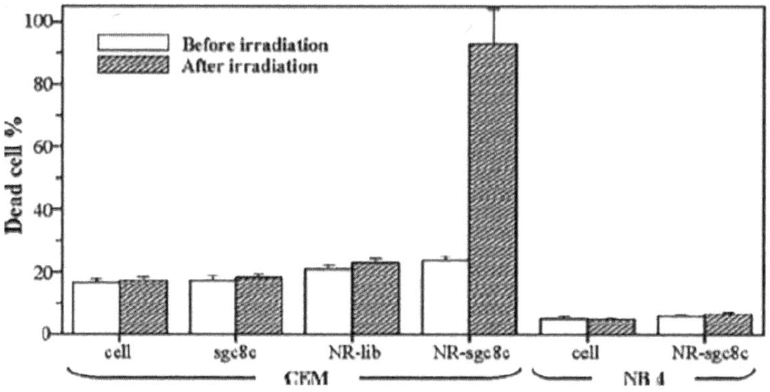

FIGURE 2.4 Comparison of the percentage of dead cells before and after irradiation for NB-4 (control cells) and CCRF-CEM (target cells) without nanorods or labelled with either aptamer sgc8c only (denoted sgc8c in figure) or nanorods only (denoted NR-lib in figure) or aptamer nanorod conjugate NR-sgc8c (denoted NR-sgc8c in figure). (Reprinted with permission from Huang, Y.-F. et al., *Langmuir*, 5, 11860–11865, 2008. Copyright 2008 American Chemical Society.)

of a radiofrequency (RF) electromagnetic field resulted in heating of the nanowires due to magnetic hysteresis. This resulted in hyperthermia-induced cell death in those cells which had internalized the wires (Choi et al. 2008). The authors proposed that the magnetic anisotropy of these particles could be further exploited to improve the heating efficiency.

More recently, it has been shown that the hyperthermia properties of magnetic particles can be tuned by synthesis of particles of different morphologies. It was calculated that cubic-shaped iron oxide NPs would have higher magnetization compared to their spherical counterparts. This was confirmed experimentally by the observation of a significant increase in the specific absorption rate (SAR) in magnetic hyperthermia when using the cubic particles compared to spherical iron oxide NPs (Bauer et al. 2016).

Similarly, alteration of the aspect ratio of iron oxide nanorods was found to yield improvements in the heating efficiency of these materials, with SAR values (862 W/g for an ac field of 800 Oe), considerably larger than those of cubic and spherical NPs of similar volume (approximately 314 and 140 W/g, respectively) (Das et al. 2016). By tuning of the applied ac field, it was possible to rapidly reach the desired temperature range at which cancer cells are more susceptible to heat than healthy cells (40°C–44°C). This suggests that the minimum amount of these NPs can be used to effect hyperthermia for cancer treatment.

2.2.4 Drug Delivery

NPs can accumulate in tumours and thus be used for the delivery of contrast agents or drugs. The enhanced permeation and retention of these materials is due to the presence of large gaps between the endothelial cells of blood vessels in the tumour. However, other factors can affect the degree of fenestrations in tumour vasculature and thus affect the accumulation of NPs in tumours. This is illustrated in Figure 2.5.

It has been shown that NP morphology affects the uptake of these materials into cells; for monodisperse poly(ethylene glycol) hydrogel particles greater than 100 nm in diameter, the highest uptake into HeLa cells was for rods, then spheres, the cylinders, then cubes (Gratton et al. 2008). It was speculated that the higher aspect ratio of the rods compared to other shapes allowed these particles a greater surface area for contact with the cell membrane, thus promoting endocytosis.

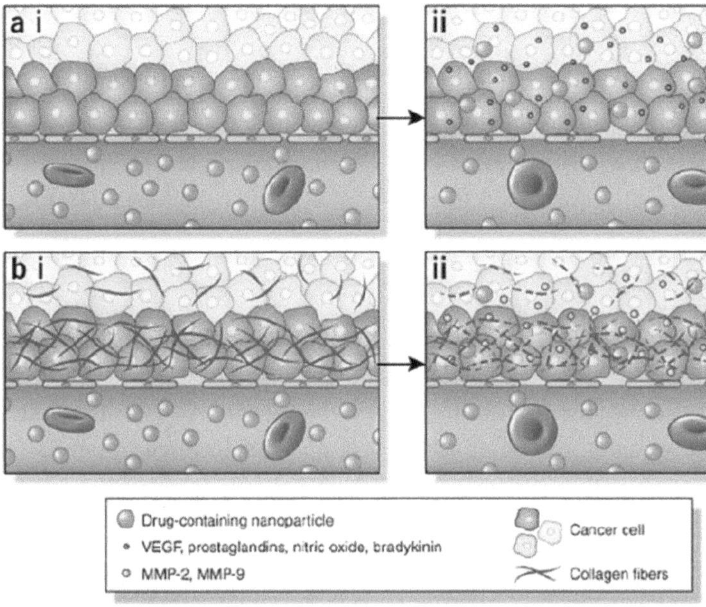

FIGURE 2.5 (a, i) Normal blood vessels typically contain narrow interendothelial junctions that inhibit the entry of particles into tissues. However, in tumour vasculature, the size of fenestrations can vary from normal blood vessels for a number of reasons, e.g., in (a, ii) vascular endothelial growth factor (VEGF) and NO have been shown to enlarge interendothelial junctions (b, i) Although tumours sometimes have a compact extracellular matrix that prevents adequate permeation of NPs into the tumour, this collagen matrix can be degraded by matrix metalloproteinases 2 and 9, as shown in (b, ii). (Reproduced from Blanco, E. et al., *Nat. Biotechnol.*, 33, 941, 2015. With permission.)

Nature has inspired the design of nanomaterials specifically tailored for the delivery of drugs. For example, silica nanocomposites consisting of larger silica NPs (c. 200 nm) coated with smaller silica particles (10 nm) were formulated to mimic the roughened structure of viruses (Niu et al. 2013). The surface topography of these particles can be seen in the SEM images shown in Figure 2.6. It was found that these materials enabled more efficient delivery of siRNA.

Antibodies are widely used for medical applications such as *in vitro* assays, *in vivo* imaging, and targeted delivery of drugs for certain diseases. Many different types of NPs have been utilized as carrier systems for these molecules, including metallic (El-sayed et al. 2006), quantum dots (Bilan et al. 2015), and polymeric (Vivek et al. 2014). It has been shown that the NP shape can play a role in enhancing the specificity of antibody interaction with target antigens (Barua et al. 2013). Polystyrene particles of three different shapes (spheres, rods, and disks) were synthesised and coated with trastuzumab. Trastuzumab is a monoclonal antibody which targets the human epidermal growth receptor 2 (HER2). The uptake of antibody-coated particles and uncoated particles into human breast cancer cells was then monitored by confocal microscopy. For uncoated particles, the spheres showed the highest uptake. However, when the coated particles were studied, it was found that the nanorods exhibited significantly higher uptake compared to the nanospheres and nanodisks. It was speculated that this higher uptake could be due to multivalent interactions of the trastuzumab with the HER2 receptors on the cell surface, facilitated by higher adhesion of the nanorods due to their increased contact area with the cell surface. Interestingly, the use of nanorods also enhanced the ability of trastuzumab to inhibit cell growth. The trastuzumab-coated nanorods induced approximately 50% inhibition at a trastuzumab concentration of 1.25 μg/mL whereas soluble trastuzumab alone

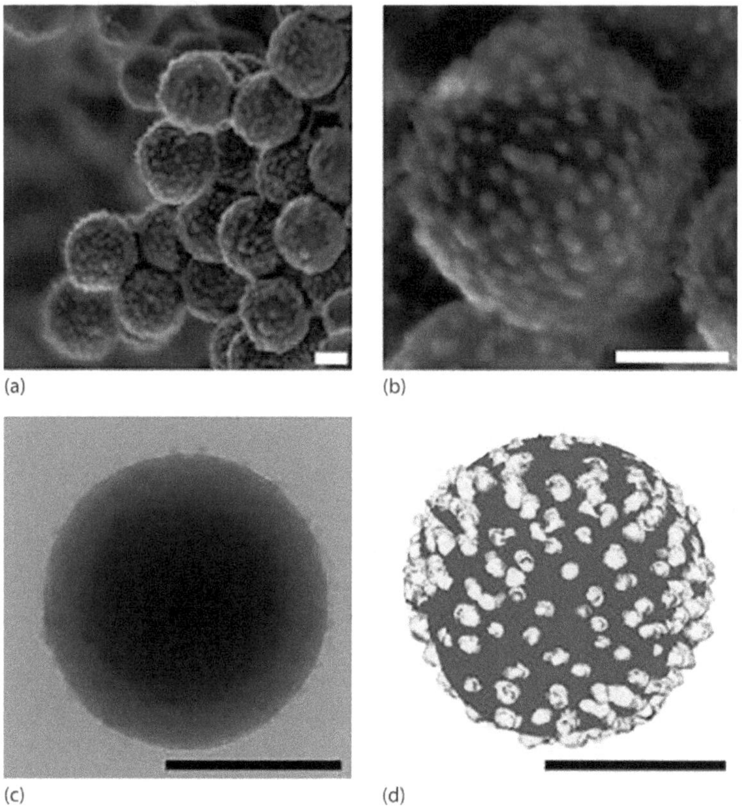

FIGURE 2.6 (a, b) SEM images of the virus mimetic surface of silica nanocomposites. (c) A zero-tilt TEM projection from a tilt series. (d) Reconstruction of the surface of a single particle. Scale bar: 100 nm. (Reproduced from Niu, Y. et al., *Adv. Mater.*, 6233–6237, 2013. With permission.)

produced only 31.9% inhibition, even though the trastuzumab concentration used was much higher (25 μg/mL).

An interesting consequence of changes in NP morphology was observed for zinc oxide NPs, where their inhibition of the activity of the β-galactosidase enzyme (GAL) was found to relate to the particle shape (Cha et al. 2015). ZnO nanopyramids showed much greater inhibition of the enzyme activity compared to spherical and plate-like particles. The nanopyramids exerted their effects via a competitive inhibition mechanism.

2.2.5 Diagnostics Applications

Gold nanorods have been used in the detection of cancer biomarkers (Truong et al. 2011) and hepatitis B (Wang et al. 2010) via their incorporation in localized surface plasmon resonance (LSPR) biosensors. The LSPR of gold nanorods is particularly responsive to changes in the dielectric constant of the surrounding medium, which can be affected by binding of various biologically significant molecules (Li et al. 2010).

To prepare such biosensors, antibodies are conjugated to the nanorods. Specific binding of target antigen to the immobilized antibody results in a peak shift of the LSPR spectrum. The degree of this peak shift is dependent on the concentration of the target analyte, thus quantitative detection can be achieved.

Triangular-shaped gold nanoplates have also been used for LSPR-based sensing. Anti-IgG functionalized Au nanoplates were reported to display a shift in λ_{max} as large as 45 nm for a 10 pg/mL solution of IgG (<1 pM) (Beeram and Zamborini 2010).

An alternative method for ultrasensitive detection of biomolecules is the use of surface-enhanced Raman scattering (SERS). SERS-active structures are generally shape-controlled metal NPs or NP aggregates where the junction between the NPs acts as a 'hot spot' for analyte detection (Hao and Schatz 2003). A SERS-based sensor for multiplexed detection of DNA was constructed by forming an array of gold nanowires conjugated to gold NP-labelled single-strand DNA (Kang et al. 2010). Gold–silver core–shell nanodumbbells with precisely engineered gaps between two particles were reported to have Raman enhancement factors of 2.7×10^{12} and hence could be used for single-molecule detection (Lim et al. 2009).

2.3 Cytotoxicity Studies Related to Various Morphologies

As outlined in the previous section, the use of shaped NPs for a variety of potential medical applications is a subject of active research. Therefore, it is essential that the possible toxicity of these materials is investigated since this will have implications for the implementation of such innovations. NP toxicity is typically assessed initially using *in vitro* toxicity tests, i.e., cytotoxicity, oxidative stress, inflammatory reactions, and genotoxicity, prior to *in vivo* studies. There are a number of ways in which coated NPs may interact with cells, as illustrated in Figure 2.7. NPs which are coated with cell-membrane receptor-specific ligands (e.g., antibodies) can bind to their target receptors and induce signal cascades within the cell without actually being internalised. Alternatively, the particles can enter the cell via exocytosis and remain encapsulated in vesicles before being ejected from the cell. In some instances, the NPs may escape from the vesicle while still within the cell and can thus interact with cellular organelles. Finally, particles may be internalized after non-specific contact with the cell membrane.

Although a lot of work has been done on the correlation between particle size and cytotoxicity (Pan et al. 2007, Gliga et al. 2014), the effect of particle shape on toxicity is not as well understood. This section will review some of the recent research on this topic.

The toxicity of silver particles of varying morphologies (silver nanowires and spherical silver NPs) towards alveolar epithelial cells (A549) was investigated (Stoehr et al. 2011). While no effects were observed for the spherical particles, incubation of the cells with the silver wires resulted in a significant reduction in cell viability and increased lactate dehydrogenase (LDH) release from the cells. However, immunotoxic effects were not observed. Therefore, it was proposed that the toxicity of the nanowires may relate to the direct mechanical interaction of the wires with the cells via the generation of punctures in the cell membrane.

Rod-shaped iron oxide NPs were also found to induce a much greater level of cytotoxic responses than spherical iron oxide NPs in a murine macrophage cell line (RAW 264.7), including increased LDH release, and higher levels of inflammatory response, ROS production, and necrosis (Han et al. 2014).

Needle- and plate-shaped hydroxyapatite NPs were found to be more cytotoxic towards BEAS-2 B cultures when compared with equivalent concentrations of sphere- and rod-shaped hydroxyapatite NPs (Zhao et al. 2013). Interestingly, however, this shape effect was not replicated with a different cell line – for RAW264.7 cells, there was no significant difference in cell death for any of the shapes.

In contrast to the previous study, when the toxicity of CeO_2 NPs of varying shape towards RAW264.7 cells was studied, the cytotoxicity was found to be shape dependent (Forest et al. 2017). LDH release and TNF-α production were significantly increased by rod-like NPs and this effect was found to be dose dependent. This phenomenon was not observed with cubic/octahedral CeO_2 NPs.

Cerium oxide nanorods and nanowires of precisely controlled lengths and aspect ratios were prepared by a hydrothermal synthetic method in order to systematically investigate the effect of aspect ratio on the viability of the human myeloid cell line THP-1 (Ji et al. 2012). While exposure of cells to CeO_2 particles of aspect ratios up to 31 did not have any effect on cell viability, the incubation of cells

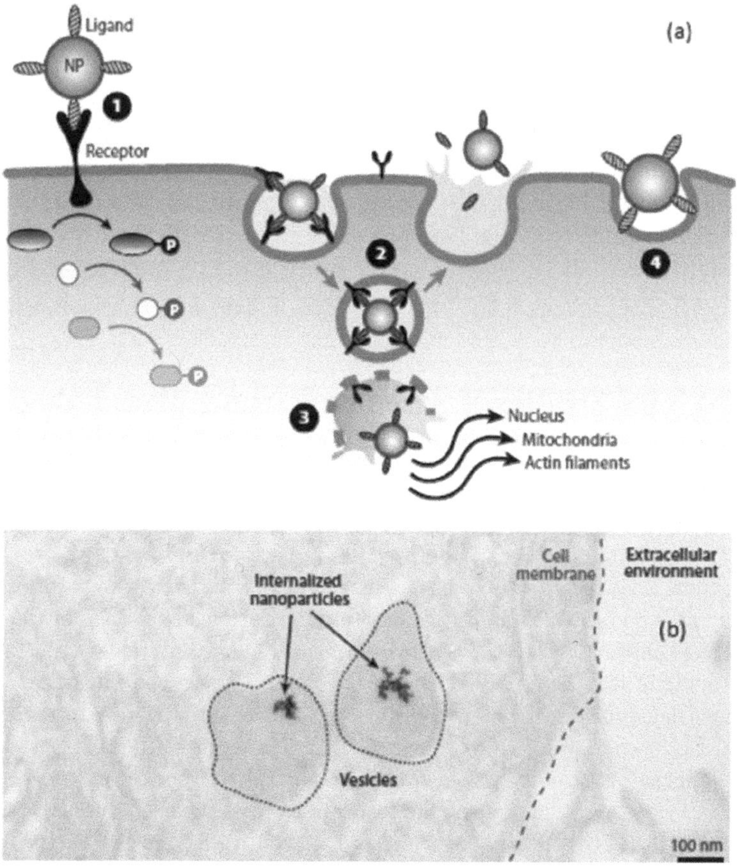

FIGURE 2.7 (a) Ligand-coated NPs interacting with cells. (1) Binding of particles to membrane receptors and induction of cell signalling cascade. (2) Exocytosis and ejection of the NPs. (3) Escape of internalized NPs from the vesicle and interaction with cell organelles. (4) Nonspecific interaction with the cell membrane and subsequently cell uptake. (b) Image of transferrin-coated gold NPs internalized by HeLa cells via intracellular vesicles. (Adapted from Albanese, A. et al., *Annu. Rev. Biomed. Eng.*, 2012. With permission.)

with particles of aspect ratios 52 and 100 for a 24-hour period resulted in a significantly higher cell death rate. Electron microscopy was used to try to understand the nature of the particles' interaction with the cells and the mechanism by which their cytotoxicity was mediated. As shown in Figure 2.8, this revealed bundling of the higher aspect ratio CeO_2 nanowires, which promoted contact with the cellular surface and in some cases, piercing of the cell membrane. Additionally, piercing of the endo-lysosomal membrane by intracellular bundles could also be observed. This can lead to rupturing of the lysosomes and a consequent detrimental cellular response. The formation of these stacking bundles of nanowires was favoured as the particle aspect ratio increased due to the magnitude of attractive forces (Van der Waals and dipole-dipole attraction) between parallel-aligned nanorods.

Analysis of the cytotoxicity of CeO_2 NPs towards *Mytilus galloprovincialis* haemocytes also indicated that particle shape plays a role in the induction of bioadverse processes in cells (Sendra et al. 2018). It was reported that rounded NPs with negative zeta potential provoked greater changes in the biomarkers of stress and immunological parameters than the NPs with a well-faceted surface with almost neutral zeta potential. However, this study did not compare the rates of uptake of the two different particle types; the different zeta potentials of the particles may have affected this. An earlier study on the effect of zeta

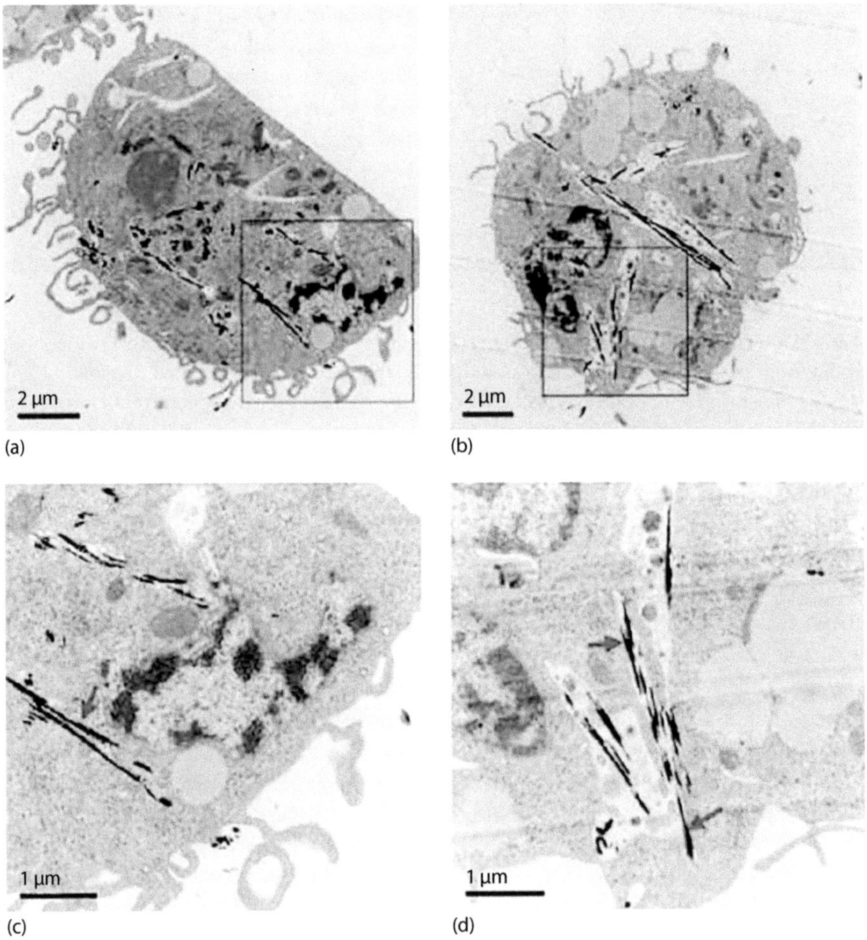

FIGURE 2.8 THP-1 cells were exposed to CeO_2 nanorods/nanowires for 24 hours at 50 μg mL^{-1} incorporation of nanorods of aspect ratio of 52 (a) and (c) (higher magnification) incorporation of nanowires of aspect ratio 100 (b) and (d) (higher magnification). The black arrows highlight the "free-floating" nanorod bundles in the cytoplasm. (Reprinted with permission from Ji, Z. et al., *ACS Nano*, 6, 5366–5380, 2012. Copyright 2012 American Chemical Society.)

potential on the uptake of gold NPs in breast cancer cells showed that positively charged gold NPs were the most internalized, followed by negatively charged NPs and then neutral NPs (Cho et al. 2009).

Incubation of neural cells (microglia and neurons) with gold NPs of varying morphologies (spheres, rods, and urchins) showed that NP geometry had a significant effect on the microglial activation status (Hutter et al. 2010). Dark-field microscopy and two-photon-induced luminescence (TPL) showed that the exposure of neural cells to gold NPs resulted in up-regulation of toll-like receptor 2 (TLR2), interleukin 1 alpha (IL-1α), granulocyte macrophage colony-stimulating factor (GM-CSF), and nitric oxide (NO) in microglia (Figure 2.9).

Gold NPs of varying morphology, i.e., gold nanocubes (AuNCs) and gold nanooctahedra (AuNOs), were synthesized and coated with either poly(acryloyl-L-valine) (L-PAV) or poly(acryloyl-D-valine) (D-PAV) and their cytotoxicity towards A549 lung cells was examined (Deng et al. 2017). While the primary aim of this study was to investigate the effect of chirality on cytotoxicity, the results showed that the particle shape also played a role. For the same enantiomer of L-PAV-coated NPs, there was a significant difference in cellular uptake of the two particle morphologies with AuNOs showing a higher

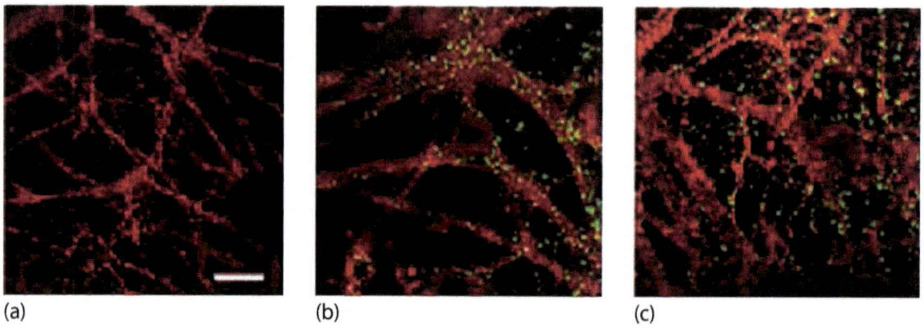

FIGURE 2.9 Two-photon luminescence imaging of (a) untreated control neurons, (b) primary hippocampal neurons incubated for 21 days with PEG-coated gold nanorods, and (c) CTAB-coated gold nanorods. Neurons were stained with MitoTracker green FM (pseudocoloured red). (Reprinted with permission from Hutter, E. et al., *ACS Nano*, 4, 2595–2606, 2010. Copyright 2010 American Chemical Society.)

amount of cell internalization than the AuNCs. The particle shape also affected the cell migration – the AuNOs had no effect on cell mobility, regardless of what enantiomer they were capped with. However, the AuNCs significantly reduced cell migration. The cytotoxicity of D-PAV-AuNOs was larger than that of D-PAV-AuNCs, even though their intracellular Au levels were similar.

The interaction between commercially available gold nanorods and A549 cells was studied by Tang and co-workers (Tang et al. 2015). The nanorods were found to enter the cells via phagocytosis. The nanorods induced a dose-dependent decrease in the viability of the A549 cells with a threshold concentration of 5 μg/mL required to provoke significant toxicity. An increase in levels of reactive oxygen species (ROS) in the cells indicated that the mechanism of the nanorod toxicity was oxidative stress.

The effect of shape on the cellular uptake of gold NPs has also been investigated by Xie et al. (2017). Three different types of methylpolyethylene glycol-coated gold NPs were prepared: stars, rods, and triangles. The uptake of these particles by RAW 264.7 cells was then analysed. Interestingly, it was found that the different particle shapes employed different mechanisms for cell entry. The star-shaped particles entered the cells via clathrin-mediated uptake, whereas the gold nanorods employed this mechanism but also took advantage of caveolae/lipid raft-mediated endocytosis. The gold nanotriangles entered the cells via multiple routes including clathrin-mediated endocytosis and a dynamin-dependent pathway.

The effects of particle morphology on cell uptake and cytotoxicity for differently shaped $NaYF_4$ NPs doped with 30 mol% Yb^{3+} and 0.5 mol% Tm^{3+} was examined by Tree-Udom et al. (Tree-Udom et al. 2015). Nanospheres, elongated nanospheres, and nanohexagonal prisms were synthesized and exposed to three different cell lines (A-375 cells, a normal skin cell line [WI-38], and HepG2). It was observed that cell uptake was most efficient for the elongated nanospheres. While all three particle morphologies exhibited minimal cytotoxicity to the cell lines, there was a significantly greater cytotoxicity associated with the elongated nanospheres. This could be attributed to the higher cellular uptake of these particles.

Statistically designed copper NPs of specified morphologies (spheres, rods, pyramids, cubes) were prepared by a thermal reduction method and their cell uptake and cytotoxicity towards NHEK and HeLa cell lines were investigated (Murugan et al. 2017). The results showed that the nanocubes had the highest uptake over a 24-hour incubation period, while the nanorods showed the lowest uptake. This was attributed to the cubes having the highest surface area of the four different particle shapes studied. It was also suggested that the sharp edges of the cubes can promote internalization for certain cell types as the cell wraps actin filaments around the particle edges. When cell viability was assessed after exposure to the NPs, a shape-dependent cytotoxic effect was observed with the lowest LC50 values for the cubes and then increasing as follows: cubes < 100 nm spheres < rods < pyramids <10 nm spheres. This effect was explained as originating from the high cellular uptake of the cubes and the higher volume of these particles, relative to the other morphologies. Analysis of glutathione activity in the cells indicated that the mechanism of the NP toxicity was oxidative stress. Once the particles enter the cells, intracellular ion release results in the production of ROS at the NP surface. This in turn leads to the formation of free radicals which can react with DNA, lipids, and proteins, with consequent oxidative damage to the cell, DNA damage, and apoptosis.

Investigation of the toxicity of zinc oxide NPs towards lung epithelial cells also showed a shape-dependent effect on cell viability (Hsiao and Huang 2011). While zinc nanorods and nanospheres (of fixed size and surface area) initially showed similar cytotoxicity after 12 hours exposure to A549 cells, after 24 hours the ZnO nanorods were more toxic than the spheres, with values of EC50 of 8.5 and 12.1 $\mu g\ mL^{-1}$, respectively. The authors commented that the contact area between a single particle and a single cell may be more important in determining cytotoxicity than the total specific surface area of the particle.

Particle aspect ratio was shown to be a factor in the effect of silica NPs on the cellular functions of A375 human melanoma cells (Huang et al. 2010). Monodisperse silica NPs of similar particle diameter, chemical composition, and surface charge but with varying aspect ratios (1, 2, and 4) were prepared. Confocal microscopy showed that the rate of cellular internalization varied with particle aspect ratio, with particles of aspect ratio 4 (long rods) exhibiting the highest rate of uptake while the particles of aspect ratio 1 (spheres) showed the lowest rate of uptake. It was proposed that this was due to the larger contact area of the rods with the cell membrane. The long rods were also found to affect the organization of cytoskeleton, with disruption of the F-actin in the cells. MTT cell proliferation assay showed that the long rods were the most cytotoxic of the three different particle types. This result was confirmed by flow cytometry evaluation of apoptosis in the cells at particle concentrations of 1 mg/mL, 0.2 mg/mL, and 0.05 mg/mL. The cells treated with the long rods were found to be the most susceptible to apoptosis.

The role of particle shape in the potential toxicity of iron oxide NPs was investigated by exposure of spherical and rod-shaped iron oxide NPs to mouse macrophage cells RAW 264.7 (Han et al. 2014). The distribution of the particles after cell internalization was found to vary with the spherical iron oxide particles aggregating in the vacuole while the nanorods were observed throughout the cytoplasm. The LDH assay showed a greater amount of cell membrane damage associated with the nanorods and higher production of TNF α and ROS compared to cells exposed to spheres only. Furthermore, the nanorods induced greater amounts of cell necrosis. It was suggested that this was a consequence of the greater extent of membrane damage and ROS production observed with the nanorods.

Interestingly, when magnetic iron oxide nanowires (diameter 200 nm, length from 1 μm to 40 μm) were prepared by controlled assembly of sub-10-nm iron oxide NPs, these particles did not display acute toxicity to NIH/3T3 mouse fibroblasts (Safi et al. 2011). It was speculated that this lack of toxicity was because the cells themselves were able to degrade the nanowires into smaller structures of length c.200 nm. Thus, the toxicity associated with higher aspect ratio particles was avoided.

Platinum NPs of defined morphologies (flowers, multipods, and spheres) were synthesized from platinum acetylacetonate precursors in diphenyl ether using hexadecylamine and adamantanecarboxylic acid as capping agents. These particles were then exposed to human umbilical vein endothelial and lung epithelial cells in order to investigate whether particle shape affected rates of cellular uptake and cytotoxicity (Elder et al. 2007). The results showed that the particles were taken up by both cell types but they did not induce oxidative stress in the cells. Indeed, it has been speculated that these particles suppress this process by scavenging of ROS. However, they did induce a mild inflammatory response in the cells.

2.4 *In Vivo* Studies

An *in vivo* study of the toxicity of gold NPs of different morphologies towards KM mice found that rod-shaped gold NPs exerted toxic effects at lower concentrations than required for cube-shaped or spherical gold NPs (Sun et al. 2016). Additionally, *in vitro* haemolysis studies showed that the rod-shaped particles exhibited higher haemolysis ratios than the other morphologies. It was suggested that these differences could be related to differences in the surface areas and surface chemistries of the different shapes.

While rod-shaped particles appear to be associated with more toxic effects than other morphologies, this is not always the case. Europium (III) hydroxide nanorods have been shown to exhibit pro-angiogenic activity and thus could potentially be used in the treatment of heart disease (Patra et al. 2008). In order to further assess the possibility of using these materials in medical applications, a study of their *in vivo* toxicity in mice was carried out (Patra et al. 2009). The results showed that the injection of the nanorods yielded minimal effects on the animals; there were no significant histological changes in the vital organs, no significant change in the weight of different organs, and no evidence of haematological toxicity. Long-term toxicity studies (more than 60 days) also showed that there were no deaths of mice using high doses of inorganic nanorods (125 $mg^{-1}\,kg^{-1}day^{-1}$).

Mannose-capped gold NPs of three different shapes (sphere, rod, and star) were analysed for their toxicity towards adult zebrafish (Sangabathuni et al. 2017). The mannose coating was utilized in order to target carbohydrate receptors on cells. The NPs were injected into zebrafish and after varying exposure times (4 hours, 24 hours, 48 hours) the fish were sacrificed and their organs (brain, eye, heart, muscles, swim bladder, and digestive system) were examined. The gold particles were found in the digestive system, heart, and swim bladder but were not observed in the brain, muscles, and eyes. The examination also revealed differences in the distribution and sequestration of the differently shaped particles. The gold nanorods showed the most rapid initial uptake into the organs but were cleared after 48 hours. However, the gold nanostars accumulated in the organs at a steadier rate and remained there for longer periods of time compared to the spheres and rods. It was proposed that the high aspect ratio of the nanorods could account for their rapid uptake and clearance while the high friction coefficient of the stars could explain their prolonged sequestration.

The morphology of silica NPs was also found to play a role in their *in vivo* toxicity (Li et al. 2015). Mesoporous silica NPs of varying aspect ratio (1, 1.75, and 5) were synthesized and then administered by injection to mice at a dose of 40 $mg\,kg^{-1}$. The exposure of the mice to the NPs was found to induce kidney damage, with the severity of tissue damage dependent on the particle shape. The most serious damage was associated with spherical silica particles while the least damage was observed for the particles of aspect ratio 5.

The effect of particle shape on the *in vivo* toxicity of Ni NPs towards zebrafish embryos was investigated (Ispas et al. 2009). Ni NPs of diameter 30, 60, and 100 nm and larger dendritic structures composed of aggregates of 60-nm particles were prepared and exposed to zebrafish embryos during embryogenesis. It was found that embryonic exposure to these materials led to the development of intestinal defects. Calculation of LD10 and LD50 values showed that the dendritic clusters exhibited the highest toxicity, exceeding the toxicities of the particles and of soluble Ni salts. It was hypothesized that the dendritic aggregates may more readily adhere and persist for a longer duration in the intestinal lumen, thus accounting for their greater toxicity.

Zebrafish embryos were also chosen as the target organism to study the *in vivo* toxicity of silver particles of varying morphologies (George et al. 2012). The results showed that triangular nanoplates of silver were the most toxic morphology compared to silver spheres and wires. Although the release of silver ions from particle dissolution has been postulated as the main mechanism of Ag NP toxicity in aquatic life-forms, this was not found to be the case in this experiment as ICP-MS analysis showed higher rates of Ag^+ release from the spheres. Instead, it appeared that the toxicity of the nanoplates was related to the high level of crystal defects (stacking faults and point defects) on their surfaces (see Figure 2.10). When these defects were passivated via coating with cysteine, this resulted in a reduction of their toxicity towards zebrafish embryos.

A range of organisms (*Lolium multiflorum*, *Danio rerio*, *Caenorhabditis elegans*, *Escherichia coli*, *Bacillus cereus*, and *Pseudomonas aeruginosa*) were used to assess the comparative toxicities of silver nanocubes, spheres, and wires (Gorka et al. 2015). This range of organisms was chosen as a representative sample of species that could be exposed to Ag in aquatic environments and included Gram-negative and Gram-positive bacteria, an aquatic nematode, and an aquatic plant. While the different particle morphologies displayed similar levels of toxicity towards most of the organisms studied, there was a

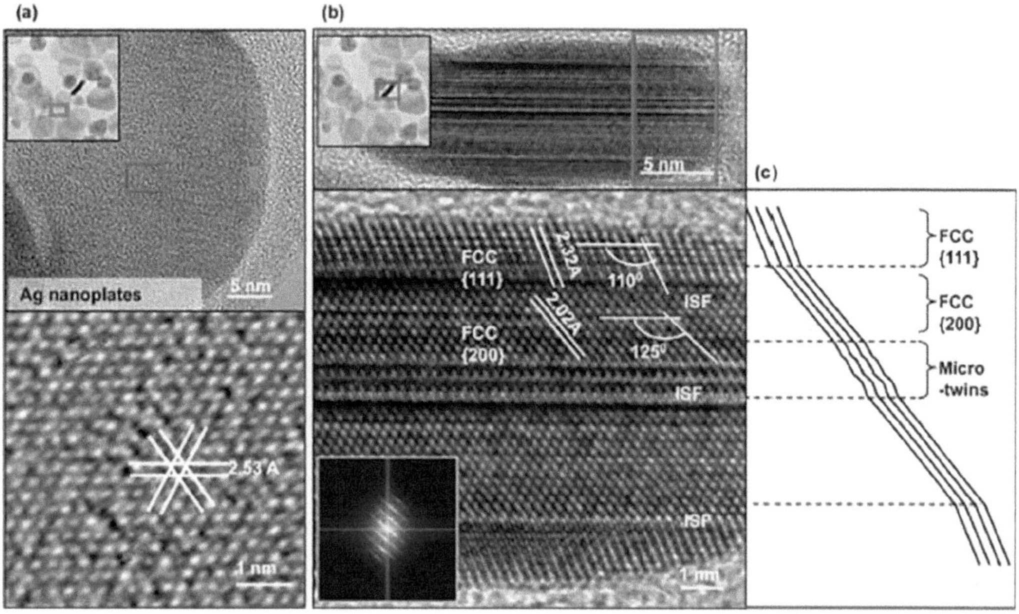

FIGURE 2.10 (a) TEM image of a flat-lying Ag nanoplate. (b) HRTEM image of vertically oriented {110} plane of an Ag nanoplate shows its structural defects. (c) Schematic representation of lattice image highlighting intrinsic stacking faults, e.g., domain boundaries and microtwins. (Reprinted with permission from George, S. et al., *ACS Nano*, 6, 3745–3759, 2012. Copyright 2012 American Chemical Society.)

notable difference observed for the plant species *Lolium multiflorum*. It was noted that there was greater inhibition of plant root growth for those plants exposed to silver spheres and wires compared to the plant exposed to nanocubes. The roots of silver nanocube-treated plants were 5.3% shorter than the control plant, while silver NP-treated plant roots were 39.6% shorter than the control (Figure 2.11).

The effect of silver NP shape on rates of skin penetration was investigated in an *in vivo* study using SKH-1 hairless mice (Tak et al. 2015). Rod-shaped, spherical, and triangular silver NPs were applied to

FIGURE 2.11 The degree of silver NP-induced inhibition of root growth of an aquatic plant was found to depend on the particle shape. (Reprinted with permission from Gorka, D.E. et al., *Environ. Sci. Technol.*, 49, 10093–10098, 2015. Copyright 2015 American Chemical Society.)

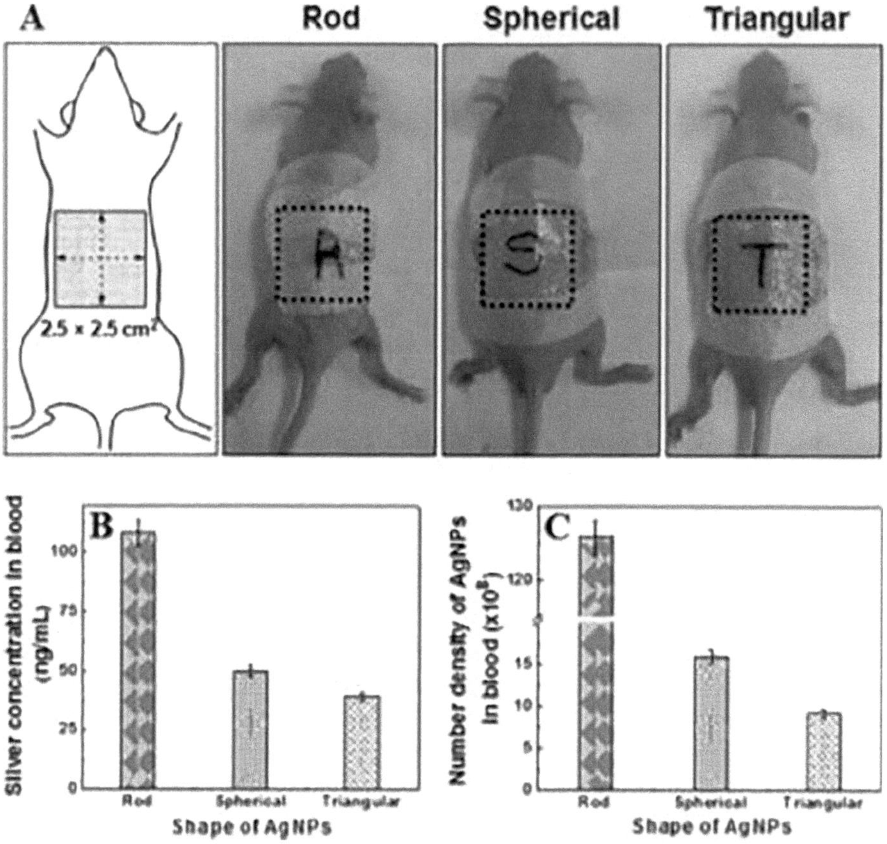

FIGURE 2.12 (A) Silver NPs of varying shapes (rods, spheres, triangles) were topically applied to the back of mice. (B) After 5 days, the concentration of silver NPs in the blood collected from their hearts were measured using ICP-MS and (C) the number density of differently shaped silver NPs in blood was also calculated. (Reproduced from Tak, Y.K. et al., *Scientific Reports*, 1–11, 2015. With permission.)

the back of the mice. Systemic circulating blood samples were then taken after the mice were treated for 5 days with silver NPs. ICP MS was used to analyse the blood samples for silver concentration. As illustrated in Figure 2.12, mice exposed to the rod-shaped particles had the highest concentration of silver in their blood at 108.57 ± 5.43 ng/mL, while animals treated with spherical silver and triangular silver had blood silver concentrations of 50.00 ± 2.50 ng/mL and 39.29 ± 1.96 ng/mL, respectively.

These findings were consistent with results from an *in vitro* analysis of skin penetration of differently shaped silver NPs. It was suggested that the lower rate of penetration of the triangular particles, compared to the spheres and rods, could be due to the differences in the predominant crystal facets in this particle morphology, given that the top basal plane has pre-dominant {111} facets, compared to predominant {100} facets in the other particle geometries. It should be noted that no significant changes were observed in the behaviour or health of the mice during the course of the study.

The importance of NP shape for tumour penetration was studied *in vivo* by intravenous injection of PEGylated fluorescent CdSe/CdS core shell spheres and nanorods into female SCID mice bearing mammary tumours (Chauhan et al. 2011). The spheres and rods had similar hydrodynamic diameters of 33–35 nm. It was observed that the nanorods penetrated the tumour more than four times as fast as the spheres. It was suggested that this was due to more rapid transport of the rods through the pores in the tumour vascular walls.

However, different results were observed when PEGylated radioactive ^{198}Au-doped NPs of various morphologies (spheres, disks, rods, cages) were injected into a tumour-bearing mouse and their tumour uptake directly imaged using the Cerenkov radiation (Black et al. 2014). In contrast to the results for the CdSe/CdS NPs, a significantly higher tumour uptake was observed for the gold spheres and disks relative to the nanorods and nanocages. This may be due to the fact that the spheres had the best blood circulation and the lowest clearance by the reticuloendothelial system of all four particle shapes.

2.5 Conclusions and Future Outlook

It is clear that shaped NPs have great potential for application in medicine, due to their distinctive physicochemical properties. As outlined earlier in this chapter, greatly increased levels of sensitivity in medical diagnostics have been achieved by taking advantage of the surface plasmon resonances of metal NPs. Control of the shape of metal and magnetic NPs has been found to be an important factor in determining their effectiveness as photothermal and hyperthermal agents for the treatment of cancer. Particle morphology has also been shown to be a critical parameter in the design of NP-based drug delivery systems.

However, while there has been extensive investigation of the roles of particle chemical composition, size, surface coating, and surface charge in relation to particle toxicity, there is still an incomplete understanding of the effect of particle morphology. A survey of the literature may lead to the tentative conclusion that an increase in NP aspect ratio leads to an increase in cytotoxicity. This effect may be related to the increased surface area of these particles for contact with cell structures or to the ability of elongated NPs to damage cell membranes or to the more efficient cellular uptake of this particle morphology such that cells are experiencing a relatively higher dose of these particles than other shapes. However, it is not clear if the changes in surface concentrations of capping agents associated with changes in particle morphology have been taken into consideration in these studies. Additionally, it should be noted that this phenomenon is not replicated across all NP types and indeed, in some instances, it is not consistent across more than one cell type when the same NPs are used. Furthermore, the role of crystal defects in inducing toxicity needs to be investigated in further detail.

It has been shown that the unique physicochemical properties of NPs can result in interference in the read out of *in vitro* assays or result in errors due to interaction with assay components, thus producing variable results (Monteiro-Riviere et al. 2009, Ferraro et al. 2016). Therefore, this factor must be taken into account when designing future studies of the toxicity of differently shaped NPs since the light absorption and/or emission of particles can change with particle morphology.

Although there has been a growth in interest in the *in vitro* toxicity of shaped NPs, there are still very few *in vivo* studies which systematically investigate this topic. There also appears to be some inconsistency between the results of *in vitro* and *in vivo* studies. Clearly, there is a need to expand current knowledge in this area before the commercialization of shaped NP-based invasive medical devices can be realistically considered. The use of three-dimensional cell culture may help to provide a more accurate picture of NP cytotoxicity (Wu et al. 2017).

References

Albanese, A., P. S. Tang, and W. C. Chan. 2012. "The Effect of Nanoparticle Size, Shape, and Surface Chemistry on Biological Systems." *Annual Review of Biomedical Engineering*. doi:10.1146/annurev-bioeng-071811-150124.

Barua, S., J.-W. Yoo, P. Kolhar, A. Wakankar, Y. R. Gokarn, and S. Mitragotri. 2013. "Particle Shape Enhances Speci Fi City of Antibody-Displaying Nanoparticles." *Proceedings of the National Academy of Sciences*. doi:10.1073/pnas.1216893110.

Bauer, L. M., S. F. Situ, M. A. Griswold, and A. C. Samia. 2016. "High-Performance Iron Oxide Nanoparticles for Magnetic Particle Imaging—Guided Hyperthermia (hMPI)." *Nanoscale* 8 (24):12162–12169. doi:10.1039/C6NR01877G.

Beeram, S. R. and F. P. Zamborini. 2010. "Purification of Gold Nanoplates Grown Directly on Surfaces for Enhanced Localized Surface Plasmon Resonance Biosensing." *ACS Nano* 4 (7):3633–3646. doi:10.1021/nn1007397.

Bilan, R., K. Brazhnik, P. Chames, D. Baty, I. Nabiev, and A. Sukhanova. 2015. "Oriented Conjugates of Single-Domain Antibodies and Fluorescent Quantum Dots for Highly Sensitive Detection of Tumor-Associated Biomarkers in Cells and Tissues." *Physics Procedia* 73:228–234. doi:10.1016/j.phpro.2015.09.162.

Black, K. C., Y. Wang, H. P. Luehmann, X. Cai, W. Xing, B. Pang, Y. Zhao et al. 2014. "Radioactive 198Au-Doped Nanostructures with Different Shapes for in Vivo Analyses of Their Biodistribution, Tumor Uptake, and Intratumoral Distribution." *ACS Nano* 8 (5):4385–4394. doi:10.1021/nn406258m.

Blanco, E., H. Shen, and M. Ferrari. 2015. "Principles of Nanoparticle Design for Overcoming Biological Barriers to Drug Delivery." *Nature Biotechnology* 33:941. doi:10.1038/nbt.3330.

Byrne, F., A. Prina-Mello, A. Whelan, B. M. Mohamed, A. Davies, Y. K. Gun'ko, J. M. D. Coey, and Y. Volkov. 2009. "High Content Analysis of the Biocompatibility of Nickel Nanowires." *Journal of Magnetism and Magnetic Materials* 321 (10):1341–1345. doi:10.1016/j.jmmm.2009.02.035.

Carbó-Argibay, E., B. Rodríguez-González, J. Pacifico, I. Pastoriza-Santos, J. Pérez-Juste, and L. M. Liz-Marzán. 2007. "Chemical Sharpening of Gold Nanorods: The Rod-to-Octahedron Transition." *Angewandte Chemie International Edition* 46 (47):8983–8987. doi:10.1002/anie.200703259.

Cha, S.-H., J. Hong, M. McGuffie, B. Yeom, J. S. VanEpps, and N. A. Kotov. 2015. "Shape-Dependent Biomimetic Inhibition of Enzyme by Nanoparticles and Their Antibacterial Activity." *ACS Nano* 9 (9):9097–9105. doi:10.1021/acsnano.5b03247.

Chauhan, V. P., Z. Popović, O. Chen, J. Cui, D. Fukumura, M. G. Bawendi, and R. K. Jain. 2011. "Fluorescent Nanorods and Nanospheres for Real-Time In Vivo Probing of Nanoparticle Shape-Dependent Tumor Penetration." *Angewandte Chemie (International Ed. in English)* 50 (48):11417–11420. doi:10.1002/anie.201104449.

Cheng, L. C., J.-H. Huang, H. M. Chen, T.-C. Lai, K.-Y. Yang, R.-S. Liu, M. Hsiao, C.-H. Chen, L.-J. Her, and D. P. Tsai. 2012. "Seedless, Silver-induced Synthesis of Star-shaped Gold/Silver Bimetallic Nanoparticles as High Efficiency Photothermal Therapy Reagent." *Journal of Materials Chemistry* 2244–2253. doi:10.1039/c1jm13937a.

Cho, E. C., J. Xie, P. A. Wurm, and Y. Xia. 2009. "Understanding the Role of Surface Charges in Cellular Adsorption versus Internalization by Selectively Removing Gold Nanoparticles on the Cell Surface with a I2/KI Etchant." *Nano Letters* 9 (3):1080–1084. doi:10.1021/nl803487r.

Choi, D. S., J. Park, S. Kim, D. H. Gracias, M. K. Cho, Y. K. Kim, A. Fung, S. E. Lee, Y. Chen, S. Khanal, S. Baral, and J. H. Kim. 2008. "Hyperthermia with Magnetic Nanowires for Inactivating Living Cells." *Journal of Nanoscience and Nanotechnology* 8 (5):2323–2327.

Das, R., J. Alonso, Z. Nemati Porshokouh, V. Kalappattil, D. Torres, M.-H. Phan, E. Garaio, J. Á. García, J. L. Sanchez Llamazares, and H. Srikanth. 2016. "Tunable High Aspect Ratio Iron Oxide Nanorods for Enhanced Hyperthermia." *The Journal of Physical Chemistry C* 120 (18):10086–10093. doi:10.1021/acs.jpcc.6b02006.

Deng, J., M. Yao, and C. Gao. 2017. "Acta Biomaterialia Cytotoxicity of Gold Nanoparticles with Different Structures and Surface-Anchored Chiral Polymers L-PAV-AuNCs." *Acta Biomaterialia* 53:610–618. doi:10.1016/j.actbio.2017.01.082.

Elder, A., H. Yang, R. Gwiazda, X. Teng, S. Thurston, H. He, and G. Oberdörster. 2007. "Testing Nanomaterials of Unknown Toxicity: An Example Based on Platinum Nanoparticles of Different Shapes." *Advanced Materials* 19 (20):3124–3129. doi:10.1002/adma.200701962.

El-sayed, I. H., X. Huang, and M. A. El-sayed. 2006. "Selective Laser Photo-Thermal Therapy of Epithelial Carcinoma Using Anti-EGFR Antibody Conjugated Gold Nanoparticles." *Cancer Letters* 239:129–135. doi:10.1016/j.canlet.2005.07.035.

Ferraro, D., U. Anselmi-Tamburini, I. G. Tredici, V. Ricci, and P. Sommi. 2016. "Overestimation of Nanoparticles-Induced DNA Damage Determined by the Comet Assay." *Nanotoxicology* 10 (7):861–870. doi:10.3109/17435390.2015.1130274.

Forest, V., L. Leclerc, J.-F. Hochepied, A. Trouvé, G. Sarry, and J. Pourchez. 2017. "Toxicology in Vitro Impact of Cerium Oxide Nanoparticles Shape on Their in Vitro Cellular Toxicity." *TIV* 38:136–141. doi:10.1016/j.tiv.2016.09.022.

Fu, A., W. Gu, B. Boussert, K. Koski, D. Gerion, L. Manna, M. Le Gros, C. A. Larabell, and A. P. Alivisatos. 2007. "Semiconductor Quantum Rods as Single Molecule Fluorescent Biological Labels." *Nano Letters* 7 (1):179–182. doi:10.1021/nl0626434.

George, S., S. Lin, Z. Ji, C. R. Thomas, L. Li, M. Mecklenburg, H. Meng et al. 2012. "Surface Defects on Plate-Shaped Silver Nanoparticles Contribute to Its Hazard Potential in a Fish Gill Cell Line and Zebrafish Embryos." *ACS Nano* 6 (5):3745–3759. doi:10.1021/nn204671v.

Gliga, A. R., S. Skoglund, I. O. Wallinder, B. Fadeel, and H. L. Karlsson. 2014. "Size-Dependent Cytotoxicity of Silver Nanoparticles in Human Lung Cells: The Role of Cellular Uptake, Agglomeration and Ag Release." *Particle and Fibre Toxicology* 11 (1):11. doi:10.1186/1743-8977-11-11.

Gorka, D. E., J. S. Osterberg, C. A. Gwin, B. P. Colman, J. N. Meyer, E. S. Bernhardt, C. K. Gunsch, R. T. DiGulio, and J. Liu. 2015. "Reducing Environmental Toxicity of Silver Nanoparticles through Shape Control." *Environmental Science & Technology* 49 (16):10093–10098. doi:10.1021/acs.est.5b01711.

Gratton, S. E., P. A. Ropp, P. D. Pohlhaus, J. C. Luft, V. J. Madden, M. E. Napier, and J. M. Desimone. 2008. "The Effect of Particle Design on Cellular Internalization Pathways." *Proceedings of the National Academy of Sciences* 105 (33):11613–11618.

Hao, E. and G. C. Schatz. 2003. "Electromagnetic Fields around Silver Nanoparticles and Dimers." *The Journal of Chemical Physics* 120 (1):357–366. doi:10.1063/1.1629280.

Hsiao, I.-L. and Y.-J. Huang. 2011. "Science of the Total Environment Effects of Various Physicochemical Characteristics on the Toxicities of ZnO and TiO_2 Nanoparticles toward Human Lung Epithelial Cells." *Science of the Total Environment* 409 (7):1219–1228. doi:10.1016/j.scitotenv.2010.12.033.

Htoon, H., J. A. Hollingworth, A. V. Malko, R. Dickerson, and V. I. Klimov. 2003. "Light Amplification in Semiconductor Nanocrystals: Quantum Rods versus Quantum Dots." *Applied Physics Letters* 82 (26):4776–4778. doi:10.1063/1.1586460.

Hu, J., L.-S. Li, W. Yang, L. Manna, L.-W. Wang, and A. P. Alivisatos. 2001. "Linearly Polarized Emission from Colloidal Semiconductor Quantum Rods." *Science* 292 (5524):2060–2063.

Huang, X., X. Teng, D. Chen, F. Tang, and J. He. 2010. "Biomaterials The Effect of the Shape of Mesoporous Silica Nanoparticles on Cellular Uptake and Cell Function." *Biomaterials* 31 (3):438–448. doi:10.1016/j.biomaterials.2009.09.060.

Huang, Y.-F., K. Sefah, S. Bamrungsap, H.-T. Chang, and W. Tan. 2008. "Selective Photothermal Therapy for Mixed Cancer Cells Using Aptamer-Conjugated Nanorods." *Langmuir* 5:11860–11865.

Hutter, E., S. Boridy, S. Labrecque, M. Lalancette-Hébert, J. Kriz, F. M. Winnik, and D. Maysinger. 2010. "Microglial Response to Gold Nanoparticles." *ACS Nano* 4 (5):2595–2606. doi:10.1021/nn901869f.

Imura, K., T. Nagahara, and H. Okamoto. 2006. "Photoluminescence from Gold Nanoplates Induced by Near-Field Two-Photon Absorption." *Applied Physics Letters* 88 (2):23104. doi:10.1063/1.2161568.

Ispas, C., D. Andreescu, A. Patel, D. V. Goia, S. Andreescu, and K. N. Wallace. 2009. "Toxicity and Developmental Defects of Different Sizes and Shape Nickel Nanoparticles in Zebrafish." *Environmental Science & Technology* 43 (16):6349–6356.

Jain, P. K., X. Huang, I. H. El-Sayed, and M. A. El-Sayed. 2008. "Noble Metals on the Nanoscale: Optical and Photothermal Properties and Some Applications in Imaging, Sensing, Biology, and Medicine." *Accounts of Chemical Research* 41 (12):1578–1586. doi:10.1021/ar7002804.

Jana, N. R., L. Gearheart, and C. J. Murphy. 2001a. "Wet Chemical Synthesis of Silver Nanorods and Nanowires of Controllable Aspect Ratio." *Chemical Communications* 7:617–618. doi:10.1039/B100521I.

Jana, N. R., L. Gearheart, and C. J. Murphy. 2001b. "Wet Chemical Synthesis of High Aspect Ratio Cylindrical Gold Nanorods." *The Journal of Physical Chemistry B* 105 (19):4065–4067. doi:10.1021/jp0107964.

Ji, G., S. Tang, B. Xu, B. Gu, and Y. Du. 2003. "Synthesis of CoFe2O4 Nanowire Arrays by Sol–gel Template Method." *Chemical Physics Letters* 379 (5):484–489. doi:10.1016/j.cplett.2003.08.090.

Ji, Z., X. Wang, H. Zhang, S. Lin, H. Meng, B. Sun, S. George, T. Xia, A. E. Nel, and J. I. Zink. 2012. "Designed Synthesis of CeO_2 Nanorods and Nanowires for Studying Toxicological Effects of High Aspect Ratio Nanomaterials." *ACS Nano* 6 (6):5366–5380. doi:10.1021/nn3012114.

Kang, T., S. M. Yoo, I. Yoon, S. Y. Lee, and B. Kim. 2010. "Patterned Multiplex Pathogen DNA Detection by Au Particle-on-Wire SERS Sensor." *Nano Letters* 10 (4):1189–1193. doi:10.1021/nl1000086.

Kim, F., S. Connor, and H. Song. 2004. "Platonic Gold Nanocrystals." *Angewandte Chemie International Edition* 3673–3677. doi:10.1002/anie.200454216.

Ledwith, D. M., A. M. Whelan, and J. M. Kelly. 2007. "A Rapid, Straight-Forward Method for Controlling the Morphology of Stable Silver Nanoparticles." *Journal of Materials Chemistry* 17 (23):2459–2464. doi:10.1039/B702141K.

Lee, J. H., J. E. Ju, B. I. Kim, P. J. Pak, E.-K. Choi, H.-S. Lee, and N. Chung. 2014. "Rod-shaped Iron Oxide Nanoparticles Are More Toxic than Sphere-shaped Nanoparticles to Murine Macrophage Cells." *Environmental Toxicology and Chemistry* 33 (12):2759–2766. doi:10.1002/etc.2735.

Li, C., C. Wu, J. Zheng, J. Lai, C. Zhang, and Y. Zhao. 2010. "LSPR Sensing of Molecular Biothiols Based on Noncoupled Gold Nanorods." *Langmuir* 26 (11):9130–9135. doi:10.1021/la101285r.

Li, L., T. Liu, C. Fu, L. Tan, X. Meng, and H. Liu. 2015. "Biodistribution, Excretion, and Toxicity of Mesoporous Silica Nanoparticles after Oral Administration Depend on Their Shape." *Nanomedicine: Nanotechnology, Biology, and Medicine* 11 (8):1915–1924. doi:10.1016/j.nano.2015.07.004.

Lim, D.-K., K.-S. Jeon, H. M. Kim, J.-M. Nam, and Y. D. Suh. 2009. "Nanogap-Engineerable Raman-Active." *Nature Materials* 9 (1):60–67. doi:10.1038/nmat2596.

Lisjak, D., and A. Mertelj. 2018. "Progress in Materials Science Anisotropic Magnetic Nanoparticles: A Review of Their Properties, Syntheses and Potential Applications." *Progress in Materials Science* 95:286–328. doi:10.1016/j.pmatsci.2018.03.003.

Mackey, M. A., M. R. Ali, L. A. Austin, R. D. Near, and M. A. El-sayed. 2014. "The Most Effective Gold Nanorod Size for Plasmonic Photothermal Therapy: Theory and In Vitro Experiments." *The Journal of Physical Chemistry B* 118(5):1319–1326.

Mohamed, M. B., C. Burda, and M. A. El-Sayed. 2001. "Shape Dependent Ultrafast Relaxation Dynamics of CdSe Nanocrystals: Nanorods vs Nanodots." *Nano Letters* 1 (11):589–593. doi:10.1021/nl0155835.

Monteiro-Riviere, N. A., A. O. Inman, and L. W. Zhang. 2009. "Limitations and Relative Utility of Screening Assays to Assess Engineered Nanoparticle Toxicity in a Human Cell Line." *Toxicology and Applied Pharmacology* 234 (2):222–235. doi:10.1016/j.taap.2008.09.030.

Murugan, K., Y. E. Choonara, P. Kumar, L. C. Toit, and V. Pillay. 2017. "Cellular Internalisation Kinetics and Cytotoxic Properties of Statistically Designed and Optimised Neo-Geometric Copper Nanocrystals." *Materials Science & Engineering C* 78:376–388. doi:10.1016/j.msec.2017.04.087.

Nangia, S. and R. Sureshkumar. 2012. "Effects of Nanoparticle Charge and Shape Anisotropy on Translocation through Cell Membranes." *Langmuir* 28(51):17666–17671. doi:10.1021/la303449d.

Niu, Y., M. Yu, S. B. Hartono, J. Yang, H. Xu, H. Zhang, J. Zhang, et al. 2013. "Nanoparticles Mimicking Viral Surface Topography for Enhanced Cellular Delivery." *Advanced Materials* 6233–6237. doi:10.1002/adma.201302737.

Oldenburg, A. L., M. N. Hansen, T. S. Ralston, and S. A. Boppart. 2009. "Imaging Gold Nanorods in Excised Human Breast Carcinoma by Spectroscopic Optical Coherence Tomography." *Journal of Materials Chemistry* 6407–6411. doi:10.1039/b823389f.

O'Neal, D. P., L. R. Hirsch, N. J. Halas, J. D. Payne, and J. L. West. 2004. "Photo-Thermal Tumor Ablation in Mice Using near Infrared-Absorbing Nanoparticles." *Cancer Letters* 209:171–176. doi:10.1016/j.canlet.2004.02.004.

Pan, Y., S. Neuss, A. Leifert, M. Fischler, F. Wen, U. Simon, G. Schmid, W. Brandau, and W. Jahnen-Dechent. 2007. "Size-Dependent Cytotoxicity of Gold Nanoparticles." *Small* 1941–1949. doi:10.1002/smll.200700378.

Patra, C. R., R. Bhattacharya, S. Patra, N. E. Vlahakis, A. Gabashvili, Y. Koltypin, A. Gedanken, P. Mukherjee, and D. Mukhopadhyay. 2008. "Pro-Angiogenic Properties of Europium(III) Hydroxide Nanorods." *Advanced Materials* 20 (4):753–756. doi:10.1002/adma.200701611.

Patra, C. R., S. S. Moneim, E. Wang, S. Dutta, S. Patra, M. Eshed, P. Mukherjee, A. Gedanken, V. H. Shah, and D. Mukhopadhyay. 2009. "In Vivo Toxicity Studies of Europium Hydroxide Nanorods in Mice." *Toxicology and Applied Pharmacology* 240 (1):88–98. doi:10.1016/j.taap.2009.07.009.

Pečko, D., M. S. Arshad, S. Šturm, S. Kobe, and K. Ž. Rožman. 2015. "Magnetization-Switching Study of Fcc Fe&–Pd Nanowire and Nanowire Arrays Studied by In-Field Magnetic Force Microscopy." *IEEE Transactions on Magnetics* 51 (10):1–4. doi:10.1109/TMAG.2015.2449773.

Peng, X., L. Manna, W. Yang, J. Wickham, E. Scher, A. Kadavanich, and A. P. Alivisatos. 2000. "Shape Control of CdSe Nanocrystals." *Nature* 404:59. doi:10.1038/35003535.

Pérez-Juste, J., I. Pastoriza-Santos, L. M. Liz-Marzán, and P. Mulvaney. 2005. "Gold Nanorods: Synthesis, Characterization and Applications." *Coordination Chemistry Reviews* 249 (17):1870–1901. doi:10.1016/j.ccr.2005.01.030.

Petryayeva, E., and U. J. Krull. 2011. "Localized Surface Plasmon Resonance: Nanostructures, Bioassays and Biosensing—A Review." *Analytica Chimica Acta* 706 (1):8–24. doi:10.1016/j.aca.2011.08.020.

Rothenberg, E., M. Kazes, E. Shaviv, and U. Banin. 2005. "Electric Field Induced Switching of the Fluorescence of Single Semiconductor Quantum Rods." *Nano Letters* 5 (8):1581–1586. doi:10.1021/nl051007n.

Safi, M., M. Yan, M.-A. Guedeau-Boudeville, H. Conjeaud, V. Garnier-Thibaud, N. Boggetto, A. Baeza-Squiban, F. Niedergang, D. Averbeck, and J.-F. Berret. 2011. "Interactions between Magnetic Nanowires and Living Cells: Uptake, Toxicity, and Degradation." *ACS Nano* 5 (7):5354–5364. doi:10.1021/nn201121e.

Sandeep, K., and N. Thomas. 2006. "Shape Control of II–VI Semiconductor Nanomaterials." *Small* 2 (3):316–329. doi:10.1002/smll.200500357.

Sangabathuni, S., R. V. Murthy, and P. M. Chaudhary. 2017. "Mapping the Glyco-Gold Nanoparticles of Different Shapes Toxicity, Biodistribution and Sequestration in Adult Zebrafish." *Scientific Reports* 1–7. doi:10.1038/s41598-017-03350-3.

Sendra, M., M. Volland, T. Balbi, R. Fabbri, M. P. Yeste, J. M. Gatica, L. Canesi, and J. Blasco. 2018. "Cytotoxicity of CeO_2 Nanoparticles Using in Vitro Assay with Mytilus Galloprovincialis Hemocytes: Relevance of Zeta Potential, Shape and Biocorona Formation." *Aquatic Toxicology* 200:13–20. doi:10.1016/j.aquatox.2018.04.011.

Shieh, F., A. E. Saunders, and B. A. Korgel. 2005. "General Shape Control of Colloidal CdS, CdSe, CdTe Quantum Rods and Quantum Rod Heterostructures." *The Journal of Physical Chemistry B* 109 (18):8538–8542. doi:10.1021/jp0509008.

Stoehr, L. C., E. Gonzalez, A. Stampfl, E. Casals, A. Duschl, V. Puntes, and G. J. Oostingh. 2011. "Shape Matters: Effects of Silver Nanospheres and Wires on Human Alveolar Epithelial Cells." *Particle and Fibre Toxicology* 1–15.

Suarasan, S., E. Licarete, S. Astilean, and A.-M. Craciun. 2018. "Colloids and Surfaces B: Biointerfaces Probing Cellular Uptake and Tracking of Differently Shaped Gelatin-Coated Gold Nanoparticles inside of Ovarian Cancer Cells by Two-Photon Excited Photoluminescence Analyzed by Fluorescence Lifetime Imaging (FLIM)." *Colloids and Surfaces B: Biointerfaces* 166:135–143. doi:10.1016/j.colsurfb.2018.03.016.

Sui, Y. C., R. Skomski, K. D. Sorge, and D. J. Sellmyer. 2004. "Nanotube Magnetism." *Applied Physics Letters* 84 (9):1525–1527. doi:10.1063/1.1655692.

Sun, Y.-N., C.-D. Wang, X.-M. Zhang, L. Ren, and X.-H. Tian. 2016. "Shape Dependence of Gold Nanoparticles on In Vivo Acute Toxicological Effects and Biodistribution Shape Dependence of Gold Nanoparticles on In Vivo Acute Toxicological Effects and Biodistribution." *Journal of Nanoscience and Nanotechnology*. doi:10.1166/jnn.2011.3094.

Tak, Y. K., S. Pal, P. K. Naoghare, and S. Rangasamy. 2015. "Shape-Dependent Skin Penetration of Silver Nanoparticles: Does It Really Matter?" *Scientific Reports* 1–11. doi:10.1038/srep16908.

Tang, Y., Y. Shen, L. Huang, G. Lv, C. Lei, X. Fan, F. Lin, Y. Zhang, L. Wu, and Y. Yang. 2015. "In Vitro Cytotoxicity of Gold Nanorods in A549 Cells." *Environmental Toxicology and Pharmacology* 39 (2):871–878. doi:10.1016/j.etap.2015.02.003.

Tree-Udom, T., J. Seemork, K. Shigyou, T. Hamada, N. Sangphech, T. Palaga, N. Insin, P. Pan-In, and S. Wanichwecharungruang. 2015. "Shape Effect on Particle-Lipid Bilayer Membrane Association, Cellular Uptake, and Cytotoxicity." *ACS Applied Materials & Interfaces* 7 (43):23993–4000. doi:10.1021/acsami.5b06781.

Truong, P. L., C. Cao, S. Park, M. Kim, and S. J. Sim. 2011. "A New Method for Non-Labeling Attomolar Detection of Diseases Based on an Individual Gold Nanorod Immunosensor." *Lab on a Chip* 11 (15):2591–2597. doi:10.1039/C1LC20085B.

Vivek, R., R. Thangam, V. NipunBabu, C. Rejeeth, S. Sivasubramanian, P. Gunasekaran, K. Muthuchelian, and S. Kannan. 2014. "Multifunctional HER2-Antibody Conjugated Polymeric Nanocarrier-Based Drug Delivery System for Multi-Drug-Resistant Breast Cancer Therapy." *ACS Applied Materials & Interfaces* 6 (9):6469–6480. doi:10.1021/am406012g.

Wang, X., Y. Li, H. Wang, Q. Fu, J. Peng, Y. Wang, J. Du, Y. Zhou, and L. Zhan. 2010. "Gold Nanorod-Based Localized Surface Plasmon Resonance Biosensor for Sensitive Detection of Hepatitis B Virus in Buffer, Blood Serum and Plasma." *Biosensors and Bioelectronics* 26 (2):404–410. doi:10.1016/j.bios.2010.07.121.

Wiley, B., Y. Sun, B. Mayers, and Y. Xia. 2005. "Shape-Controlled Synthesis of Metal Nanostructures: The Case of Silver," 454–463. doi:10.1002/chem.200400927.

Wu, Z., R. Guan, M. Tao, F. Lyu, G. Cao, M. Liu, and J. Gao. 2017. "Assessment of the Toxicity and Inflammatory Effects of Different-Sized Zinc Oxide Nanoparticles in 2D and 3D Cell Cultures." *RSC Advances* 7 (21):12437–12445. doi:10.1039/C6RA27334C.

Xie, X., J. Liao, X. Shao, Q. Li, and Y. Lin. 2017. "The Effect of Shape on Cellular Uptake of Gold Nanoparticles in the Forms of Stars, Rods, and Triangles." *Scientific Reports*:1–9. doi:10.1038/s41598-017-04229-z.

Yuan, F., H. Chen, J. Xu, Y. Zhang, Y. Wu, and L. Wang. 2014. "Aptamer-Based Luminescence Energy Transfer from Near-Infrared-to-Near-Infrared Upconverting Nanoparticles to Gold Nanorods and Its Application for the Detection of Thrombin." *Chemistry—A European Journal* 20 (10):2888–2894. doi:10.1002/chem.201304556.

Zeng, S., K.-T. Yong, and I. Roy. 2011. "A Review on Functionalized Gold Nanoparticles for Biosensing Applications." *Plasmonics* 491–506. doi:10.1007/s11468-011-9228-1.

Zhao, X., S. Ng, B. C. Heng, J. Guo, L. Ma, T. T. Tan, K. W. Ng, and S. C. Loo. 2013. "Cytotoxicity of Hydroxyapatite Nanoparticles Is Shape and Cell Dependent." *Archives of Toxicology* 87 (6):1037–1052. doi:10.1007/s00204-012-0827-1.

3

Environmental and Cytotoxicity Risks of Graphene Family Nanomaterials

Saoirse Dervin

3.1 Introduction

A global endeavour to safeguard environmental and human health is evolving at a rapid pace. The list of environmental tribulations faced by humanity may be huge, but one solution to tackling our planet's threats is remarkably small. The use of nanotechnology, often referred to as a 'key technology of the twenty-first century', will facilitate the design and development of innovative products and new market potentials, as well as radical medical developments which will revolutionize our future lives. Nanotechnology has the potential to contribute significantly to planetary protection through several remarkable pathways. Nanomaterials (NMs) have facilitated the development of numerous novel green products and technologies, also capable of minimizing the generation of adverse by-products by improved manufacturing processes. NMs have also improved technologies for decontamination and remediation of waste sites, polluted water sources, and air supplies and improved energy technologies.

Nanotechnology is undoubtedly one of the fastest growing areas of research and technology. Engineered NMs have rapidly integrated society's daily lives in the form of electronic devices, robotics, food packaging, therapeutics, medical devices biosensors, and more (Akinwande et al., 2014; Bora and Dutta, 2014; Hussein, 2015; Kumar et al., 2014; Nikalje, 2015; Peng et al., 2018). This expeditious surge in nanoscale products and technologies has consequently intensified the likelihood of environmental and human exposure (Guo and Mei, 2014; Nel et al., 2006; Xia et al., 2009). Although NMs provide great benefits in many industries, there is still a great deal unknown about their impact on the environment and human health. NMs are often considered a 'two-edged sword'. The distinct properties of the nanoscale structures attributed to revolutionary product and process design, and innovative technologies are the exact properties which may also pose environmental and cytotoxic threats.

Carbon nanostructures are the forerunner of nanotechnology and have considerably influenced the advancement of contemporary science and technology (Gupta et al., 2015). In the last two decades, a new generation of low dimensional carbon nanostructures has gained mounting interest because of their profound properties and versatile utilization. Compared with zero-dimensional fullerenes and one-dimensional nanotubes, a newly emerging two-dimensional member of the carbon family, graphene, exhibits exceptional properties including unique electronic conductivity, outstanding mechanical strength, good thermal stability, amplified external surface area, ease of modification, chemical inertness, high mobility of charge carriers, extraordinary electrocatalytic activities, and improved optical transmittance (Choi et al., 2010; Hu and Zhou, 2013; Jastrzębska and Olszyna, 2015; Lee et al., 2008; Pretti et al., 2014). The single-atom layer of sp^2-hybridized carbon, packed densely in a honeycomb crystal lattice, has triggered a 'gold rush', since its fascinating discovery by Nobel Laureates Novoselov and Geim in 2004 (Dervin et al., 2016; Novoselov et al., 2004). The scientific community has subsequently applied this material to a variety of strategic disciplines, such as nanoelectronics (Westervelt, 2008), supercapacitors (Vivekchand et al., 2008), battery electrodes (Paek et al., 2008), solar cells (Yin et al., 2014), field effect transistors (Lee et al., 2013), photonics, printable inks (Wang et al., 2009), heterogeneous catalysis (Garg et al., 2014), antibacterial agents (Dikin et al., 2007), conducting polymers, sensors and biosensors (Shao et al., 2010), transport barriers, cell imaging (Chen et al., 2011a; Wang et al., 2013b), drug delivery, tissue engineering (Cha et al., 2014; Schinwald et al., 2012; Yang et al., 2011b), structural composite materials, and many more (Novoselov et al., 2012; Ou et al., 2016; Pretti et al., 2014; Sanchez et al., 2011; Syrgiannis, 2017). The astounding zero-band gap energy, small electrical resistivity, fast heat dissipation, and fast electron transfer properties of graphene envisaged the 2D carbon nanostructures ideal application in nanoelectronics, where it was initially used as an atomic thin coating for microchips (Pumera, 2014). In this form, an immobilized component, graphene, is unlikely to elicit a grave environmental or cytotoxic impact. As the graphene 'gold rush' proceeded, utilization of the 2D atomic structure within the sectors swiftly advanced and the possibility of environmental and human exposure dramatically increased (Figure 3.1) (Sanchez et al., 2011). Due to the heightened possibility of exposure, the primary objective of the scientific community has been directed toward investigating the biosafety of graphene-based materials (Card et al., 2011; Paul and Sharma, 2011; Sanchez et al., 2011). Fundamental knowledge of graphene's interactions with biological molecules, cells, tissue structures, organs, and organisms is crucial to safe graphene product design. Accordingly, the nature of and risks associated with graphene exposure, including its chemical behaviour, environmental fate, life cycle, ecosystem effects, biological responses, and related diseases, require in-depth exploration (Hu and Zhou, 2013; Sanchez et al., 2011).

FIGURE 3.1 Schematic illustration of the life cycle and exposure events of GFNs. (From Park, M.V.D.Z. et al., *ACS Nano.*, 11, 9574–9593, 2017.)

3.2 The Graphene Family

Graphene is the precursor to a broader family of ultrathin carbon allotropes. In addition to single-layer graphene, several graphene-derived, layered, carbonaceous NMs have also gained significant scientific attention (Bianco et al., 2013). The collectively termed graphene family of NMs, analogous to carbon nanotubes (CNTs), varies in the layer number, stacking order and orientation, surface functionalities and interface states, lateral dimensions, structural defects, and impurities (Figure 3.2) (Pretti et al., 2014). Graphene family NMs (GFNs) include single-layer or monolayer graphene, graphite, graphane, perfluorographane, epitaxial graphene, bilayer graphene, twisted bilayer graphene or turbostratic bilayer graphene, twisted few-layer graphene, trilayer graphene, few-layer graphene, graphene nanoplate (or graphene nanoplatelet), graphite oxide, graphene oxide (GO), and reduced graphene oxide (rGO) (as well as carbon materials derived from GFNs; atomically-thin predecessors which can be twisted, wrinkled, folded, stacked or assembled into a myriad of diverse three-dimensional (3D) designs (ISO/TS 80004-13:2017) (Bianco, 2013; Bianco et al., 2013; Cui et al., 2012; Li et al., 2012a; Pollard and Clifford, 2017). The wealth of advances in the field of graphene research has led to rapid development of the graphene family. In the nascent period of the wonder material, graphene was a term which explicitly referred to a single, flat sheet of carbon atoms arranged in a honeycomb lattice but has since become a generic idiom used to describe a variety of GFNs (Bianco et al., 2013; Wick et al., 2014). The ambiguous use of the term in both publications and patents has created a misperception and has steered the incorrect use of multiply-defined graphene acronyms (Bianco, 2013; Bianco et al., 2013; Pollard and Clifford, 2017). If a clear and

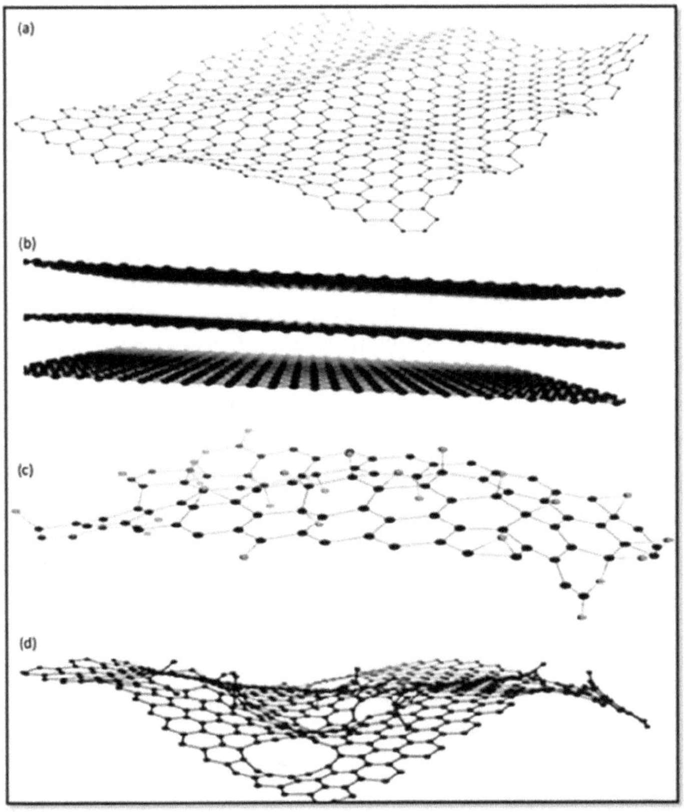

FIGURE 3.2 Structural variations of GFNs [(a) graphene, (b) FLG, (c) GO and (d) rGO]. (From Bianco, A., *Angew. Chem. Int. Ed.*, 52, 4986–4997, 2013.)

mutual perception of graphene-related terms and definitions is not shared by both industry and academia, this thriving industry may flounder.

A lack of well-defined terminology can also create momentous difficulty when attempting to apprehend the potential environmental and cytotoxic impacts of GFNs. If the toxicity of a specific material is explored to understand the related safety implications, but the material itself is not correctly recognized or classified, and the specific material properties not appreciated, the observed toxic responses become senseless (Pollard and Clifford, 2017). It is fundamental that the paramount relationship between the physicochemical characteristics of any NM type and its safety hazards should be acknowledged (Bussy et al., 2012; Petros and DeSimone, 2010). The significance of this relationship has previously been highlighted by the CNT experience (Bussy et al., 2012). Though the safety hazards of CNTs have been studied for a considerable period, the dangers of GFNs are equivocal as a consequence of early-stage development (Bussy et al., 2012). In addition, contradictory conclusions regarding the safety of GFNs have widely been issued, again emphasizing the unmistakable need for standardization of graphene terms and definitions, in addition to clearly defined standard protocol for the characterization and nanotoxicity testing of all NM types (Ferrari et al., 2015; Pollard and Clifford, 2017). Evidently, a generalization of GFNs and their toxicity would be evasive in view of GFNs diversity in structure, chemistry, dimensions, and manufacturing processes (Bianco et al., 2013; Bussy et al., 2012; Syrgiannis, 2017). Distinguishing multiple forms of graphene will accentuate the different characteristics of each member of the graphene family of NMs and associate the toxic effects with the material specific properties of each family member (Bianco, 2013). Prior to the physical isolation of graphene in 2004, the International Union of Pure and Applied Chemistry (IUPAC) in 1995 first defined the term 'graphene layer' (Fitzer et al., 1995). The first description explicably pertained to a graphene layer stacked within a graphitic structure, rather than a freestanding, single layer of graphene. The term 'graphene' was also classified in a terminology standard by the International Organization for Standardization (ISO) in 2010 (https://www.iso.org/obp/ui/#iso:std:iso:ts:80004:-13:ed-1:vl:en, 2017). In both cases, however, graphene is identified as a single layer of graphite, with graphite thus being characterized as a stacked structure of multiple graphene layers. Despite attempts, these definitions did not provide a clear classification framework for GFNs (Pollard and Clifford, 2017).

Since 2013 scientific publications have uncovered the academic and industrial confusion caused by a lack of clear-cut graphene terminology, with many authors recommending widespread nomenclature and progressive classification frameworks. The first nomenclature for 2D carbon forms was published by Bianco et al. in late 2013. The incentive for this editorial was rationalized by the statement 'precise names promote precise ideas' (Li et al., 2013b; Zhu et al., 2011). Preceding the endeavour for a more logical graphene nomenclature, the EU Graphene Flagship project also proposed a nomenclature and classification framework. The classification approach recommended by the Graphene Flagship, which was published in an article by Wick et al. (2014), addressed the nomenclature discrepancies throughout previously disseminated GFN nanotoxicity studies. The advised framework suggests GFNs are defined by three physicochemical descriptors, including number of graphene layers, the lateral size, and the carbon to oxygen ratio (C/O ratio), which are fundamental properties most relevant to prevalent GFNs and ignite biosafety concern (Jastrzębska et al., 2012; Park et al., 2017; Wick et al., 2014).

Subsequently, an extensive list of graphene acronyms, established within the framework of Graphene Flagship, was developed in a further attempt to regulate graphene nomenclature (Ferrari et al., 2015). In 2017, however, the International Organization for Standardization (ISO) published its first international standard for graphene and related 2D materials (https://www.iso.org/obp/ui/#iso:std:iso:ts:80004:-13:ed-1:vl:en, 2017). Together, the UK's National Physical Laboratory, the Institute of Physics, which publishes Physics World, and the UK's National Graphene Institute (NGI), all of which involved experts from 37 different countries, derived 99 terms and definitions classifying the different 2D materials types and material manufacturing and characterization techniques, in addition to the distinct material properties of GFNs (Kulvinder Singh, 2017; Pollard and Clifford, 2017). Formally referred to as ISO/TS 80004-13:2017, this standard formalizes the terms and definitions of each graphene family member, as outlined as follows (Table 3.1):

TABLE 3.1 International Standardized Terminology for Graphene Family Nanomaterials

Nomenclature	Material Description	Notes to Entry
Graphene/graphene layer/single-layer graphene/monolayer graphene	A single layer of carbon atoms with each atom bound to three neighbours in a honeycomb structure	1. It is an important building block of many carbon nano-objects. 2. As graphene is a single layer, it is also sometimes called monolayer graphene or single-layer graphene and abbreviated as 1LG to distinguish it from bilayer graphene (2LG) and few-layered graphene (FLG). 3. Graphene has edges and can have defects and grain boundaries where the bonding is disrupted.
Graphite (Gt)	An allotropic form of the element carbon, consisting of layers stacked parallel to each other in a three-dimensional, crystalline, long-range order	1. Adapted from the definition in the IUPAC Compendium of Chemical Terminology. 2. There are two primary allotropic forms with different stacking arrangements: hexagonal and rhombohedral.
Graphane	A single-layer material consisting of a two-dimensional sheet of carbon and hydrogen with the repeating unit of $(CH)_n$	1. Graphane is the full hydrogenated form of graphene with carbon bonds in the sp^3 bonding configuration.
Perfluorographane	Single layer material consisting of a two-dimensional sheet of carbon and fluorine with each carbon atom bonded to one fluorine atom with the repeating unit of $(CF)_n$	1. Perfluorographane has carbon bonds in the sp^3 bonding configuration. 2. Perfluorographane is sometimes referred to as fluorographene.
Epitaxial graphene	<graphene> A graphene layer grown on a silicon carbide substrate	1. Graphene can be grown by epitaxy on other substrates, for example, Ni (111), but these materials are not termed epitaxial graphene. 2. This specific definition applies only in the field of graphene. In general, the term 'epitaxial' refers to the epitaxial growth of a film on a single crystal substrate.
Bilayer graphene (2LG)	A two-dimensional material consisting of two well-defined stacked graphene layers	1. If the stacking registry is known, it can be specified separately, for example, as 'Bernal stacked bilayer graphene'.
Twisted bilayer graphene/ turbostratic bilayer graphene (tBLG/t2L)	A two-dimensional material consisting of two well-defined graphene layers that are turbostratically stacked, with a relative stacking angle, also known as commensurate rotation, rather than *Bernal* (hexagonal) rhombohedral	
Twisted few-layer graphene (t(*n*+*m*) LG)	A two-dimensional material consisting of a few-layers of graphene of *n* Bernal stacked layers which are situated with a relative stacking angle upon *m* Bernal stacked layers	

(Continued)

TABLE 3.1 (*Continued*) International Standardized Terminology for Graphene Family Nanomaterials

Nomenclature	Material Description	Notes to Entry
Trilayer graphene (3LG)	A two-dimensional material consisting of three well-defined stacked graphene layers	1. If the stacking registry is known, it can be specified separately, for example, as 'twisted trilayer graphene'.
Few-layer graphene (FLG)	A two-dimensional material consisting of three to ten well-defined stacked graphene layers	
Graphene nanoplate/ graphene nanoplatelet (GNP)	A nanoplate consisting of graphene layers	1. GNPs typically have thickness of between 1 and 3 nm and lateral dimensions ranging from approximately 100 nm to 100 µm.
Graphite oxide (GtO)	Chemically modified graphite prepared by extensive oxidative modification of the basal planes	1. The structure and properties of graphite oxide depend on the degree of oxidation and the synthesis method.
Graphene oxide (GO)	A chemically modified graphene prepared by oxidation and exfoliation of graphite, causing extensive oxidative modification of the basal plane	1. Graphene oxide is a single-layer material with a high oxygen content, typically characterized by C/O atomic ratios of approximately 2,0 depending on the method of synthesis.
Reduced graphene oxide (rGO)	A reduced oxygen content form of graphene oxide	1. This can be produced by chemical, thermal, microwave, photo-chemical, photo-thermal, or microbial/bacterial methods or by exfoliating reduced graphite oxide. 2. If graphene oxide was fully reduced, then graphene would be the product. However, in practice, some oxygen-containing functional groups will remain and not all sp3 bonds will return to sp2 configuration. Different reducing agents will lead to different carbon to oxygen ratios and different chemical compositions in reduced graphene oxide. 3. It can take the form of several morphological variations such as platelets and worm-like structures.

Source: Nanotechnologies—Vocabulary—Part 13: Graphene and related two-dimensional (2D) materials. https://www.iso.org/obp/ui/#iso:std:iso:ts:80004:-13:ed-1:v1:en, 2017, ISO/TS **80004-13:2017**

The recent standardization of GFN nomenclature has a twofold gain: in essence, the graphene industry can advance with an in-depth and shared understanding (Bianco et al., 2013). Furthermore, the devised terminology standard will elucidate the relationship between physicochemical characteristics of GFNs and their environmental and cytotoxic risks (Wick et al., 2014).

3.2.1 Properties of Graphene Determining Environmental and Cytotoxic Risks

A strong prerequisite of nanotoxicity is consideration of the potential toxic criteria of the NM itself. Graphene embraces various forms, one impediment which renders rationalization of GFN toxicity effects difficult (Zhu et al., 2013). In addition, the tuneable physiochemical properties of GFNs are often vigorously modified under diverse biological and environmental systems (Zhu et al., 2013). This dynamic change in GFN properties poses novel challenges to a growing understanding of bio-physicochemical interactions at the nano-bio interface. To partly address this problem, scientific research and classification strategies have identified a number of potentially hazardous GFN characteristics likely to be responsible for environmental and cytotoxic risks (Bianco and Prato, 2015).

3.2.1.1 Size and Shape

The properties of nanoscale substances are often directly derived from size; indeed, the hazardous potential of NMs is also presumed to be a consequence of size. It is often supposed that a decrease in size leads to an increase in surface area, thus increasing the number of atoms on the NMs' surface which are readily available to interact with biological surroundings (Oberdörster et al., 2005; Park et al., 2017). The lateral dimensions of NMs are also considered to dictate cellular uptake, biodistribution, accumulation, and clearance of NMs, in addition to many other particles' size-dependent biological responses (Sanchez et al., 2011). NMs <100 nm in size can infiltrate the cell membrane, <40 nm can permeate the cell nucleus, and smaller than <35 nm can traverse the blood-brain barrier (Jennifer and Maciej, 2013). The range of GFN lateral sizes can stretch from the nanoscale (10 nm) to microscale (>20 mm). As with many NMs, the size of GFNs significantly shapes the key textural properties of graphene derivatives (Hu and Zhou, 2013; Russier et al., 2013; Sanchez et al., 2011; Yue et al., 2012). The lateral dimensions of GFNs also determine crucial parameters key to cellular interactions, such as layer number and measure of deformability. Laterally larger plates are generally more capable of being reshaped than smaller plates of a uniform layer number (Sanchez et al., 2011). The cellular uptake of sheet-like nanostructures is not yet fully recognized. However, size-sensitive uptake is evident when the size of GFNs ranges from 100 to 500 nm, the smallest sheets are likely to cause the most severe toxicity; below 40 nm, however, smaller sheets may be the safer (Ou et al., 2016). Cells can adhere and disperse across the available surfaces of larger GFN sheets but successfully internalize smaller GFN sheets or experience a form of frustrated uptake with potential biological risk repercussions (Sanchez et al., 2011).

The dimensions and shape of GFNs may also define the degree of bioavailability in an organism, for instance the scope of airborne GFN deposition and clearance in the lung (Park et al., 2017). Determining the amount of airborne GFNs capable of adverse inhalation and deposition is a crucial parameter in the consideration of GFN toxicity often unseen. Park et al., in accordance with the Multiple-Path Particle Dosimetry (MPPD) model (www.ara.com), recently highlighted that the physiochemical properties of materials devising a mass median aerodynamic diameter (MMAD) less than 10 μm can be inhaled, those less than 4 μm moving as far as the lower respiratory tract (Park et al., 2017). GFNs with an area diameter of 25 μm and a thickness of 0.1 μm can generate a MMAD < 3 μm and deposit in the alveoli (Schinwald et al., 2012). The MMAD of GFNs should thus be determined during early stages of initial screening and preliminary toxicity investigation. The flake-like shape and diverse physiochemical properties of GFNs may, however, promote aerodynamic behaviour atypical of spherical particles. For this reason, the outcome generated using the MPPD model, which does not directly consider the dissimilar nature of sheet-like nanostructures, should carefully be construed (Park et al., 2017).

The mechanisms of prolonged NM retention and blood clearance are also dependent on size, but besides this are heavily influenced by shape (Longmire et al., 2008). The challenging clearance of fibres and nanostructures from the lung has dominated the agenda of particle and nanotoxicity since discovering the cardiovascular system is a target for inhaled particles (Donaldson and Seaton, 2012). Lung macrophages cannot effectively engulf or clear materials with a length of ~10 μm and an aspect ratio of >3 μm (Park et al., 2017). The large surface dimensions of graphene-like materials and their potentially sharp edges may cause toxic effects comparable with fibres. Furthermore, the shape and geometry of certain GFNs may vary in diverse environmental or biological systems. Biological interaction can cause flakes or sheets to twist, generating fibre-like nanostructures. It is also commonly acknowledged that GFNs composed of sharp edges can penetrate cell membranes (Li et al., 2013c).

Additionally, size and shape establish the extent to which GFNs intersect biological barriers including the skin, gastrointestinal tract, lung epithelium, blood-brain barrier, and placenta. There is, however, limited understanding of biodistribution and persistence of GFNs in the body, perhaps due to the challenging detection of GFNs in carbon matrices (Guo et al., 2013). Radio-labelling has been proposed as one approach for GFN detection. Examining carbon-14 radio-labelled graphene flakes, introduced into the lung of mice, detected small fractions in the liver (1%) and spleen (0.18%) (Mao et al., 2016). As an excessive dose was administered, however, the degree of translocation witnessed may be an overvaluation of the extent occurring at lower, more realistic dose levels of airborne GFNs. On the other hand, a high dosage of NMs is often associated with lower uptake in organs such as the gut, most probably owing to the formation of aggregates at extreme doses, so the aggregated NMs are subsequently too large to cross the intestinal barrier (van der Zande et al., 2014). Lower concentrations circumvent the likelihood of aggregation, hence distributing smaller materials which are presumed to cross biological barriers with ease. Restricted knowledge is available regarding GFN particle size-dependent translocation, though evidence has shown that translocation is also determined by characteristics other than size, such as surface chemistry and structure (Braakhuis et al., 2014, 2015). A recent study which modelled the interaction of GO with lung surfactant revealed that the distinct 2D formation of GO sheets cause the graphene derivative to lie down quickly upon contact with the lung surfactant film, suggesting the oxidized form of graphene is unlikely to translocate (Hu et al., 2015a). The lack of detailed understanding surrounding the mechanism underlying these size- and shape-dependent effects may also be due to the difficulty associated with the specific size determination of GFNs, as structures aggregate and vary with the dispersion medium *in vivo*. Size distribution is a crucial parameter for GFN nanotoxicity because the exposure dose is directly related to their mass and size (Hu and Zhou, 2013). Furthermore, the toxicity, cellular uptake, biodistribution, localization, organ uptake, and etiopathology of GFNs are greatly affected by size distribution (Cho et al., 2007, 2009; Dan et al., 2012; De Jong et al., 2008). GFNs possess a wide size variation, stretching from tens of nanometres to several micrometres. Respectively, varying size distributions produce inhomogeneous physicochemical properties (Hu and Zhou, 2013). Accordingly, GFN toxicity and pharmacokinetics assessments starved of size distribution consideration are unreasonable. It has previously been suggested, however, that an optimum size range for uptake in terms of producing a biological exists (Jiang et al., 2008; Patlolla and Vobalaboina, 2005). GFNs of narrow size distribution and uniform properties are critical to precisely define the risks posed to environmental and human health by GFNs. Precise control of NM size and size distribution will considerably influence the rational assessment of GFN environmental and cytotoxicity threats. Smaller GFNs have also demonstrated more oxidative stress mediated damage than larger structures; in the presence of compact graphene sheets adherent human skin fibroblasts extensively generated more reactive oxygen species (ROS) than in the presence of less densely packed GO (Fojtů et al., 2017).

3.2.1.2 Layer Number

Currently, little is understood about the possible differences in the biological behaviour of GFNs with varying layer numbers and lateral sizes. The number of layers in a GFN is, however, an important consideration as layer number subsequently determines thickness, specific surface area, and bending stiffness/elasticity (Wick et al., 2014).

It is presumed that the adsorptive capacity of GFNs for biological molecules will grow as the structure's layer number decreases; in other words, the greater the layer number, the lower the adsorptive capacity for biological molecules (Sanchez et al., 2011; Wick et al., 2014).

The role of stiffness in the toxic potential of sheet-like nanostructures has not yet been fully elucidated. The lessons learned from the safety assessment of CNTs and fibres, however, have taught us that the stiffness of material significantly influences pathological responses. Thirty-one thinner GFNs such as monolayer graphene or GO are more susceptible to deformation by weak forces than multilayer materials (Bagri et al., 2010; Guo et al., 2011; Patra et al., 2009). An increase in GFN stiffness thus provides a platform to potentially form rigid bodies during cellular interactions.

3.2.1.3 Surface Chemistry and Charge

The graphene family includes NMs with considerably different chemical compositions and C/O ratios and various types of oxygen-containing functional groups. For instance, the surface of pristine graphene is relatively inert and hydrophobic, generally possessing a water contact angle near 90°, offering relatively low solubility. Additionally, biological interactions tend to occur at the edge of graphene sheets or adjacent to defect sites (Hu and Zhou, 2013). GO, on the other hand, is an oxygen-rich derivative which contains a multitude of oxygen functionalities such as carbonyl, epoxy, and hydroxy groups, established along the planar surfaces and edges of graphene sheets during the oxidation and exfoliation of graphite. The surface of GO comprises hydrophobic islands with hydrophilic regions (typical water contact angle 40°–50°), exhibiting a diverse extent of basal reactivity, and contains negative charges on edge-sites associated with the presence of carboxylate groups (Cote et al., 2009). This oxygen-rich, hydrophilic nature greatly improves the solubility and dispersity of GO, in comparison to pristine graphene (Hu and Zhou, 2013). Due to a relatively high solubility and ease of functionalization, many biomedical developments have integrated the use of GO NMs as opposed to alternative GFNs. To ensure the long-term stability of GO in physiological solutions such as saline or culture medium, however, further surface functionalization is often required (Feng and Liu, 2011; Sun et al., 2008). Typical GO compositions correspond to C/O ratios ranging from 4 to 2. GO can also be reduced to produce rGO via thermal, chemical, or electrochemical methods. Oxygen removal during reduction produces basal vacancy defects and increases the C/O ratio to ~12, though ratios as large as have been reported, yielding intermediate hydrophilicity and basal reactivity (Bagri et al., 2010).

When assessing the environmental and cytotoxic risks of GFNs, it is critical to consider the surface chemistry and charge, but the innumerable GFN production processes must also be appreciated. Given that oxygen coverage is subject to the degree of oxidation during production, an assortment of production methods will yield a variety of inconsistent final products (Dreyer et al., 2010). Thus far, the specific toxic mechanism of GFN surface chemistry and charge has yet to be rationalized, but research studies have recently begun to highlight the need for more in-depth consideration of these imperative parameters and the substantial impact they can have on GFN toxicity.

An examination of the toxicological behaviour of GO NMs prepared by different oxidative routes (Staudenmaier, Hofmann, Hummers, and Tour) in lung epithelial cells indicated that GFN toxicity is governed by the amount and type of oxygen functionalities present. GO NMs prepared via the Hummers and Tour method exhibited a greater oxygen content, lower C/O ratio, and greater degree of toxicity than those prepared following the Staudenmaier and Hofmann methods (Chng and Pumera, 2013). Similarly, a further study comparing the toxicity of GO nanoribbons (GONRs) and nanoplatelets (GONPs) indicated that the toxicity profile of each of the materials was determined by the striking difference in the amount of the C=O groups present (Khim Chng et al., 2014), hence correlating the toxicity profile of GO NMs to enhanced oxygen content and diminished C/O ratio.

In contrast, a limited amount of evidence has suggested that GO is less toxic than rGO due to enhanced high-oxygen content, and thus improved hydrophilicity, dispersity, and stability (Chng et al., 2014; Jarosz et al., 2016; Jin et al., 2014; Sasidharan et al., 2012). Chatterjee et al. proposed that the distinct hydrophilicity of GO might contribute to the internalization and uptake of the oxidized carbon nanostructures. The hydrophobic surface of rGO, however, causes adsorption and aggregation at the cell surface, limiting uptake (Chatterjee et al., 2014). More recently, Majeed et al. suggested that an increase in oxidative functionalization of graphene reduces its toxic impact. Their assessment of the role of surface chemistry in GFN cytotoxicity demonstrated that pristine graphene was the most toxic. The extent of toxicity decreased as the oxygen content of the graphene materials increased. Increased surface oxygen also encouraged long-term dispersity in aqueous physiological solutions. The group proposed that rather than ingesting poorly dispersed toxic aggregates, the aforementioned high level of dispersity may have resulted in the internalization of individual, oxidized graphene flakes, ~1 nm thick (Majeed et al., 2017).

Like many other GFN characteristics of emerging significance, the biological relevance of surface charge remains unsolved. The effects of this parameter on the environmental and cytotoxic effects of GFNs may be rationalized by electrostatic interactions that trigger biomolecule and ion adsorption (Hu and Zhou, 2013). In general, neutral surfaces are considered biocompatible and cationic surfaces are considered more toxic than anionic surfaces. Cationic surfaces are also more expected to provoke haemolysis and platelet aggregation (Goodman et al., 2004). This may be due to their affinity towards negative phospholipids or proteins (Hu and Zhou, 2013). As such, the high haemolytic activity of GFNs has been ascribed to the interaction of negatively charged surface oxygen with positively charged phosphatidylcholine on the outer cell membranes of red blood cells (Liao et al., 2011). The electrostatic disruption modifies the morphology of cells and causes significant lysis (Liao et al., 2011). Additionally, the surface charge of GFNs affects cellular uptake and internalization (Jiang et al., 2010; Wang et al., 2013a; Yue et al., 2011). Owing to the strong electrostatic repulsion between negatively charged GO and corresponding cell surface, GO internalization in non-phagocytic cells is deemed negligible (Yue et al., 2012). However, cellular internalization of negatively charged NMs is reportedly made possible through the availability of cationic sites on the cell surface. Negatively charged NMs bind to positively charged groups, such as amino groups on the cell surface, and are taken up by scavenger receptors (Chatterjee et al., 2014; Jarosz et al., 2016; Yue et al., 2011).

The surface charge of NMs has previously demonstrated crucial importance in modulating membrane potential (Arvizo et al., 2010). Inherent cellular membrane potential is regulated by ion permeability, changes in ionic concentrations of intracellular and extracellular environments, and electrical and agonist stimuli. Compounding this, the membrane potential itself can attune intracellular pathways such as calcium flux, the cell cycle, proliferation, and apoptosis, all of which are processes that determine normal cell structure and function and disease progression (Monteith et al., 2007). Atomically thin GO sheets can induce an aggregatory response in human platelets like that evoked by thrombin, a potent physiological platelet agonist (Singh et al., 2011). Intravenous administration of GO into mice also elicited substantial pulmonary thromboembolism. The pro-thrombotic character of GO was attributed to surface charge as rGO had a considerably lesser effect on platelet aggregation. GO-induced platelet aggregation is thus motivated by the release of intracellular Ca^{2+}, an essential moderator of intracellular signalling concerning platelet function and activation of non-receptor Src family protein tyrosine kinases. Interestingly, graphene surfaces modified with amine functional groups are not equipped with thrombo-toxic properties. Stimulatory effects on human platelets were not exhibited, nor was pulmonary thromboembolism exhibited in mice; rather, a cyto-protective action was exhibited (Singh et al., 2012).

3.2.1.4 Surface Functionalization

Due to electrostatic charges and nonspecific protein binding many of the graphene family members tend to aggregate in biological fluids (Guo and Mei, 2014; Pan et al., 2012). To address the issue of poor dispersity and provide adequate suspension in physiological solutions, stabilizing agents or surfactants have previously been employed. The problematic toxicity of the latter, however, has been uncovered by previous experience with CNTs (Sanchez et al., 2011). On the other hand, the development of functionalized GFNs has provided improved solubility and biocompatibility and reduced environmental and cytotoxic impacts (Guo and Mei, 2014). Surface chemical functionalization alters the surface chemistry and charge of GFNs, modulating nanotoxicity (Rafiee et al., 2010). Surface functionalization or modification also commonly affects the blood retention and organ translocation of NMs, either placating or provoking toxicity (Bettinger et al., 2009). For example, hydrophilic carboxyl-functionalized graphene, with an increased degree of oxidation, exhibited intracellular internalization in comparison to the ill consequences of hydrophobic, non-functionalized graphene (Sasidharan et al., 2011).

The functionalization of GFNs can be accomplished through one of two routes: covalent conjugation or noncovalent physisorption (Liu et al., 2013a; Pan et al., 2012). Unlike the unclear effects of many other GFN parameters, covalent functionalization with compounds such as aliphatic and aromatic amines (Arvidsson et al., 2013; Liu et al., 2013a; Singh et al., 2012), amino acids (Liu et al., 2013a), block copolymers (Seo et al., 2011) like pluronic and (Hu et al., 2012) tetronic, in addition to polyethylene glycol (PEG) (Li et al., 2014; Yang et al., 2010a; Yang et al., 2010b), PEGylated poly-L-lysine (PLL) (Zhang et al., 2012), lactobionic acid-polyethylene glycol (LA-PEG) (Wang et al., 2013a), polyethylenimine (PEI) (Wang et al., 2013a), poly(ε-caprolactone) (Wojtoniszak et al., 2012), polyvinyl alcohol (Yang et al., 2011c), amine terminated biomolecules (Liu et al., 2013a), proteins (Lee et al., 2011), silanes (Liu et al., 2013a), enzymes (Liu et al., 2013a), ssDNA (Seo et al., 2011), carboxyl groups (Ou et al., 2016), dextran groups (Sahu et al., 2012; Zhang et al., 2011a), and chitosan (Wu et al., 2015) have been widely reported to largely decrease the toxicity and improve the biocompatibility of graphene (Guo and Mei, 2014; Sasidharan et al., 2011). Noncovalent functionalization is often regarded as more versatile than covalent methods of surface modification. This route exploits electrostatic binding, π–π interaction, van der Waals forces, and hydrophobic interactions (Liu et al., 2013a; Pan et al., 2012). Multi-functionalization is also an emerging trend. For instance, graphene demonstrated a wide distribution in the zebrafish, but when coated with polylactic acid and fluorescein o-methacrylate for imaging purposes, the functionalized atomic structure did not considerably impact the survival rate of zebrafish (*Danio rerio*) embryos (Gollavelli and Ling, 2012). Similarly, self-assembled graphene complexes modified with both PEG and anticancer drugs such as curcumin or doxorubicin did not alter the development of zebrafish from the embryo to the larval stages (Park et al., 2017).

Though GFN toxicity is usually diminished as a result of surface functionalization, it is not always eliminated. Furthermore, chemical modification can also antagonize GFN environmental and cytotoxic risks. For instance, the reported IC50 for the human breast cancer line MCF-7 of covalently PEGylated GO and noncovalently PEGylated rGO was ~80 mg/L and ~99 mg/L, respectively. Even though PEGylation was accomplished by two distinct methods of surface interactions, the toxic impact of the resulting materials was quite similar (Robinson et al., 2011). Moreover, unmodified GO and carboxyl-functionalized graphene elicited plasma membrane damage, induced oxidative stress, and altered metabolic pathways and cell structure (Lammel et al., 2013). Furthermore, GO nanoribbons water-solubilized with 1,2-distearoyl-sn-glycero-3-phosphoethanolamine-N-amino-PEG (PEG-DSPE) demonstrated a significant dose- and time-dependent reduction in the viability HeLa, MCF-7, NIH 3T3 mouse fibroblasts, and Sloan Kettering breast cancer cells (SKBR3) (Chowdhury et al., 2013).

3.2.1.5 Surface Area and Aggregation State

The biological interactions of NMs and their potential toxicity are largely representative of their surface characteristics and large surface area to mass ratio (Nel et al., 2009). In fact, surface area is considered the most the biologically effective dose metric for nanotoxicity (Schmid and Stoeger, 2016). As the specific surface area increases, a greater proportion of atoms are displayed on the surface, enhancing surface reactivity, adsorption properties, and potential toxicity.

In general, the surface areas of GFNs tend to decrease as the number of graphitic layers within the structure grows. The remarkable theoretical surface area of pristine, monolayer graphene, however, is at least one order of magnitude beyond the surface area of most biologically studied NMs. Due to the nanostructures' atomic thickness, sp^2-bonded carbon atoms are exposed to their surroundings from both sides of the crystalline lattice, yielding ultrahigh surface areas of up to 2600 m^2/g. Monolayer GO often exhibits contending surface areas but is determined by atomic-scale roughness (Sanchez et al., 2011).

Consequently, the measurements and stability of GFN surface area in biological solutions remains somewhat ambiguous. Due to van der Waals attractions on GFN inter-sheet surfaces, aggregation or restacking is inevitable. Unlike spherical nanoparticles that make only point contacts during aggregation, sheet-like nanostructures restack and pack face-to-face. Review sheet-to-sheet aggregation thus significantly compromises the unique properties possessed by individual sheets, such as high specific surface area (Yang et al., 2011d). Face-to-face packing also occurs during processes such as filtration, centrifugation, or drying, instigating stacking between interlayer spaces and rendering sheet surfaces unreachable for biological interactions. Furthermore, the characterization of GFNs is most often completed after the NMs have been dried, that is by measuring the properties of dry powders (Montes-Navajas et al., 2013). However, the environmental and biological applications of GFNs are based on the use of liquid phase, physiological solutions. Due to the unavoidable stacking and agglomeration of sheets during solvent removal, the surface area and other properties of dry GFN powders will undoubtedly differ from those of suspended GFNs. It will thus be difficult to effectively measure the surface area of GFNs in environmental or biological systems using conventional techniques such as filtration, drying, or vapor adsorption methods.

3.2.1.6 Structural Imperfections and Chemical Impurities

Structural and chemical defects could inadvertently or inexorably disrupt the reactive microenvironment and bio-response pathways of GFNs (Banhart et al., 2010; Stadler et al., 2011). Graphene sheets possess liquid-like topography, a trait responsible for the extreme flexibility of the honeycomb lattice. Furthermore, the atomic structure is not seamlessly flat, sheets typically exhibit nanometre-scale wrinkles, ripples, crumples, and folds (Deng and Berry, 2016; Meyer et al., 2007). Additionally, the solid-liquid adhesion energy of GFNs ranging from 1–5 sheets (0.45 J–0.31 J m^{-2}) are larger than those of typical micromechanical structures (Koenig et al., 2011). Together these characteristics influence the intense adhesion of graphene sheets on bacterial and cellular membranes, which then induces growth, or an envelopment of bacteria, resulting in noninvasive damage (Mohanty et al., 2011).

The edges of graphene sheets are generally sharp and decorated with highly reactive dangling bonds (Yan et al., 2012). The sharp edges of graphene-like sheets pose threats of physical damage reportedly due to physical disruption of cellular, bacteria, and viral membranes (Akhavan and Ghaderi, 2010; Liu et al., 2011). Both the intense chemical and mild biological methods of GO reduction promote nano-hole formation within sheets (Jiao et al., 2011; Kotchey et al., 2011). When the initial hydroxyl and epoxy groups set to be cleaved from the graphene plane are close to one another, carbon atoms are also easily removed alongside, hence the formation of nano-holes (Bagri et al., 2010; Erickson et al., 2010). The carbonyl-rich nano-holes certainly must affect the solubility, conductivity, stiffness, and affinity sites of GFNs in physiological solutions, but as of yet this matter requires further consideration (Hu and Zhou, 2013). The chemical purity of NMs is also an important consideration. Residual synthetic contaminants such as metals are often responsible for detected toxicity, masking the true nature of the NM itself and giving rise to conflicting nanotoxicity data (Nezakati et al., 2014; Yin et al., 2015). NM impurities also often cause nonspecific interference in toxicological assays (Sanchez et al., 2011; Wick et al., 2014). Furthermore, impurities provide a new platform for biomolecule adsorption. A new biological character is thus produced upon interactions with environmental or biological systems, one which likely would not emerge in the presence of high-purity NMs (Monopoli et al., 2012).

Both *in vitro* and *in vivo* studies have evidenced that heavy metal impurities like iron and nickel reside as trace contaminants after the manufacture of other carbon-based NMs like single- and multi-walled CNTs. The presence of the remaining metal catalysts is, to some extent, responsible for CNT toxicity (Aldieri et al., 2013). Unlike CNTs, however, metal catalysts are not typically used for the assembly of GFNs, which therefore do not typically contain residual metal catalysts. However, GO production methods use a variety of oxidizing and intercalation reagents such as permanganate, nitrate, sulphate, chromate, peroxide, and persulfate, in addition to reductants like hydrazine and borohydride. Accordingly, salts and acidic or metallic impurities can reside if adequate purification of the final product is not accomplished. Due to the gelation triggered by physical interactions of giant molecular disks during washing, GO purification can be a tedious process and the complete removal of impurities is often challenging (Kim et al., 2010a). For instance, conventional GO NMs often contain residuals of highly mutagenic Mn^{2+} and Fe^{2+}, so the nonspecific release of these ions will likely result in enhanced environmental and cytotoxicity. For instance, highly pure GO with sound stability and dispersity did not garner significant toxic effects *in vitro* and following intraperitoneal injection no obvious inflammatory responses or granuloma formations were witnessed, in comparison to the adverse effects generated by conventional GO (Ali-Boucetta et al., 2013). Furthermore, by reason of soluble acidic impurities, inefficiently washed GO acted as an antibacterial agent yet highly purified GO neither inhibited nor stimulated *E. coli* growth (Roberts, 2016).

3.2.1.7 π Interactions and Electronic Conductivity

The π-π stacking interactions between aromatic residues and carbon-based NM largely dictate NM-protein interactions (Zuo et al., 2011). In fact, π-π stacking interactions are considered to dominate NM-protein interactions and modulate both adsorption kinetics and thermodynamics. Graphene's unique sp^2 hybridization exhibits in-plane σ bonds, the rigid backbone of the 2D atomic structure. In addition to π orbits perpendicular to the plane controlling inter-sheet interactions and as such are responsible for the aggregation of GFNs in physiological solutions. The GFN-related π bonding system impedes covalent interaction with environmental and biological systems but facilitates versatile reactions capable of forming complexes with organic compounds through π–π or H–π bonds (Balapanuru et al., 2010; Kim et al., 2011; Wu et al., 2011). The π-stacking interactions therefore encourage the random adsorption of biocomponents such as nucleic acids, proteins, enzymes, bacteria, and viruses on GFN surfaces (Hu and Zhou, 2013). For example, extensive nonspecific adsorption of serum albumin, proteins, and lectins on graphene-like surfaces have been widely recognized (Chen et al., 2011b; Haque et al., 2012; Salihoglu et al., 2012). π-Stacking interactions can also lead to a collapse of secondary and tertiary protein structures (Zuo et al., 2011). Consequently, these interactions between biomolecules and GFNs should not be overlooked (Pan et al., 2011). This nonspecific binding can produce competitive adsorption and trigger unexpected and/or adverse bio-responses.

The π-bond structure and high surface area of graphene-like materials yield electron mobility beyond 15,000 $cm^2/(V \cdot s)$ at room temperature, values considerably greater than those of CNTs (Shin et al., 2016). As such, graphene as an electronic acceptor for the elimination of electron pairs has been widely noted (Lightcap et al., 2010; Matsumoto et al., 2011). By this means, the superlative electronic conductivity of GFNs may lead to peculiar environmental and biological behaviours (Hu and Zhou, 2013). The high electrical conductivity of GFNs has been shown to stimulate signal channels and the corresponding proteins via electron transfer, which in turn affects cell growth, activity, and signalling. However, GFNs, like CNTs, may also disturb cell function, though this theory requires further exploration (Hu and Zhou, 2013; Shin et al., 2016).

3.2.1.8 Interrelationship Between the Parameters

As is evident from the previous consideration of physicochemical properties responsible for determining the environmental and cytotoxic risks of GFNs, the various characteristics of graphene-like materials are often interdependent. Accordingly, the fundamental characteristics of GFNs should be considered in terms of the interrelationship between the parameters and the health and environmental risks of any NM. For example, GFN size affects electron density and conductivity, charge density, and solubility (Matsumoto et al., 2011; Stadler et al., 2011). Graphene sheets must be of an appropriate size, thickness, and lateral dimension to suitably interface with biological systems. Gathering evidence also suggests that GFN toxicity is regulated by surface oxygen content (Das et al., 2013; Sanchez et al., 2011; Shi et al., 2012).

Surface chemistry also drastically effects the bio-responses and environmental behaviours of GFNs. Theoretically, smaller-sized graphene sheets with relatively more edges exhibit stronger hydrophilicity. The pH of physiological solutions should also be considered when evaluating the environmental fate and toxic risks of GFNs. In addition to size, the pH of GFN surroundings can affect the degree of ionization at the edge of graphene sheets where carbonyl groups likely exist and thus transform surface chemistry (Kim et al., 2010b). Moreover, numerous reports have evidenced the dose-dependent toxicity of GFNs in animals and cells (Ou et al., 2016). In addition, these parameters are also affected by reaction medium and time (Chang et al., 2011; Maiorano et al., 2010). Thus, the importance of GFN systematic properties and related microenvironment is noteworthy for ecological and human health sustainability as their interdependent characteristics modulate GFNs interactions with organism and thus their environmental and cytotoxic fates (Hu and Zhou, 2013).

3.3 Environmental Risks of GFNs

It is reasonable to assume that GFNs will inevitably be released into the environment during production, use, and/or disposal (Hammes et al., 2013). The discharge and potential toxicity of GFNs in the environment is thus an urgent issue for concern. Upon release, GFNs will interact with a variety of physiochemical and biological factors present in waters, sediments, and soils. Members of the graphene family will likely be exposed to aquatic and terrestrial ecosystems where they may bioaccumulate, be further transported, chemically transformed, and/or eventually become ingested by organisms, therefore entering the food chain (He et al., 2017). The nature of the exposed 2D carbon NM (graphene family member, e.g., GO, rGO, etc.), exposure dose, and the specific physiochemical characteristics of the graphene family, strongly influence the potential environmental threats posed by GFNs, in addition to the length of exposure phase, growth medium, administration route, and target species (Jastrzębska et al., 2012; Jiang et al., 2009; Khosravi-Katuli et al., 2017; Sun et al., 2016).

Currently, the environmentally relevant concentrations of GFNs have not yet been reported. However, the environmental fate of GFNs is considered comparable to conventional NMs, like functionalized CNTs (Deng et al., 2017). Accordingly, the predicted environmental concentrations of GO, the most extensively studied GFN in terms of environmental fate, is thought comparable to multi-walled CNTs (MWCNTs), carbon-based nanostructures that present environmentally relevant concentrations ranging from 0.001–1000 μg/L (Lanphere et al., 2014; Nouara et al., 2013; Zhang et al., 2017a, 2017b). Furthermore, despite their flat nature, GO nanosheets (GONS) predominantly follow the same theories of stability and transport as many other 'carbonaceous-oxide' NMs due to their ionic response and the presence of natural organic matter (Lanphere et al., 2014). GO tends to become less stable in groundwater, which ordinarily has more hardness and a lower concentration of natural organic matter. GO eventually sediments and can be removed in subsurface environments. In contrast, in surface waters that contain more organic material and less hardness GO remains stable and is transported farther (Lanphere et al., 2014).

Despite the increasing use and production, there is still an incomplete understanding of the environmental fate of GFNs (De Marchi et al., 2018). This may be due to several inherent difficulties including the lack of specific quantification analysis, which may hinder GFN environmental determination, the problematic concentration measurement of GFNs, and the use of non-standardized experimental techniques and parameters. Furthermore, most studies restrict their focus to water and soil ecosystems; there are currently no available studies detailing the release or fate of GFNs in the atmosphere (Park et al., 2017). It is therefore imperative that the scientific community shift focus to understanding the interactions of GFNs within diverse ecosystems and their subsequent impact on residing organisms (Fojtů et al., 2017).

3.3.1 Toxicity of GFNs in Aquatic and Terrestrial Organisms

In order to understand the environmental fate of the graphene family, several studies have explored the impact of GFN exposure to individual representatives of various trophic levels, including microorganisms, plants, and multicellular aquatic or terrestrial organisms. The environmental impact of GFNs is first addressed by examining exposure to simple organisms located toward the bottom of the food chain, such as bacteria and algae. These microorganisms provide nourishment for invertebrates such as crustaceans, nematodes, and earthworms (Peralta-Videa et al., 2011). Few studies, however, have performed in-depth examinations of GFNs' interaction with soil environments and the subsequent impact on terrestrial organisms. Soil systems are generally very complex, so research deliberating the release fate and transport of GFNs in a soil system should be considered (Jastrzębska and Olszyna, 2015).

In an aquatic system, crustaceans like Daphnia magna are often the ultimate recipients of an ever-increasing quantity and range of environmental contaminants. Accordingly, the exposure of Daphnia magna to GFNs has been broadly examined. In turn, aquatic invertebrates are consumed by fish. Many studies concerning the toxic impact of GFNs at this trophic level are based on a model animal such as zebrafish (Chen et al., 2012b; Souza et al., 2017; Wang et al., 2015b; Zhang et al., 2017a, 2017b; Zhou et al., 2012b). What is more, many of these investigations have concentrated heavily on the bioaccumulation and distribution of GFNs in a variety of tissues and organs, in addition to the excretion of the planar NMs, yet the effect of administration route lacks consideration (Jastrzębska and Olszyna, 2015). The current fate and impact of GFN exposure to aquatic and terrestrial organisms have been summarized in Figure 3.3.

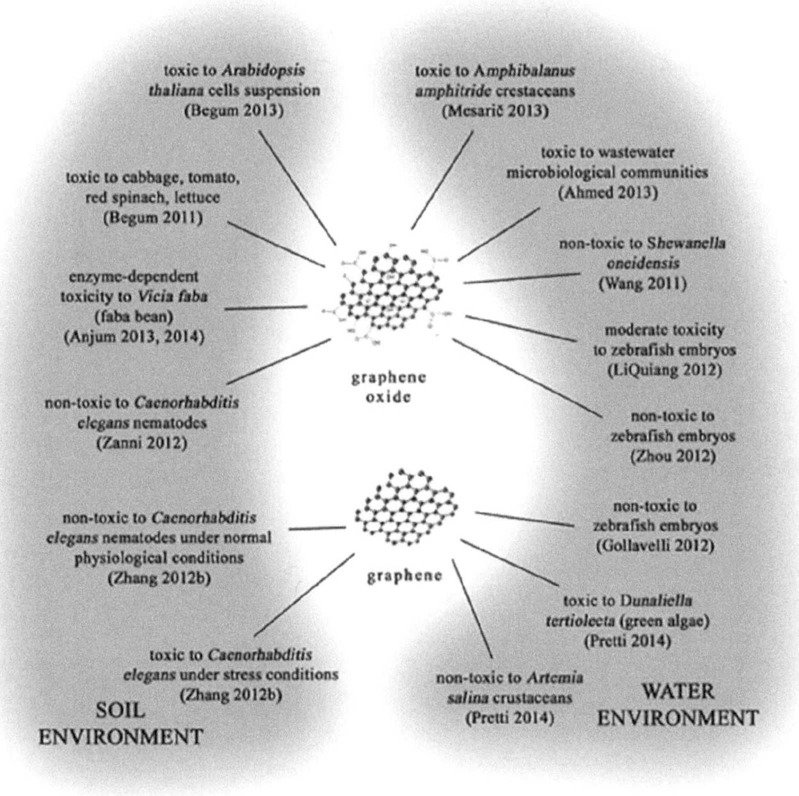

FIGURE 3.3 Schematic summary of the impact of GFN exposure to aquatic and terrestrial organisms. (From Jastrzębska, A.M. and Olszyna, A.R., *J. Nanopart. Res.*, 17, 40, 2015.)

3.3.1.1 Interactions of GFNs with Microorganisms

Concern has been raised for GFN-induced toxic effects on nonpathogenic microorganisms that play an important role in the environment. Bacteria are model organisms, ever-present in the environment and receptive to diverse factors (Bianco, 2013). To date, GFN-induced bacteria toxicity has been extensively considered, yet conflicting reports prevail. Several studies describe GFNs, particularly GO, as outstanding biocompatible NMs (Fojtů et al., 2017). At the same time, GFNs have been reported to exhibit promising antibacterial properties capable of provoking substantial *in vitro* and *in vivo* toxicity in a number of diverse microorganisms (Akhavan and Ghaderi, 2010; Hu et al., 2010). This reported combination of biocompatibility and simultaneous antimicrobial activity renders GFNs ideal candidates for biomedical applications (Fojtů et al., 2017). However, other investigations have suggested that toxic responses due to GFN exposure are negligible (Ruiz et al., 2011).

Thus far, many mechanisms of GFN bacterial toxicity have been suggested (Figure 3.4) (Cai et al., 2011; Hu et al., 2010; Liu et al., 2011). It is generally accepted that graphene sheets first entrap bacteria; the sharp edges of the sheets subsequently penetrate the cytoplasmic membrane which then facilitates GFNs in altering physiological activity of bacteria, thus ultimately leading to bacterial inactivation. Having assessed the antibacterial activity of Gt, GtO, GO, and rGO toward *E. coli*, Liu and co-workers suggested that the antimicrobial mechanism of GFNs involved three distinct actions, which were comparable to the previously proposed cytotoxicity mode of CNTs. To be precise, cells first deposit on GFNs, the sharp edges of the GFN sheets then induce membrane stress, followed by subsequent superoxide anion-independent oxidation.

The induced membrane stress ultimately contributed to physical disruption and leakage of cellular content, including RNA. Furthermore, GO exhibited superior antibacterial activity followed by rGO, Gt, and GtO (Liu et al., 2011). GO and rGO nanowalls deposited on stainless steel substrates also incited RNA and cytoplasmic material effluence from *E. coli* and *S. aureus* strains via cell membrane disruption caused by direct contact with the extremely sharp edges of the nanowall structures (Akhavan and Ghaderi, 2010). The authors suggested that both GO and rGO nanowalls directly

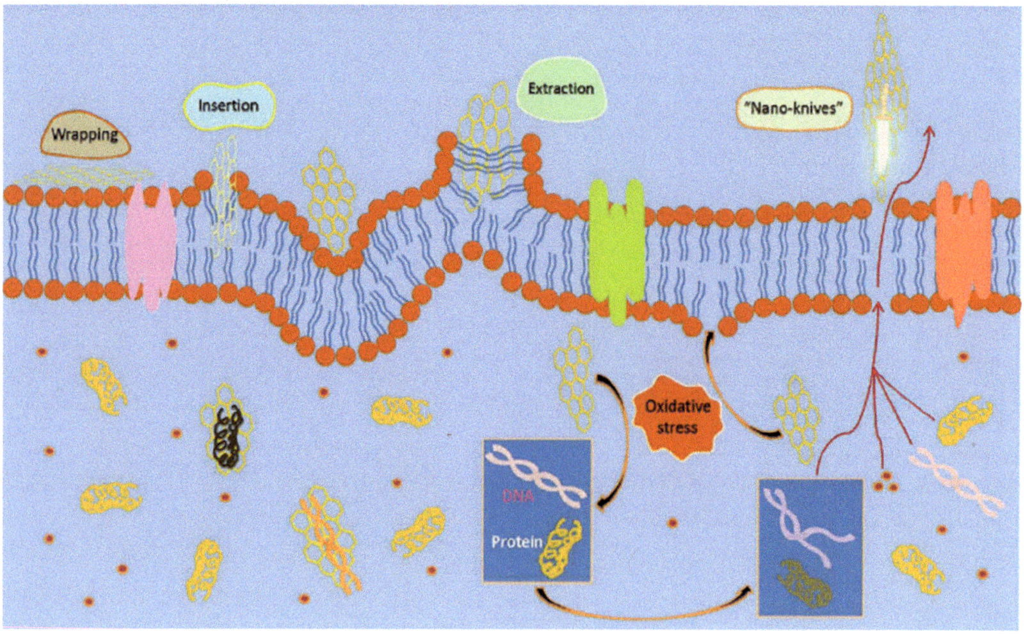

FIGURE 3.4 Schematic illustration of GFN antibacterial mechanisms. (From Zou, X. et al., *J. Am. Chem. Soc.*, 138, 2064–2077, 2016.)

interacted with the negatively charged bacterial membrane, thus improving the mutual charge transfer which caused further membrane damage, eventually leading to total membrane disruption. This proposed mechanism of toxicity was supported by the superior toxicity of rGO nanowalls. The reduction of GO leads to the sharpening of sheet edges, thus enhancing physical interactions. Considering also that Gram-positive bacteria lack an outer membrane toxicity induced by physical interaction with the sharper sheets of rGO also rationalizes the increased toxic responses from outer-membrane lacking Gram-positive bacteria *S. aureus*. Furthermore, the reduction of GO also improves the possibility of electron transfer (Akhavan and Ghaderi, 2010). As of yet the molecular mechanism of antibacterial activity is not well understood. GO and bacteria form a redox system via electron transfer and a mediator, GO acts as an electron acceptor and bacteria as a donor, but the proposed electron transfer pathways are conflicting. Salas et al. first revealed that GO can act as a terminal electron acceptor for and is thus reduced by heterotrophic, metal-reducing, environmental bacteria such as the genus *Shewanella* (Salas et al., 2010). These bacteria are known as exoelectrogens as they engage in extracellular electron transfer (EET) with solid compounds such as iron oxide (Fredrickson et al., 2008; Hau and Gralnick, 2007; Nealson and Scott, 2006). The pathway uncovered by Salas and coworkers indicated that electrons flow from the inner-membrane quinine pool to the periplasmic protein and are mediated by cytochromes MtrA, MtrB, and MtrC/OmcA, while CymA, a protein involved in anaerobic respiration, was not essential for electron transfer (Myers and Myers, 1997). However, a later report from Jiao et al. highlighted CymA as an indispensable component of the GO-bacteria redox system (Jiao et al., 2011). More recently, after analyzing computer simulation data, Tu et al. realized a potentially novel molecular mechanism for GFN-induced bacterial membrane degradation. Initially Tu and co-workers observed a membrane cutting and insertion mechanism, similar to that described in the case of GO and rGO nanowalls. Further examination uncovered subsequent extraction of lipid molecules which was directed by the complicated and collective movements of phospholipids along the graphene nanosheet. Lipid molecule fluctuations were first propelled by seemingly short-range van der Waals attractions. Once extracted, strong hydrophobic interactions between graphene and the lipid tails motivate co-operative phospholipid climbing. Both the insertion/cutting mode and destructive extraction of phospholipids degrade the *E. coli* cell membrane, thus decreasing cell viability (Tu et al., 2013). Other studies have reported the oxidative stress induced antibacterial activity of GO and rGO (Gurunathan et al., 2012). However, it is too early to draw a solid conclusion about the bacterial toxicity of graphene. Published studies concerned with the antimicrobial activity of GFNs are not consistent (Fojtů et al., 2017). Liu et al. reported that the greatest toxicity towards *E. coli* was induced by GO (Liu et al., 2011). In agreement, Ahmed and Rodrigues revealed high toxic effects of GO, demonstrating the adverse effects of the oxidized graphene form on bacterial metabolic activity and bacterial viability in addition to the biological removal of nutrients (Ahmed and Rodrigues, 2013). The antibacterial activity of GONS was also demonstrated toward both Gram-negative and Gram-positive bacteria *E. coli* and *S. iniae*. The study indicated that the generation of ROS strongly influenced the antibacterial mechanism and demonstrated that GONS proved more toxic toward outer-membrane lacking, Gram-positive bacteria (Krishnamoorthy et al., 2012; Seabra et al., 2014).

Compared with GO, rGO has limited reactive groups and may therefore exhibit weaker affinity for bacteria, which renders a slighter degree of toxicity (Hu et al., 2010). The superior antibacterial activity of GO was demonstrated towards a representative phytopathogenic bacterium, *X. oryzae pv. oryzae*, commonly responsible for rice infections. GO demonstrated greater antibacterial activity than both rGO, with the former and latter dispersed in different media and bismerthiazol, a common bactericide. GO-induced bacterial inactivation was likely due to physical interaction with sharp sheet edges and the generation of ROS (Chen et al., 2013).

Other studies, however, have asserted that rGO is more toxic than GO. Akhavan and Ghaderi exposed rGO nanowalls to be more toxic than its oxidized counterpart (Chen et al., 2013; Sanchez et al., 2011). A similar trend was reported by Hu et al., but studies by both Hu et al. and Feng and Liu also recognized that the interaction of bacteria with GO sheets triggered bacterial membrane disruption and glutathione oxidation (Feng and Liu, 2011; Hu et al., 2010).

Despite the specific mechanism of GFNs antimicrobial activity, wastewater treatment plants will be one of the ultimate depositories for graphene-based wastes. Efficient waste treatment such as biological wastewater treatment processes rely heavily on the purposes of diverse microbial communities like the bacteria and protozoa found in the activated sludge. The presence of graphene-based contaminants in the wastewater may adversely affect the functions of these microorganisms yet only a handful of studies have explored the impact of GFNs on microbial wastewater communities (Dasari Shareena et al., 2018; Suárez-Iglesias et al., 2017). Ahmed and Rodrigues reported that the presence of GO significantly compromised the bacterial community found in activated sludge by diminishing metabolic activity, viability, and nutrient removal capacity. Authors observed a significant increase in ROS production in the presence of increased GO concentrations, in addition to the accumulation of GO within the activated sludge floc matrix. GO also deteriorated effluent and sludge quality. The final effluent exhibited increased levels of BOD, ammonia nitrogen, and phosphate, as well as greater turbidity. Meanwhile, the sludge possessed less dewaterability and required higher suction times. Such demerits generate regulatory infringements and increase the cost of sludge disposal (Ahmed and Rodrigues, 2013). Combarros et al. reported the effect of GO on *P. putida* growth and physiological states which consisted of viable, damaged, and dead cell subpopulations. In this particular study the term viable refers to cells with an intact membrane, damaged cells are described as those with a disrupted membrane but the capability to retain metabolic activity and finally, dead cells refer to those that are non-viable and inactive. GO exposure inhibited the growth of *P. putida* at 0.05 mg/mL and elicited a continuous reduction in the number of viable cells, in both simulated industrial and urban wastewater as the exposure dose increased. GO did not, however, significantly increase the number of dead cells but did induce an elevation of damaged cells. The evidence suggests that the sharp edges of GO act as blades to disrupt membrane integrity but preserve metabolic activity (Combarros et al., 2016). In a more recent study, the acute toxicity of graphene was investigated in a simulated biological wastewater treatment process. The study mimicked accidental spills and high concentration releases of graphene-based contaminants into wastewater treatment plants in the presence of ammonia oxidizing bacteria and phosphate accumulating organisms. The presence of graphene significantly reduced the metabolic activity of the microbial community, which subsequently diminished the removal of BOD, nitrogen, and phosphorous. Additionally, graphene considerably altered the structure of the sludge microbial community structure. A considerable number of bacteria genera that participate in the wastewater treatment process were no longer detected by deep sequencing analyses when graphene was present (Nguyen et al., 2017). Yang et al. also examined the removal of GO and FLG from wastewater by wastewater biomass. At biomass concentrations of 1,000 mg/L and a settling time of 30 min at least 65% of GO, with an initial concentration of 25 mg/L was removed while a biomass of 50 mg/L removed 16% of 1 mg/L FLG (Yang et al., 2015).

Besides the antibacterial effects of the GFNs, antifungal activity has also been reported. rGO demonstrated antifungal activity against nonpathogenic fungus *Aspergillus oryzae* and pathogenic fungi *Aspergillus niger* and *Fusarium oxysporum*. The reported IC50 values for rGO were 50, 100, and 100 μg/mL against *F. oxysporum*, *A. niger*, and *A. oryzae*, respectively. The unproblematic attachment of rGO to the external cell wall by means of glycoprotein hydroxyl oxygen species likely motivated superior toxicity against *F. oxysporum*. While GFN toxicity to pathogenic fungi may be useful, antifungal activity against nonpathogenic microorganisms that are important for the metabolism or the environment is a cause for concern (Sawangphruk et al., 2012). GO also triggered substantial ecological implications in the white rot fungus Phanaerochaete chrysosporium. Exposure doses of up to up to 4000 mg/L induced morphology changes, ultrastructure disruption and complete loss of the decomposition activity but lower concentrations stimulated growth of the fungus (Xie et al., 2016).

GFNs have also exhibited growth inhibition, ROS generation, and oxidative stress, in addition to cell division permeability in different species of green algae and cyanobacteria. Guo et al. measured the effect of graphene on the alga, *C. pyrenoidosa*'s cell survival rates in the presence of dissolved organic matter. As the exposure dose of graphene increased the survival rate of cells decreased. Wang et al. also demonstrated a concentration-dependent growth inhibition of unicellular green alga S. *obliquus* in the presence of graphene (Wang et al., 2016). Toxicity was induced via ROS mediated oxidative damage. Pretti et al. examined the toxicity of pristine graphene monolayer flakes (PGMF) and graphene nanopowder grade C1 (GNC1) in unicellular marine alga *D. tertiolecta* (Pretti et al., 2014). In comparison with bulk graphite (GRP), both PGMF and GNC1 demonstrated an enhanced degree of *D. tertiolecta* growth inhibition. Moderate particle dimension-dependent toxicity (PGMF>GNC1>GRP) was induced by the GFNs. Cell swelling and a loss of flagella were also observed. The authors thus proposed that the toxic effects elicited by pristine graphene were likely due to physical interaction and a loss of cell membrane integrity (Pretti et al., 2014). GO also provoked membrane damage and ROS generation in green alga *R. subcapitata* and inhibited cell division in *C. vulgaris* (Hu et al., 2015b; Nogueira et al., 2015).

Different effects by GO and combined systems of GO and Cd^{2+} (concentrations between 1–50 mg/L and 0.2–0.7 mg/L, respectively) were observed in cyanobacteria, *M. aeruginosa*, suggesting that GFN release in aquatic systems may lead to a potential enhancement of background contaminants toxicity, even at low non-toxic concentrations (Tang et al., 2015). GO alone at low concentrations below 10 mg/L had no significant toxicity. GO with Cd^{2+} easily clung to cell walls and entered the algal cells, yet surprisingly significant visible damage did not occur. Rather, mortality was due to the subsequent uptake of Cd^{2+}. The toxicity of the combined system intensified concentrations of GO increased, evidenced by the changes in ROS and MDA levels.

GONS and quantum dots also proved toxic to the model green alga *C. vulgaris*. Ouyang et al. (Ouyang et al., 2015) studied the possible synergistic envelopment–internalization effects of the GO-derived nanostructures using metabolomics. Larger GONS intensively entrapped single-celled *C. vulgaris*, which in turn reduced cell permeability. In contrast, smaller GO quantum dots (GOQD) stimulated intense shrinkage of *C. vulgaris* cell's plasma membrane and improved cell permeability via remarkable internalization reactions such as plasmolysis, GFN uptake and increased oxidative stress, in addition to the inhibition of cell division and chlorophyll biosynthesis. The phytotoxicity of rGO in microalgae *S. obliquus* also inhibited chlorophyll a and b levels which suppressed microalgae growth (Du et al., 2016b).

3.3.1.2 Interactions of GFNs with Invertebrates

Invertebrates are considered to be the ultimate receivers of environmental contaminants. Though the scale of studies concerning the impact of GFN environmental exposure to invertebrates is expanding, comprehensive knowledge of the toxic threats posed to these organisms by GFNs is still poor. By this time, toxicity in invertebrates has reportedly been caused by graphene-based composites including pristine graphene, FLG, GO, or rGO hybrids with polymeric materials such as PVP and polyvinylpyrrolidone-PVP or metal oxides. Mechanisms of previously observed toxicity include accumulation and transfer from adults to neonates, organ accumulation and retention, oxidative stress, and altered physiological activity. FLG exposed to shrimp *Litopenaeus vannamei* via diet elicited alterations in the redox state of cells. FLG induced a rise in ROS generation, altered the activity of antioxidant enzymes including glutamate-cysteine ligase and glutathione-S-transferase, diminished glutathione levels, and moderated the total antioxidant capacity. The stacked family member also engendered histological changes and lipid peroxidation in hepatopancreas tissues, all of which indicate an oxidative stress situation. In addition to unicellular green alga *D. tertiolecta*, Pretti et al. also exposed brine shrimp *A. salina* to PGMF and GNC1 (Pretti et al., 2014). GFN aggregates accumulated in the shrimp gut, generating an oxidative situation. Subsequently, an altered pattern of oxidative stress biomarkers was revealed, which ensued a substantial escalation of catalase and glutathione peroxidase activities. The members of the graphene family also induced enhanced levels of membrane lipid peroxidation. However, at lower exposure doses GFNs did not significantly affect *A. salina* and acute toxicity was not witnessed.

More recently, Mesarič et al. (2015) explored the influence of GO on *A. Salina* larval stages. GO aggregates attached to larvae on gills and appendages and to the surface larvae, affecting swimming behaviour. Contrary to Pretti et al., Mesarič and co-workers exposed nauplius stage larvae to more concentrated doses of GO concentrations and as a result, observed acute mortality at maximum concentrations. Mesarič et al. (2013) also previously investigated the toxic effects of single-layer GO toward *Amphibalanus amphitrite* larvae. Authors examined the effects of monolayer GO on the settlement of cypris larvae and on the mortality and swimming behaviour of the nauplius larvae. A dose-dependent anti-settlement effect on cyprids was generated by monolayer GO. This effect was made reversible, however, by exposing the cyprids to clean fresh water for 24 hours after rinsing. Increased doses and exposure times of GO also decreased the swimming speed of nauplii and eventually increased mortality.

Earthworms and nematodes are important components of terrestrial ecosystems due to their beneficial effects on both soil structure and function. Consequently, interactions between GFNs and terrestrial environments may disrupt biological activities. In addition, earthworms and nematodes are permanently exposed to their surrounding via their exterior epidermis and also through their intestinal tract due to continuous soil ingestion (Sanchez-Hernandez, 2006). Furthermore, terrestrial worms are in contact with both aqueous and solid-phase substrates (Römbke et al., 2005). These features have rendered appropriate sentinel organisms for the determination of GFN impact on terrestrial environments.

Thus far, only a few studies examining GFN toxicity toward soil organisms have been conducted. The free-living nematode *C. elegans* is often exploited as a model organism in the assessment of GFNs' toxic impact in soil. Zanni et al. demonstrated the absence of acute toxicity of graphite nanoplatelets *in vivo* on *C. elegans*, graphite nanoplatelets of various concentrations did not increase nematode mortality (Zanni et al., 2012). Authors also observed the transition of nanoplatelets from the intestine to the gonads; however, the investigations of the chronic exposure of nanoplatelets on the nematodes' lifespan showed lack of their reproductivity inhibition. However, a homogeneous distribution of graphite nanoplatelets was found along the inside of the living nematodes body which failed to effect nematode lifespan. Zhang et al. also confirmed the lack of toxicity of functionalized GO (GO/PP), modified with PEGylated poly-L-lysine (PLL-PEG) on *C. elegans* while assessing the effects on the aging process. The uptake of GO/PP was witnessed under normal physiological conditions but no changes in the mean longevity of the nematodes were observed, indicating GO did not exert toxic effects. The possibility of body wall damage by GO/PP nanosheets was also excluded. Furthermore, there were no significant differences in the crawling characteristics or movement rates of the nematodes. Elevated GO doses did, however, instigate increased ROS generation, but no oxidative damage occurred. Interestingly, GO/PP nanosheets significantly impaired the stress-resistance capacity of *C. elegans* under oxidative and temperature induced stress conditions. Due to oxidative stress created by juglone, a widely used oxidative stress inducing peroxidant, the mean lifespan of the nematodes treated with GO diminished. In addition, when subjected to heat stress, the cumulated survival percentage of *C. elegans* treated with GO/PP reduced by over 48% (Zhang et al., 2012). Wu et al. (2013) considered the acute and prolonged exposure of GO *in vivo* on *C. elegans*. The authors found that prolonged GO damaged the functions of both primary and secondary organs such as the intestine and the neuron and reproductive organs. Zhang et al. (2017a) also demonstrated negligible toxicity of GO on cosmopolitan oligochaete, *Tubifex tubifex*. However, GO exposure did cause a significant reduction in burrowing activity. Nevertheless, GO modified biochemical performances in the polychaeta *Diopatra neapolitana*. The presence of GO modulated energy-related responses and caused cellular damage, despite the enhanced activities of antioxidant and biotransformation enzymes in the exposed individuals (De Marchi et al., 2017). A recent study assessing the environmental fate of GO was conducted on a further model organism, the house cricket (*Acheta domesticus*). The *in vivo* toxicity study revealed that injection of GO into the haemolymph provoked oxidative stress, yet the organism easily coped with the derived stress and was capable of recovery within a short time.

3.3.1.3 Interactions of GFNs with Fish

Most studies highlighting the impact of GFNs on aquatic organisms are currently devoted to freshwater fish species, in which the embryos of zebrafish (*Danio rerio*) are among one of the most commonly employed models for aquatic toxicity assessments. Furthermore, GFNs have widely been reported to cause growth inhibition and hatching delay in *D. rerio* embryos (Chen et al., 2012b; Souza et al., 2017; Wang et al., 2015b; Zhang et al., 2017a, 2017b; Zhou et al., 2012b). Zebrafish embryonic growth development was impaired by the presence of GO, whose presence leads to DNA modification, protein carbonylation, and significant ROS generation, particularly superoxide radicals. Notably, toxic responses witnessed developed in a non-monotonic manner. GO further disturbed collagen and matrix metalloproteinase (MMP)-related genes which subsequently modified the cardiac and skeletal development of zebrafish. Additionally, inhibition of amino acid metabolism and the ratios of unsaturated fatty acids to saturated fatty acids contributed to GO-induced toxicity (Zhang et al., 2017b). Envelopment, hypoxia, and hatching delay also results from the adherence of GO to the chorion of zebrafish embryos (Chen et al., 2012b). In addition, due to GO exposure, biochemical responses from zebrafish embryos including excessive ROS generation and oxidative stress, DNA damage, and apoptosis in both mitochondria and the circulatory system have also been corroborated (Chen et al., 2012b; Souza et al., 2017; Wang et al., 2015b; Zhang et al., 2017a, 2017b; Zhou et al., 2012b). Chen et al. also established that embryos exposed to GO exhibited morphological malformation and degradation such as bending of the spine and minor tail deformities, in addition to opaqueness in the yolk which suggests the presence of apoptotic tissue (Chen et al., 2012b). GO exposure also caused histopathology damage in *D. rerio,* increasing the number of gill cells in early apoptotic and necrotic stages. GO also elicited damage to hepatocytes yielding non-uniformly shaped cells, pyknotic nuclei, the formation of vacuole, cell fracture, and necrosis (Souza et al., 2017). Moreover, GO and GQDs exposed to *D. rerio* embryos induced bioaccumulation in different tissues and organs, modulated excretion rates, oxidative stress, and mortality as well as embryonic malformations and larval hyperactivity (Chen et al., 2012b; Souza et al., 2017; Wang et al., 2015b; Zhang et al., 2017a, 2017b; Zhou et al., 2012b). On the other hand, graphene coated with PAA and FMA proved biocompatible with zebrafish and failed to induce any significant abnormalities or affect the survival rate of fish embryos (Gollavelli and Ling, 2012). GO also generated bioaccumulation and ROS in another fish species, *Poeciliopsis lucida* (Lammel and Navas, 2014).

3.3.1.4 Interactions of GFNs with Plants

Plants are essential constituents of both aquatic and terrestrial ecosystems. Accordingly, comprehension of GFN interactions with plants is crucial to understand the fate of graphene family members in the environment, but much progress has yet to be made (Khodakovskaya et al., 2011). The phytotoxic effect of graphene has been shown to considerably constrain the growth of cabbage, tomato, red spinach, and lettuce (Begum et al., 2011). The effects of graphene on root and shoot growth, biomass, shape, cell death, and ROS were examined to reveal that graphene negatively affects the root morphology of tomato and red spinach plants yielding a loosely or completely detached epidermis. The number and size of graphene-treated plants leaves reduced in a dose-dependent manner. Similarly, authors also detected a dose-dependent increase in ROS and cell death. Furthermore, graphene-treated plants exhibited visible evidence of necrotic lesions which likely represent graphene-induced oxidative stress necrosis. Under the same conditions, insignificant toxicity was observed in lettuce seedlings. In their later work, Begum and Fugetsu investigated an alternative plant model, *Arabidopsis thaliana* T87 cell suspension, which has been extensively studied in many biological fields including nanotoxicity (Begum and Fugetsu, 2013). Graphene exerted harmful effects on *Arabidopsis thaliana* T87 cells, including nuclei fragmentation, membrane damage, mitochondrial dysfunction, an increase in ROS and accumulation, all leading to cell death. ROS was suggested as a key mediator in the cell death signalling pathway given the fact the graphene exposure caused a 3.3-fold increase in ROS generation. By reason of the observed endocytosis-like structures, the authors also proposed cellular uptake due to endocytosis and confirmed the cellular translocation of graphene. In addition, GO toxicity and translocation in *A. thaliana* plants was explored in the presence of normal and stressful conditions (Wang et al., 2014; Zhao et al., 2015). Four weeks of GO exposure to growing *A. thaliana* plants did not affect germination or the development of seed sprouting. GO was found to largely accumulate in the cell compartments of the cotyledon cells. In the seedlings, GO accumulated in the root system only which suggests that *A. thaliana* can strongly cope with the translocation of GO from root to stem or leaves. Like in the previously discussed case of nematodes, when coupled with a preexisting stress such as drought or salt, GO reduced the stress-resistance capacity of A. thaliana plants, inducing more severe oxidative stress and membrane ion leakage, increasing the potential for GO translocation from roots to leaves.

Anjum et al. (2013, 2014) demonstrated concentration-dependent differential sensitivity of a common *faba* bean food crop, *V. faba* toward GO. The model plant system was exposed to various GO concentrations (0, 100, 200, 400, 800, and 1,600 mg/L) which revealed that *V. faba*. was tolerant to 400 and 800 mg/L, but sensitive to doses of 100, 200 and 1600 mg/L. In their first work, the authors studied the significance of the glutathione redox system in the toxic responses of *V. faba* to GO, which revealed a decreased glutathione redox ratio, and reduced glutathione pool, as well as considerably increased activities of glutathione-regenerating and glutathione-metabolizing enzymes at toxic doses (Anjum et al., 2013). On the other hand, the exposure of *V. faba* to 400 and 800 mg/L increased the ratio of the reduced glutathione and oxidized glutathione and reduced both the glutathione pool and glutathione reductase activity but decreased activities of glutathione-metabolizing enzymes, increasing GO tolerance and *V. faba* health. In their later work, Anjum et al. studied the impact of GO exposure to germinating *V. faba* seedlings under similar conditions to before (Anjum et al., 2014). Again, 100, 200, and 1600 mg/L of GO produced negative impacts, decreasing growth parameters and the activity of H_2O_2-decomposing enzymes, in addition to an increase in the levels of electrolyte leakage, as well as H_2O_2, lipid, and protein oxidation. In the presence of 400 and 800 mg/L GO the electrolyte levels of *V. faba* decreased as well as H_2O_2, lipid and protein oxidation. Furthermore, the H_2O_2-decomposing APX and CAT activity, proline and seed relative water content increased (Jastrzębska and Olszyna, 2015). The increased *V. faba* toxic responses were likely due to increased oxidative stress and a reasonably damaged glutathione metabolism (Syrgiannis, 2017). Ecotoxicological effects of graphene-based materials The tolerance of *V. faba* to 400 and 800 mg/L GO is likely due to the elevated glutathione regeneration combined with a lowered glutathione consumption. Furthermore, hydrated graphene ribbons (HGRs), graphene, and GO were responsible for inducing specific metabolic pathways that distinctively modulate the metabolism of plant model *Triticum aestivum* (Hu and Zhou, 2014). Both graphene and GO inhibited the germination of *Triticum aestivum* while the HGRs unexpectedly relieved oxidative stress, as well as promoting root elongation and aged seed germination rates. Being the more hydrated of the graphene derivatives, the positive effects of HGRs was likely due to molecular features as the authors also reported the presence of nitrogen and oxygen in the functionalized graphene form (Syrgiannis, 2017). It was also reported that the presence of GO intensifies the phytotoxicity of arsenic in wheat (Hu et al., 2014). In general, the GFN-related plant nanotoxicity mechanisms include biochemical responses such as ROS generation or oxidative stress, size-dependent interactions like physical cell disruption and affinity-based interactions (Dietz and Herth, 2011). However, several issues such as the unclear mechanisms of uptake, translocation, long-term exposure, and bio-responses must still be addressed in an attempt to fully understand the interactions of GFNs and plant systems (Hu and Zhou, 2013).

The interactions of graphene family members with plants are illustrated in Figure 3.5.

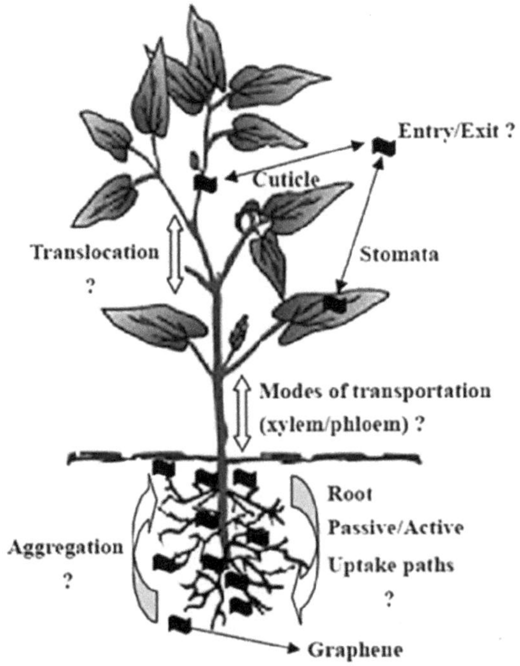

FIGURE 3.5 Schematic illustration of GFN-plant interactions. (From Hu, X. and Zhou, Q., *Chem. Rev.*, 113, 3815–3835, 2013.)

3.4 Cytotoxic Risks of GFNs

To safeguard and advance the biomedical implementation of GFNs it is imperative we understand their behaviour and interactions in human systems. Humans could be exposed to GFNs in a variety of scenarios such as on-site product manufacture and design, product usage, exposure to environmentally prevalent GFN-contaminated industrial waste, etc., via various exposure routes (Figure 3.1) (Dasari Shareena et al., 2018; Syama and Mohanan, 2016). Graphene-like NMs have been reported to generate acute and chronic injuries in tissues subsequent to administration and accumulate in organs such as the lung, liver, spleen, etc. For instance, GFN-based aerosols could be inhaled, subsequently depositing 2D NMs in the respiratory tract. Inhaled GFNs can easily penetrate tracheobronchial airways and transfer to the lower lung airways, which will likely result in the subsequent growth of granulomas and lung fibrosis (Sanchez et al., 2011; Su et al., 2016). GFNs dispersed in air may therefore present a threat to individuals exposed on a daily basis such as those in an industrial workplace setting (Pretti et al., 2014).

A significant number of cytotoxicity works specify that GFNs do not cause toxic effects. Indeed, many well-established cell lines have presented biocompatibility of graphene family members *in vitro*, yet many others have highlighted the adverse effects caused by GFNs. Due to the rapid development of nanotechnology and next-generation materials standard cytotoxicity testing does not yet satisfy the strict screening criteria necessary to safely translate new NMs into clinical use. Many gaps have yet to be filled in terms of GFN risk-related knowledge. Thus, the European Scientific Committee on Emerging and Newly Identified Health Risks has recently included graphene in the category of hazard substances (Bianco and Prato, 2015; Reina et al., 2017).

Although the human health risks of GFNs have been extensively studied at both a cellular and animal model level, the understanding of the potential biocompatibility and toxic effects of GFNs still lacks a fundamental clarity, hastily required (Dasari Shareena et al., 2018). To date countless studies and reviews have summarized the latest potential biological applications of GFNs for drug delivery, gene delivery, biosensors, tissue engineering, and neurosurgery, considered the unique properties responsible for such possibilities, the relationship and interactions of GFNs with biological systems as a result of such astonishing properties and ultimately the influence of this interrelationship and behaviour on the biocompatibility of GFNs in microbial, mammalian, and plant cells in addition to animals such as crustaceans, worms, nematodes, zebrafish, and mice, yet uncertain determination and rationalization of the exact mechanisms of GFN biocompatibility and toxicity prevail (Jastrzębska et al., 2012; Ou et al., 2016).

For the most part, studies indicate that GFNs exhibit commendable mammalian cell biocompatibility and simultaneous antimicrobial activity. Remarkably, this anomaly is rarely deliberated; research describing the possible mechanism underlying selective toxicity has yet to be publicized. There is little reason as to why GFNs should be selectively toxic to prokaryotic cells and not hazardous to eukaryotic cells, such as mammalian lines that are generally considered more sensitive to ambient factors. In contrast, outlying studies aforementioned claim that GFNs are bacteriostatic or do not possess antibacterial activity. In effect, the scaffolding effect of GFNs stimulates nonspecific cellular attachment and growth (Ruiz et al., 2011).

The toxicity mechanisms of GFNs proposed in previous studies predominantly include inflammatory response, DNA damage, apoptosis, autophagy, necrosis, etc. (Ou et al., 2016). Unfortunately, the determination of GFN cytotoxicity lacks currently uniformity (Jastrzębska et al., 2012). The determinants of GFN toxicity span far beyond the chief toxicological criteria of simple molecules such as dose, exposure type, and time. In addition to such parameters the size, shape, surface chemistry, and agglomeration state of graphene-related NMs significantly alters toxic responses. Consequently, the safety evaluation of GFNs is a complex and difficult task, in comparison with simple molecules, due to the various atomic compositions of each of the distinct graphene family members. For this reason, a unique mechanism of GFN toxicity is unlikely, but rather a more complex scenario of mechanisms (Reina et al., 2017). Nevertheless, direct or indirect ROS leading mediated oxidative stress in target cells is often the main mechanism proposed for the toxicity of engineered NMs, including GFNs (Oberdörster et al., 2005; Stone et al., 2009). At this point it should also be mentioned that, to maintain a balanced internal environment cellular homeostasis processes employ antioxidant enzymes, such as superoxide dismutase, catalase, or glutathione peroxidase to reduce or eliminate ROS as required to retain stable levels. When ROS levels cannot be reduced by antioxidant enzymes, macromolecules such as polyunsaturated fatty acids in membrane lipids can be altered, and in the end protein and DNA destruction occurs (Sanchez et al., 2011). The extremely high hydrophobic surface areas associated with GFNs facilitate significant interactions with membrane and induce toxicity via direct physical damage or indirectly due to the adsorption of biological molecules (Jastrzębska et al., 2012; Stone et al., 2009).

3.4.1 Human Exposure

At present, many proposed GFN products and applications are still under development. Therefore, the greatest concern of human exposure to GFNs is currently occupational respiratory exposure during the handling of GFNs in production facilities. Potential exposure of workers is likely a concern throughout all phases of product manufacture and design, while direct consumer exposure will only occur during the use and disposal of finished products (Park et al., 2017). GFNs can enter the human body via inhalation, ingestion, dermal penetration, injection, or implantation (Sanchez et al., 2011). As discussed, the major exposure route for GFNs in an occupational setting is airway exposure, thus inhalation and intratracheal instillation are administration techniques commonly used to simulate human exposure to GFNs in mice. Though inhalation routes are more realistic, in a laboratory setting instillation is both more effective and efficient (Lee et al., 2017; Li et al., 2013a; Schinwald et al., 2012, 2014; Wang et al., 2013a). Previous reports exhibited GFN deposition and accumulation in the lungs of mice, which were retained for more than 3 months, demonstrating slow clearance rates (Sydlik et al., 2015).

Another widely exercised route of administration is intravenous injection. Using this course of exposure graphene has been described to move through the body of mice in 30 min, gathering in both the liver and bladder at a working concentration (Ou et al., 2016). However, GO derivatives orally administered to adult mice had limited intestinal adsorption and were rapidly excreted (Fu et al., 2015; Yang et al., 2013). Interestingly, in comparison to micron-sized GO, the subcutaneous injection of nano-sized GO in the neck region caused less mononuclear cells to infiltrate adipose tissue (Yue et al., 2012). Furthermore, intraperitoneal injection caused agglomeration of GO near the injection site while aggregates settled in close propinquity to the liver and spleen (Kurantowicz et al., 2015; Yang et al., 2013).

After exposure, GFNs can reach various locations within the human body through blood circulation or via the penetration of biological barriers such as the blood-air barrier, blood-testis barrier, blood-brain barrier, and blood-placental barrier, which can thus result in varying degrees of distribution, uptake, and retention in cells, tissues, and organs (Ou et al., 2016). The lung pulmonary surfactant (PS) film is the first line of host defence facing inhaled particles. Inhaled GFNs, namely GO, reportedly abolish both the ultrastructure and biophysical properties of the defensive film through the creation of pores, specifically due to surfactant inhibition (Hu et al., 2015a). Graphene has also been shown to reach the lower part of the respiratory tract in the rats which caused an increase of lavage markers and microgranulomas, indicative of inflammatory response (Ma-Hock et al., 2013). Schinwald et al. also highlighted that graphene nano-platelets consisting of FLG deposit beyond the ciliated airways subsequent to inhalation rendering similar adverse effects to the former (Schinwald et al., 2012).

The blood-brain barrier is a tightly protective reticulum surrounding the brain to prohibit the entrance of foreign substances. The tight junctions between adjacent capillary endothelial cells possess a semi permeable property to modulate the role of the blood-brain barrier allowing only small molecules and lipids to cross (Alam et al., 2010). The intricate arrangement of the blood-brain barrier, consisting of numbers of membrane receptors and highly selective carriers, only exerts subtle influence on blood circulation and the brain microenvironment compared to the peripheral vascular endothelium (Abbott et al., 2010). The breakdown of the blood-brain barrier, which also comprises an intricate arrangement of membrane receptors and highly selective carriers, induces cell damage, neurodegeneration, and brain inflammation. The relevant mechanisms of the blood-brain barrier involved in GFN nanotoxicity are not yet well understood but research has begun to make some progress. However, studies focusing on the transport of graphene through the blood-brain barrier and the possibility of subsequent neurotoxicity are rare; a greater insight is required before conclusions can be drawn. Some literature has reported that graphene derivatives such as GO and PEGylated-graphene were detected in the brain following intravenous injection, yet Wang et al. suggested that GO failed to enter the brain due to impediment by the blood-brain barrier (Yang et al., 2011a; Zhang et al., 2011b). Matrix-assisted laser desorption/ionization (MALDI) mass spectrometry imaging (MSI) also revealed that rGO pervaded the paracellular pathway and transferred into the inter-endothelial cleft by means of a transitory decrease in paracellular tightness (Mendonça et al., 2015).

Studies of GFN skin permeation are very rare. Thus far, most studies examining the dermal exposure of NMs administer the potential toxins to subcutaneous tissues to evaluate bio-responses and cellular toxicity (Johnston et al., 2010). Though the mechanisms of GFN skin permeation are not yet known, single-walled CNTs revealed the possibility of dermal penetration subsequent to skin deposition. SWCNT skin penetration increased epidermal thickness and the accumulation of dermal fibroblasts (Murray et al., 2009). Furthermore, Hu et al. proposed a possible route of GFN skin permeation: graphene is absorbed through sweat pores, is transferred from the epidermis into the dermis and subcutaneous fatty tissues, and finally penetrates the veins and arteries, though this theory requires substantiation (Hu and Zhou, 2013).

The placenta system, inclusive of the maternal blood sinus, mononuclear trophoblast, syncytiotrophoblast, endothelium, and foetal blood capillary, joins the growing foetus to the uterine wall. The placental barrier is thus vital to pregnancy preservation, as it exercises crucial metabolic function, facilitates the exchange of nutrients and metabolic wastes, and secretes hormones (Ou et al., 2016). Considerable focus has been paid to the potential crossing of NMs through the placenta and thus disturbed foetal growth. A recent review has advised that the placenta fails to provide a sufficient barrier against nanoparticle transfer to the foetus (Ema et al., 2016b). Additional reports have also demonstrated passage nanoparticles cross the placental barrier, which resulted in significant influence on embryo development (Du et al., 2016a; Ema et al., 2016a; Warheit et al., 2015; Zhou et al., 2015). GFNs could impact foetal health and development via direct translocation through the placenta or cause indirect disorder of placenta functions (Hu and Zhou, 2013). But investigations of the possibility, mechanism, and antagonistic effects of GFNs across the placenta are lacking. Thus, the latter and the pathway for NM to embryo transfer should be evaluated.

3.4.2 Biodistribution and Clearance of GFNs

The accumulation, distribution, and clearance of GFNs from biological systems are influenced by administration routes, physicochemical properties, and particle agglomeration. As previously highlighted, evidence has demonstrated that the route of GFN administration strongly dictates the subsequent path of distribution. For instance, intratracheally instilled FLG passing through the air-blood barrier mainly accumulated in the lung (Mao et al., 2016). Intravenously administered GO passed through the body via blood circulation and resided in the lung, liver, spleen, and bone marrow. GO accumulation and retention in the lung elicited inflammatory cell permeation, granuloma growth, and pulmonary oedema (Zhang et al., 2011b). After intraperitoneal injection, GO derivatives functionalized with PEG accumulated in the reticuloendothelial systems (RES) of the liver and spleen. In contrast, exposure to PEGylated GO and FLG via oral administration did not occasion absorption or tissue uptake in the gastrointestinal tract (Yang et al., 2013). The toxicological criteria of GFNs, such as their size, dose, and surface chemistry also influence distribution profiles. For example, smaller GO sheets, 10–30 nm in size predominantly distributed in the liver and spleen, in comparison to larger sheets (10–800 nm) that accumulated in the lungs (Wang et al., 2010; Yang et al., 2010a; Zhang et al., 2011b). Furthermore, GO are often found wedged in arteries and capillaries near the injection site if the administered material is larger than the size of the exposed blood vessels. In addition, the GO accumulation in the lungs increased in a size- and dose-dependent manner (Liu et al., 2012). Biofunctionalization and biomolecule adsorption on the surface of GO also alter biodistribution (Yang et al., 2013; Yue et al., 2012; Zhang et al., 2011a).

The mechanisms of GFN clearance vary within different organs. In the lungs, the excretion of NMs occurs through mucociliary self-clearance by alveolar macrophages, or translocation through the epithelial layer (Ruge et al., 2012). Intratracheally instilled FLG was cleared from the body in the form of faeces after 28 days of exposure (Mao et al., 2016). The accumulation of GO coated with PEG was gradually cleared from the liver and spleen by both renal and faecal excretion. The small GONS, ~8 nm in size, could infiltrate renal tubules, transfer into urine, and rapidly be removed from the body, whereas larger sheets of 200 nm were confined by physical splenic filtration. However, the excretion paths of GFNs require further consideration renal and faecal routes appear to be the chief means of elimination. Various routes of GFN biodistribution and clearance have been reported thus far. The breakdown and evacuation of NMs from biological systems require extended periods. However, the assessment of GFN toxicity is currently limited to short term studies. Consequently, the effects of long-term cellular, tissue, and organ accumulation are vague. Thus, to guarantee biosafety, long-term and more in-depth studies of GFN distribution and excretion must be performed in the presence of diverse cellular and animal models (Ou et al., 2016).

3.4.3 Uptake and Location of GFNs in Cells

Diverse cell lines have exerted different modes of GFN uptake and location. Furthermore, GFNs undergo intracellular ingestion via a number of routes (Li et al., 2012b; Zhang et al., 2010). Again, the physicochemical parameters such as the size, shape, coating, charge, hydrodynamic diameter, isoelectric point, and pH gradient play an important role on GO cellular membrane penetration (Sydlik et al., 2015). For instance, GQDs can transfuse cell membranes directly without the means of energy-dependent pathways (Peng et al., 2010; Wang et al., 2015a). Larger biofunctionalized GO nanoparticles (PCGO) reaching 1 μm in size (PCGO) (~1 μm) were ingested via phagocytosis. On the other hand, smaller PCGO nanoparticles entered cells through clathrin-mediated endocytosis (Mu et al., 2012). GO cellular interactions can also induce either cell membrane adherence, lipid bilayer insertion, or cellular internalization (Xu et al., 2016). Comparably, cellular membrane interactions with the hydrophobic, graphitic domains of both rGO and PEGylated rGO (PrGO) promoted the adherence to the lipid bilayer cell membrane (Kostarelos and Novoselov, 2014; Sasidharan et al., 2011). Reports have also demonstrated that GO sheets are phagocytosed by macrophages due to interactions with the plasma membrane (Ou et al., 2016).

3.4.4 Cytotoxicity Mechanisms of GFNs

The predominant mechanisms of GFN cytotoxicity are highlighted in Figure 3.6. Although the toxicity of GFNs has been widely examined, the exact underlying mechanisms are ambiguous though physical

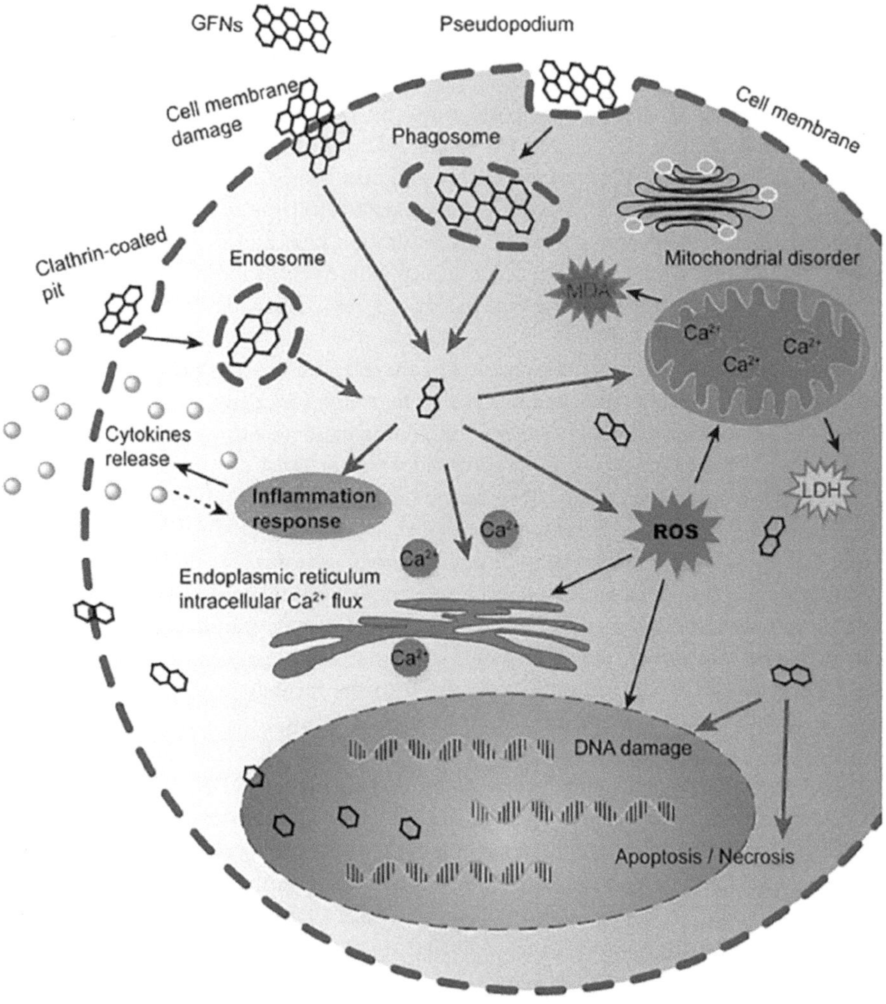

FIGURE 3.6 Schematic illustration of the cytotoxic effects of GFNs. (From Ou, L. et al., *Part. Fibre Toxicol.*, 13, 57, 2016.)

disruption, ROS-mediated oxidative stress, mitochondrial and DNA damage, inflammatory response, apoptosis, autophagy, and necrosis are the main toxic effects of GFNs discussed in literature. As previously mentioned, physical disruption of cell membranes immensely contributes to cytotoxic responses due to GFN exposure (Gurunathan et al., 2014; Qi et al., 2014; Yuan et al., 2011). For instance, pristine graphene attached to the surfaces of murine macrophage cells, RAW 264.7, and induced abnormal cell membrane stretching (Sasidharan et al., 2012). The formidable hydrophobic reactions between graphene and the cell membrane elicited the morphological expansion of F-actin filopodial and cytoskeletal dysfunction.

Oxidative stress develops when increasing levels of ROS cannot be modulated by the activity of antioxidant enzymes (Burton and Jauniaux, 2011). ROS represent second messengers in intracellular signalling cascades that induce macromolecular damage, such as membrane lipid disruption, protein damage, DNA destruction, and mitochondrial dysfunction (Chen et al., 2016; Chong et al., 2014; Waiwijit et al., 2014). Furthermore, ROS generation, often induced by GFNs such as GO, is considered the first phase in the mechanisms of carcinogenesis, ageing, and mutagenesis (Li et al., 2012b). GFN-induced ROS generation and mediated oxidative stress played a meaningful role in GO-induced acute lung injury (Li et al., 2013a). GO-generated oxidative stress also caused apoptosis and DNA damage in human lung fibroblasts (Wang et al., 2013a). Furthermore, intracellular ROS generated due to the exposure to pristine graphene induced oxidative stress which, in turn activated metabolic pathways and stimulated apoptosis (Li et al., 2012b).

Mitochondria are central to apoptotic regulation and are also the energy manufacture centres involved in several signalling pathways in (Li et al., 2012b). Both GO- and carboxyl-functionalized graphene depolarized the mitochondrial membrane and diminished the degree of mitochondrial cells present in HepG2 cells (Lammel et al., 2013). GO also donated electrons to site I/II of the mitochondrial electron transport chain, increased mitochondrial respiration rates and ROS generation and activated inflammatory and apoptotic pathways (Duch et al., 2011). Therefore, in addition to the widely renowned mechanisms of plasma membrane damage and oxidative stress, GFNs can alter mitochondrial activity and cause apoptosis (Park et al., 2015).

DNA damage can instigate the development of cancer and affect the health of the forthcoming generation should potential GO mutagenicity arise in reproductive cells (Ciccia and Elledge, 2010; Liu et al., 2013b). The unique collection of properties, including nanoscale dimensions, substantial surface area, and surface chemistry of GFNs, could potentially trigger DNA destruction. GO has reportedly caused severe DNA damage, including chromosomal disruption, DNA strand fragmentation, mutations, and oxidative DNA injury in a number of established cell lines (Chatterjee et al., 2016; Ivask et al., 2015; Wang et al., 2015a). The breakdown of the nuclear membrane during mitosis also affords GO with an opportunity for DNA disturbance, even in the absence of cell nucleus interactions (Magdolenova et al., 2014; Ren et al., 2010; Wang et al., 2015a). An elevated degree of mutagenesis was elicited from GO in comparison with an increased dose of cyclophosphamide, a classic mutagen (Liu et al., 2013b). GO induces chromosomal fragmentation in addition to the formation of DNA adducts and point mutations via oxidative stress or inflammatory responses through the activation of metabolic and apoptotic pathways (Chatterjee et al., 2014; Jarosz et al., 2016; Liu et al., 2013b). Furthermore, the strong intermolecular forces of GFNs relentlessly distort the end base pairs of DNA. The π stacking interactions of carbon rings within the honeycomb lattice and the hydrophobic DNA base pairs motivate segments of DNA to 'stand up' or 'lay on' the surface of graphene (Zhao, 2011). Considering that the AT or GC base planes are perpendicular to the graphene surface the ending base planes must rotate 90° to induce face-face contact. Thus, if both of its ends of the DNA segment form stable structures with the graphene surface, severe deformation ensues. After breaking, the ending bases are flexible and free to stack effectively with the carbon surface.

Intratracheal instillation or intravenous administration of GFNs prompts considerable inflammatory responses including inflammatory cell intrusion, pulmonary oedema, and granuloma development (Li et al., 2013a; Zhang et al., 2011b). During inflammatory response due to clot formation, platelets are the important components responsible for the attack of pathogens. GO has been shown to directly stimulate thrombi-rich platelet activation in an attempt to block lung vessels subsequent to intravenous injection (Fujimi et al., 2006; Singh et al., 2012). Subcutaneous injection of GO also triggered inflammatory responses and tissue injury due to cytokine and chemokine secretion, inclusive of IL-6, IL-12, TNF-α, MCP-1, IFN-g, and Th1/Th2 (Chen et al., 2012a; Reshma et al., 2016; Wen et al., 2015; Zhou et al., 2012a). Graphene (Zhou et al., 2012a) and rGO (Chatterjee et al., 2014) also reduced an inflammatory response by binding to toll-like receptors (TLRs) and activating the NF-κB signalling pathway in cells (Lawrence, 2009).

GFNs also evoke pro-apoptotic cellular effects (Jaworski et al., 2013; Matesanz et al., 2013; Reshma et al., 2016; Vallabani et al., 2011). Apoptosis is regarded as the gene-regulated self-destruction of a cell via a complicated scenario of mechanisms (Hengartner, 2000; Li et al., 2012b). Following inhalation, GO and rGO have reportedly caused apoptosis and an inflammatory response in the lungs of mice (Duch et al., 2011). Similarly, rGO-induced apoptosis was caused by the death-receptor and canonical mitochondrial pathway (Chatterjee et al., 2014). Another investigation highlighted three possible routes that may trigger apoptosis due to exposure of three distinctly functionalized GFNs. The direct interaction of GO with protein receptors and consequential activation of the B-cell lymphoma-2 (Bcl-2) educed ROS-dependent apoptosis. Carboxylate GO bound to protein receptors which prompted the communication of a submissive apoptosis signal to nuclear DNA, this time actuating a ROS-independent route. On the other hand, polyethylenimine-modified GO triggered apoptosis by means of severe T lymphocyte damage (Ding et al., 2014). Autophagy or non-apoptotic cell death is the self-destruction of damaged cells prior to new cell formation. Under stress conditions autophagy processes surge but activation is first required in the form of Beclin 1-containing autophagosomes, various autophagy-associated proteins, and microtubule-related proteins such as light chain 3 and p62 (Levine et al., 2011). The accumulation of autophagosomes has been correlated with an exposure to a variety of NMs (Chen et al., 2015; Halamoda Kenzaoui et al., 2012; Hussain and Garantziotis, 2013). What is more, autophagy processes are capable of removing extracellular organisms and dismantle the organisms in the cytosol (Chen et al., 2014). Studies demonstrated that both GO and GQDs evoked autophagosome accumulation, the adaption of LC3-I to LC3-II and inhibition of autophagic substrate p62 protein degradation (Markovic et al., 2012; Wan et al., 2013). In addition, GO instantaneously activated TLR4 and TLR9 reactions in macrophages (Chen et al., 2012a; Yue et al., 2012) and colon carcinoma cells (Chen et al., 2014). Thus, the route of autophagy processes is associated with phagocytosis by TLR macrophage signalling (Chen et al., 2014; Sanjuan et al., 2007). Necrosis is a form of cell death induced by disease or injury. Elevated doses of graphene have previously been reported to provoke apoptosis and necrosis (Lalwani et al., 2016; Li et al., 2012b). GO too induced macrophagic necrosis through TLR4 signalling activation (Qu et al., 2013).

3.5 Conclusions and Outlook

At present, GFN nanotoxicity is an emergent field; thus, literature and our common understanding are too limited to draw solid conclusions regarding the environmental and cytotoxicity risks of graphene family members (Sanchez et al., 2011). Though juvenile, reports concerning GFN bio-interactions are emerging rapidly. Evidence suggests that GFNs may be detrimental to aquatic, terrestrial, and human health. Opposite opinions regarding the environmental and cytotoxic risks of GFNs have been uncovered. Some researchers propose that GFNs elicit negligible toxic effects though others report adverse biological responses and toxicity (Ou et al., 2016). These inconsistencies may be caused by a few factors including varying physicochemical characteristics of GFNs, the use of a variety of cellular or animal models and varying experimental techniques and parameters (Ou et al., 2016). Accordingly, a generalization of GFN nanotoxicity should be avoided. The environmental and cytotoxic risks of these newly emerging NMs are also contingent on specific applications and development

(Bianco, 2013). A comparison between GFNs and CNTs has recently provided us with a certain extent of clarity in terms of GFN toxicity modulation. If the lessons to be learned from CNTs are heeded, GFN nanotoxicity may evade the previous blunders encountered by CNTs (Bianco, 2013; Bussy et al., 2012).

This new family of NMs have the potential to become hazardous to both the environmental and human health but adverse effects can be alleviated through sufficient design, engineering, and manipulation of GFN physiochemical properties (Bianco, 2013). To correctly evaluate the environmental and cytotoxic risks of the graphene family, the development of a standardized, multilevel approach for nanotoxicity testing is required. The modification and expansion of traditional evaluation methods should also be considered when defining standard nanotoxicity assays. Furthermore, the use of human cells should be implemented when assessing the human health risks of GFNs. In addition, GFN toxicity testing lacks long-term test evaluations. These efforts and a continuing advancement of the field from interdisciplinary communities will greatly contribute to the growing knowledge on the environmental and cytotoxic fate of GFNs (Hu and Zhou, 2013).

Acknowledgements

S. Dervin would like to thank IT Sligo for providing financial support.

References

Abbott, N. J., A. A. K. Patabendige, D. E. M. Dolman, S. R. Yusof, and D. J. Begle. 2010. Structure and function of the blood–brain barrier, *Neurobiology of Disease*, **37(1)**: 13–25.

Ahmed, F. and D. F. Rodrigues. 2013. Investigation of acute effects of graphene oxide on wastewater microbial community: A case study, *Journal of Hazardous Materials*, **256–257**: 33–39.

Akhavan, O. and E. Ghaderi. 2010. Toxicity of graphene and graphene oxide nanowalls against bacteria, *ACS Nano*, **4(10)**: 5731–5736.

Akinwande, D., N. Petrone, and J. Hone. 2014. Two-dimensional flexible nanoelectronics, *Nature Communications*, **5**: 5678.

Alam, M. I., S. Beg, A. Samad, S. Baboota, K. Kohli, J. Ali, A. Ahuja, and M. Akbar. 2010. Strategy for effective brain drug delivery, *European Journal of Pharmaceutical Sciences*, **40(5)**: 385–403.

Aldieri, E., I. Fenoglio, F. Cesano, E. Gazzano, G. Gulino, D. Scarano, A. Attanasio, G. Mazzucco, D. Ghigo, and B. Fubini. 2013. The role of iron impurities in the toxic effects exerted by short multiwalled carbon nanotubes (MWCNT) in murine alveolar macrophages, *Journal of Toxicology and Environmental Health, Part A*, **76(18)**: 1056–1071.

Ali-Boucetta, H., D. Bitounis, R. Raveendran-Nair, A. Servant, J. Van den Bossche, and K. Kostarelos. 2013. Purified graphene oxide dispersions lack *in vitro* cytotoxicity and *in vivo* pathogenicity, *Advanced Healthcare Materials*, **2(3)**: 433–441.

Anjum, N. A., N. Singh, M. K. Singh, I. Sayeed, A. C. Duarte, E. Pereira, and I. Ahmad. 2014. Single-bilayer graphene oxide sheet impacts and underlying potential mechanism assessment in germinating faba bean (Vicia faba L.), *Science of the Total Environment*, **472**: 834–841.

Anjum, N. A., N. Singh, M. K. Singh, Z. A. Shah, A. C. Duarte, E. Pereira, and I. Ahmad. 2013. Single-bilayer graphene oxide sheet tolerance and glutathione redox system significance assessment in faba bean (Vicia faba L.), *Journal of Nanoparticle Research*, **15(7)**: 1770.

Arvidsson, R., S. Molander, and B. A. Sandén. 2013. Review of potential environmental and health risks of the nanomaterial graphene, *Human and Ecological Risk Assessment: An International Journal*, **19(4)**: 873–887.

Arvizo, R. R., O. R. Miranda, M. A. Thompson, C. M. Pabelick, R. Bhattacharya, J. D. Robertson, V. M. Rotello, Y. Prakash, and P. Mukherjee. 2010. Effect of nanoparticle surface charge at the plasma membrane and beyond, *Nano Letters*, **10(7)**: 2543–2548.

Bagri, A., C. Mattevi, M. Acik, Y. J. Chabal, M. Chhowalla, and V. B. Shenoy. 2010. Structural evolution during the reduction of chemically derived graphene oxide, *Nature Chemistry*, **2(7)**: 581.

Balapanuru, J., J. X. Yang, S. Xiao, Q. Bao, M. Jahan, L. Polavarapu, J. Wei, Q. H. Xu, and K. P. Loh. 2010. A graphene oxide–organic dye ionic complex with DNA-sensing and optical-limiting properties, *Angewandte Chemie International Edition*, **49(37)**: 6549–6553.

Banhart, F., J. Kotakoski, and A. V. Krasheninnikov. 2010. Structural defects in graphene, *ACS Nano*, **5(1)**: 26–41.

Barbolina, I., Woods, C. R., Lozano, N., Kostarelos, K., Novoselov, K. S., and Roberts, I. S. 2016. Purity of graphene oxide determines its antibacterial activity, *2D Materials*, **3(2)**: 025025.

Begum, P. and B. Fugetsu. 2013. Induction of cell death by graphene in Arabidopsis thaliana (Columbia ecotype) T87 cell suspensions, *Journal of Hazardous Materials*, **260**: 1032–1041.

Begum, P., R. Ikhtiari, and B. Fugetsu. 2011. Graphene phytotoxicity in the seedling stage of cabbage, tomato, red spinach, and lettuce, *Carbon*, **49(12)**: 3907–3919.

Bettinger, C. J., R. Langer, and J. T. Borenstein. 2009. Engineering substrate topography at the micro-and nanoscale to control cell function, *Angewandte Chemie International Edition*, **48(30)**: 5406–5415.

Bianco, A. and M. Prato. 2015. Safety concerns on graphene and 2D materials: A flagship perspective, *2D Materials*, **2(3)**: 030201.

Bianco, A., H.-M. Cheng, T. Enoki, Y. Gogotsi, R. H. Hurt, N. Koratkar, T. Kyotani, M. Monthioux, C. R. Park, J. M. D. Tascon, and J. Zhang. 2013. All in the graphene family—A recommended nomenclature for two-dimensional carbon materials, *Carbon*, **65**: 1–6.

Bianco, A. 2013. Graphene: Safe or toxic? The two faces of the medal, *Angewandte Chemie International Edition*, **52(19)**: 4986–4997.

Bora, T., and J. Dutta. 2014. Applications of nanotechnology in wastewater treatment—A review, *Journal of Nanoscience and Nanotechnology*, **14(1)**: 613–626.

Braakhuis, H. M., M. V. Park, I. Gosens, W. H. De Jong, and F. R. Cassee. 2014. Physicochemical characteristics of nanomaterials that affect pulmonary inflammation, *Particle and Fibre Toxicology*, **11(1)**: 18.

Braakhuis, H. M., S. K. Kloet, S. Kezic, F. Kuper, M. V. D. Z. Park, S. Bellmann, M. van der Zande, S. Le Gac, P. Krystek, R. J. B. Peters, I. M. C. M. Rietjens, and H. Bouwmeester. 2015. Progress and future of in vitro models to study translocation of nanoparticles, *Archives of Toxicology*, **89(9)**: 1469–1495.

Burton, G. J. and E. Jauniaux. 2011. Oxidative stress, *Best Practice & Research Clinical Obstetrics & Gynaecology*, **25(3)**: 287–299.

Bussy, C., H. Ali-Boucetta, and K. Kostarelos. 2012. Safety considerations for graphene: Lessons learnt from carbon nanotubes, *Accounts of Chemical Research*, **46(3)**: 692–701.

Cai, X., S. Tan, M. Lin, A. Xie, W. Mai, X. Zhang, Z. Lin, T. Wu, and Y. Liu. 2011. Synergistic antibacterial brilliant blue/reduced graphene oxide/quaternary phosphonium salt composite with excellent water solubility and specific targeting capability, *Langmuir*, **27(12)**: 7828–7835.

Card, J. W., T. S. Jonaitis, S. Tafazoli, and B. A. Magnuson. 2011. An appraisal of the published literature on the safety and toxicity of food-related nanomaterials, *Critical Reviews in Toxicology*, **41(1)**: 20–49.

Cha, C., S. R. Shin, X. Gao, N. Annabi, M. R. Dokmeci, X. Tang, and A. Khademhosseini. 2014. Controlling mechanical properties of cell-laden hydrogels by covalent incorporation of graphene oxide, *Small*, **10(3)**: 514–523.

Chang, Y., S.-T. Yang, J.-H. Liu, E. Dong, Y. Wang, A. Cao, Y. Liu, and H. Wang. 2011. In vitro toxicity evaluation of graphene oxide on A549 cells, *Toxicology Letters*, **200(3)**: 201–210.

Chatterjee, N., H.-J. Eom, and J. Choi. 2014. A systems toxicology approach to the surface functionality control of graphene–cell interactions, *Biomaterials*, **35(4)**: 1109–1127.

Chatterjee, N., J. Yang, and J. Choi. 2016. Differential genotoxic and epigenotoxic effects of graphene family nanomaterials (GFNs) in human bronchial epithelial cells, *Mutation Research/Genetic Toxicology and Environmental Mutagenesis*, **798–799**: 1–10.

Chen, G.-Y., C.-L. Chen, H.-Y. Tuan, P.-X. Yuan, K.-C. Li, H.-J. Yang, and Y.-C. Hu. 2014. Graphene oxide triggers toll-like receptors/autophagy responses in vitro and inhibits tumor growth in vivo, *Advanced Healthcare Materials*, **3(9)**: 1486–1495.

Chen, G.-Y., C.-L. Meng, K.-C. Lin, H.-Y. Tuan, H.-J. Yang, C.-L. Chen, K.-C. Li, C.-S. Chiang, and Y.-C. Hu. 2015. Graphene oxide as a chemosensitizer: Diverted autophagic flux, enhanced nuclear import, elevated necrosis and improved antitumor effects, *Biomaterials*, **40**: 12–22.

Chen, G.-Y., H.-J. Yang, C.-H. Lu, Y.-C. Chao, S.-M. Hwang, C.-L. Chen, K.-W. Lo, L.-Y. Sung, W.-Y. Luo, H.-Y. Tuan, and Y.-C. Hu. 2012a. Simultaneous induction of autophagy and toll-like receptor signaling pathways by graphene oxide, *Biomaterials*, **33(27)**: 6559–6569.

Chen, J., X. Wang, and H. Han. 2013. A new function of graphene oxide emerges: Inactivating phytopathogenic bacterium Xanthomonas oryzae pv. Oryzae, *Journal of Nanoparticle Research*, **15(5)**: 1658.

Chen, L., P. Hu, L. Zhang, S. Huang, L. Luo, and C. Huang. 2012b. Toxicity of graphene oxide and multi-walled carbon nanotubes against human cells and zebrafish, *Science China Chemistry*, **55(10)**: 2209–2216.

Chen, M., J. Yin, Y. Liang, S. Yuan, F. Wang, M. Song, and H. Wang. 2016. Oxidative stress and immunotoxicity induced by graphene oxide in zebrafish, *Aquatic Toxicology*, **174**: 54–60.

Chen, M.-L., J.-W. Liu, B. Hu, M.-L. Chen, and J.-H. Wang. 2011a. Conjugation of quantum dots with graphene for fluorescence imaging of live cells, *Analyst*, **136(20)**: 4277–4283.

Chen, Y., H. Vedala, G. P. Kotchey, A. Audfray, S. Cecioni, A. Imberty, S. Vidal, and A. Star. 2011b. Electronic detection of lectins using carbohydrate-functionalized nanostructures: Graphene versus carbon nanotubes, *Acs Nano*, **6(1)**: 760–770.

Chng, E. L. K. and M. Pumera. 2013. The toxicity of graphene oxides: Dependence on the oxidative methods used, *Chemistry—A European Journal*, **19(25)**: 8227–8235.

Chng, E. L. K., Z. Sofer, and M. Pumera. 2014. Cytotoxicity profile of highly hydrogenated graphene, *Chemistry—A European Journal*, **20(21)**: 6366–6373.

Cho, M., W.-S. Cho, M. Choi, S. J. Kim, B. S. Han, S. H. Kim, H. O. Kim, Y. Y. Sheen, and J. Jeong. 2009. The impact of size on tissue distribution and elimination by single intravenous injection of silica nanoparticles, *Toxicology Letters*, **189(3)**: 177–183.

Choi, H. S., W. Liu, P. Misra, E. Tanaka, J. P. Zimmer, B. I. Ipe, M. G. Bawendi, and J. V. Frangioni. 2007. Renal clearance of quantum dots, *Nature Biotechnology*, **25(10)**: 1165.

Choi, W., I. Lahiri, R. Seelaboyina, and Y. S. Kang. 2010. Synthesis of graphene and its applications: A review, *Critical Reviews in Solid State and Materials Sciences*, **35(1)**: 52–71.

Chong, Y., Y. Ma, H. Shen, X. Tu, X. Zhou, J. Xu, J. Dai, S. Fan, and Z. Zhang. 2014. The in vitro and in vivo toxicity of graphene quantum dots, *Biomaterials*, **35(19)**: 5041–5048.

Chowdhury, S. M., G. Lalwani, K. Zhang, J. Y. Yang, K. Neville, and B. Sitharaman. 2013. Cell specific cytotoxicity and uptake of graphene nanoribbons, *Biomaterials*, **34(1)**: 283–293.

Ciccia, A. and S. J. Elledge. 2010. The DNA damage response: Making it safe to play with knives, *Molecular Cell*, **40(2)**: 179–204.

Combarros, R. G., S. Collado, and M. Díaz. 2016. Toxicity of graphene oxide on growth and metabolism of Pseudomonas putida, *Journal of Hazardous Materials*, **310**: 246–252.

Cote, L. J., F. Kim, and J. Huang. 2009. Langmuir–Blodgett assembly of graphite oxide single layers, *Journal of the American Chemical Society*, **131(3)**: 1043–1049.

Cui, Y., S. N. Kim, R. R. Naik, and M. C. McAlpine. 2012. Biomimetic peptide nanosensors, *Accounts of Chemical Research*, **45(5)**: 696–704.

Dan, M., P. Wu, E. A. Grulke, U. M. Graham, J. M. Unrine, and R. A. Yokel. 2012. Ceria-engineered nanomaterial distribution in, and clearance from, blood: Size matters, *Nanomedicine*, **7(1)**: 95–110.

Das, S., S. Singh, V. Singh, D. Joung, J. M. Dowding, D. Reid, J. Anderson, L. Zhai, S. I. Khondaker, W. T. Self, and S. Seal. 2013. Oxygenated functional group density on graphene oxide: Its effect on cell toxicity, *Particle & Particle Systems Characterization*, **30(2)**: 148–157.

Dasari Shareena, T. P., D. McShan, A. K. Dasmahapatra, and P. B. Tchounwou. 2018. A review on graphene-based nanomaterials in biomedical applications and risks in environment and health, *Nano-Micro Letters*, **10(3)**: 53.

De Jong, W. H., W. I. Hagens, P. Krystek, M. C. Burger, A. J. Sips, and R. E. Geertsma. 2008. Particle size-dependent organ distribution of gold nanoparticles after intravenous administration, *Biomaterials*, **29(12)**: 1912–1919.

De Marchi, L., C. Pretti, B. Gabriel, P. A. A. P. Marques, R. Freitas, and V. Neto. 2018. An overview of graphene materials: Properties, applications and toxicity on aquatic environments, *Science of the Total Environment*, **631–632**: 1440–1456.

De Marchi, L., V. Neto, C. Pretti, E. Figueira, L. Brambilla, M. J. Rodriguez-Douton, F. Rossella, M. Tommasini, C. Furtado, A. M. V. M. Soares, and R. Freitas. 2017. Physiological and biochemical impacts of graphene oxide in polychaetes: The case of diopatra neapolitana, *Comparative Biochemistry and Physiology Part C: Toxicology & Pharmacology*, **193**: 50–60.

Deng, S. and V. Berry. 2016. Wrinkled, rippled and crumpled graphene: An overview of formation mechanism, electronic properties, and applications, *Materials Today*, **19(4)**: 197–212.

Deng, Y., J. Li, M. Qiu, F. Yang, J. Zhang, and C. Yuan. 2017. Deriving characterization factors on freshwater ecotoxicity of graphene oxide nanomaterial for life cycle impact assessment, *The International Journal of Life Cycle Assessment*, **22(2)**: 222–236.

Dervin, S., D. D. Dionysiou, and S. C. Pillai. 2016. 2D nanostructures for water purification: Graphene and beyond, *Nanoscale*, **8.33**: 15115–15131.

Dietz, K.-J. and S. Herth. 2011. Plant nanotoxicology, *Trends in Plant Science*, **16(11)**: 582–589.

Dikin, D. A., S. Stankovich, E. J. Zimney, R. D. Piner, G. H. Dommett, G. Evmenenko, S. T. Nguyen, and R. S. Ruoff, 2007, Preparation and characterization of graphene oxide paper, *Nature*, **448(7152)**: 457–460.

Ding, Z., Z. Zhang, H. Ma, and Y. Chen. 2014. In vitro hemocompatibility and toxic mechanism of graphene oxide on human peripheral blood T lymphocytes and serum albumin, *ACS Applied Materials & Interfaces*, **6(22)**: 19797–19807.

Donaldson, K. and A. Seaton. 2012. A short history of the toxicology of inhaled particles, *Particle and Fibre Toxicology*, **9**: 13–13.

Dreyer, D. R., S. Park, C. W. Bielawski, and R. S. Ruoff. 2010. The chemistry of graphene oxide, *Chemical Society Reviews*, **39(1)**: 228–240.

Du, J., S. Wang, H. You, R. Jiang, C. Zhuang, and X. Zhang. 2016a. Developmental toxicity and DNA damage to zebrafish induced by perfluorooctane sulfonate in the presence of ZnO nanoparticles, *Environmental Toxicology*, **31(3)**: 360–371.

Du, S., P. Zhang, R. Zhang, Q. Lu, L. Liu, X. Bao, and H. Liu. 2016b. Reduced graphene oxide induces cytotoxicity and inhibits photosynthetic performance of the green alga Scenedesmus obliquus, *Chemosphere*, **164**: 499–507.

Duch, M. C., G. R. S. Budinger, Y. T. Liang, S. Soberanes, D. Urich, S. E. Chiarella, L. A. Campochiaro, A. Gonzalez, N. S. Chandel, M. C. Hersam, and G. M. Mutlu. 2011. Minimizing oxidation and stable nanoscale dispersion improves the biocompatibility of graphene in the lung, *Nano Letters*, **11(12)**: 5201–5207.

Ema, M., K. S. Hougaard, A. Kishimoto, and K. Honda. 2016b. Reproductive and developmental toxicity of carbon-based nanomaterials: A literature review, *Nanotoxicology*, **10(4)**: 391–412.

Ema, M., M. Gamo, and K. Honda. 2016a. Developmental toxicity of engineered nanomaterials in rodents, *Toxicology and Applied Pharmacology*, **299**: 47–52.

Erickson, K., R. Erni, Z. Lee, N. Alem, W. Gannett, and A. Zettl. 2010. Determination of the local chemical structure of graphene oxide and reduced graphene oxide, *Advanced Materials*, **22(40)**: 4467–4472.

Feng, L. and Z. Liu. 2011. Graphene in biomedicine: Opportunities and challenges, *Nanomedicine*, **6(2)**: 317–324.

Ferrari, A. C., F. Bonaccorso, V. Fal'ko, K. S. Novoselov, S. Roche, P. Bøggild, S. Borini et al. 2015. Science and technology roadmap for graphene, related two-dimensional crystals, and hybrid systems, *Nanoscale*, **7(11)**: 4598–4810.

Fitzer, E., K. H. Kochling, H. P. Boehm, and H. Marsh. 1995. Recommended terminology for the description of carbon as a solid (IUPAC Recommendations 1995), *Pure and Applied Chemistry*, **3**: 473.

Fojtů, M., W. Z. Teo, and M. Pumera. 2017. Environmental impact and potential health risks of 2D nanomaterials, *Environmental Science: Nano*, **4(8)**: 1617–1633.

Fredrickson, J. K., M. F. Romine, A. S. Beliaev, J. M. Auchtung, M. E. Driscoll, T. S. Gardner, K. H. Nealson et al. 2008. Towards environmental systems biology of Shewanella, *Nature Reviews Microbiology*, **6**: 592.

Fu, C., T. Liu, L. Li, H. Liu, Q. Liang, and X. Meng. 2015. Effects of graphene oxide on the development of offspring mice in lactation period, *Biomaterials*, **40**: 23–31.

Fujimi, S., M. P. MacConmara, A. A. Maung, Y. Zang, J. A. Mannick, J. A. Lederer, and P. H. Lapchak. 2006. Platelet depletion in mice increases mortality after thermal injury, *Blood*, **107(11)**: 4399–4406.

Garg, B., T. Bisht, and Y.-C. Ling. 2014. Graphene-based nanomaterials as heterogeneous acid catalysts: A comprehensive perspective, *Molecules*, **19(9)**: 14582–14614.

Gollavelli, G. and Y.-C. Ling. 2012. Multi-functional graphene as an in vitro and in vivo imaging probe, *Biomaterials*, **33(8)**: 2532–2545.

Goodman, C. M., C. D. McCusker, T. Yilmaz, and V. M. Rotello. 2004. Toxicity of gold nanoparticles functionalized with cationic and anionic side chains, *Bioconjugate Chemistry*, **15(4)**: 897–900.

Guo, F., F. Kim, T. H. Han, V. B. Shenoy, J. Huang, and R. H. Hurt. 2011. Hydration-responsive folding and unfolding in graphene oxide liquid crystal phases, *ACS Nano*, **5(10)**: 8019–8025.

Guo, X. and N. Mei. 2014. Assessment of the toxic potential of graphene family nanomaterials, *Journal of Food and Drug Analysis*, **22(1)**: 105–115.

Guo, X., S. Dong, E. J. Petersen, S. Gao, Q. Huang, and L. Mao. 2013. Biological uptake and depuration of radio-labeled graphene by Daphnia magna, *Environmental Science & Technology*, **47(21)**: 12524–12531.

Gupta, A., T. Sakthivel, and S. Seal. 2015. Recent development in 2D materials beyond graphene, *Progress in Materials Science*, **73**: 44–126.

Gurunathan, S., J. Han, J. H. Park, and J. H. Kim. 2014. An in vitro evaluation of graphene oxide reduced by Ganoderma spp. in human breast cancer cells (MDA-MB-231), *International Journal of Nanomedicine*, **9**: 1783–1797.

Gurunathan, S., J. W. Han, A. A. Dayem, V. Eppakayala, and J.-H. Kim. 2012. Oxidative stress-mediated antibacterial activity of graphene oxide and reduced graphene oxide in Pseudomonas aeruginosa, *International Journal of Nanomedicine*, **7**: 5901.

Halamoda Kenzaoui, B., C. Chapuis Bernasconi, S. Guney-Ayra, and L. Juillerat-Jeanneret. 2012. Induction of oxidative stress, lysosome activation and autophagy by nanoparticles in human brain-derived endothelial cells, *Biochemical Journal*, **441(3)**: 813.

Hammes, J., J. A. Gallego-Urrea, and M. Hassellöv. 2013. Geographically distributed classification of surface water chemical parameters influencing fate and behavior of nanoparticles and colloid facilitated contaminant transport, *Water Research*, **47(14)**: 5350–5361.

Haque, A.-M. J., H. Park, D. Sung, S. Jon, S.-Y. Choi, and K. Kim. 2012. An electrochemically reduced graphene oxide-based electrochemical immunosensing platform for ultrasensitive antigen detection, *Analytical Chemistry*, **84(4)**: 1871–1878.

Hau, H. H. and J. A. Gralnick. 2007. Ecology and biotechnology of the genus *Shewanella*, *Annual Review of Microbiology*, **61(1)**: 237–258.

He, K., G. Chen, G. Zeng, M. Peng, Z. Huang, J. Shi, and T. Huang. 2017. Stability, transport and ecosystem effects of graphene in water and soil environments, *Nanoscale*, **9(17)**: 5370–5388.

Hengartner, M. O. 2000. The biochemistry of apoptosis, *Nature*, **407**: 770.

Nanotechnologies—Vocabulary—Part 13: Graphene and related two-dimensional (2D) materials. https://www.iso.org/obp/ui/#iso:std:iso:ts:80004:-13:ed-1:v1:en, 2017, ISO/TS 80004-13:2017.

Hu, H., J. Yu, Y. Li, J. Zhao, and H. Dong. 2012. Engineering of a novel pluronic F127/graphene nanohybrid for pH responsive drug delivery, *Journal of Biomedical Materials Research Part A*, **100(1)**: 141–148.

Hu, Q., B. Jiao, X. Shi, R. P. Valle, Y. Y. Zuo, and G. Hu. 2015a. Effects of graphene oxide nanosheets on the ultrastructure and biophysical properties of the pulmonary surfactant film, *Nanoscale*, **7(43)**: 18025–18029.

Hu, W., C. Peng, W. Luo, M. Lv, X. Li, D. Li, Q. Huang, and C. Fan. 2010. Graphene-based antibacterial paper, *Acs Nano*, **4(7)**: 4317–4323.

Hu, X. and Q. Zhou. 2013. Health and ecosystem risks of graphene, *Chemical Reviews*, **113(5)**: 3815–3835.

Hu, X. and Q. Zhou. 2014. Novel hydrated graphene ribbon unexpectedly promotes aged seed germination and root differentiation, *Scientific Reports*, **4**: 3782.

Hu, X., J. Kang, K. Lu, R. Zhou, L. Mu, and Q. Zhou. 2014. Graphene oxide amplifies the phytotoxicity of arsenic in wheat, *Scientific Reports*, **4**: 6122.

Hu, X., S. Ouyang, L. Mu, J. An, and Q. Zhou. 2015b. Effects of graphene oxide and oxidized carbon nanotubes on the cellular division, microstructure, uptake, oxidative stress, and metabolic profiles, *Environmental Science & Technology*, **49(18)**: 10825–10833.

Hussain, S. and S. Garantziotis. 2013. Interplay between apoptotic and autophagy pathways after exposure to cerium dioxide nanoparticles in human monocytes, *Autophagy*, **9(1)**: 101–103.

Hussein, A. K. 2015. Applications of nanotechnology in renewable energies—A comprehensive overview and understanding, *Renewable and Sustainable Energy Reviews*, **42**: 460–476.

Ivask, A., N. H. Voelcker, S. A. Seabrook, M. Hor, J. K. Kirby, M. Fenech, T. P. Davis, and P. C. Ke. 2015. DNA melting and genotoxicity induced by silver nanoparticles and graphene, *Chemical Research in Toxicology*, **28(5)**: 1023–1035.

Jarosz, A., M. Skoda, I. Dudek, and D. Szukiewicz. 2016. Oxidative stress and mitochondrial activation as the main mechanisms underlying graphene toxicity against human cancer cells, *Oxidative Medicine and Cellular Longevity*, **2016**.

Jastrzębska, A. M. and A. R. Olszyna. 2015. The ecotoxicity of graphene family materials: Current status, knowledge gaps and future needs, *Journal of Nanoparticle Research*, **17(1)**: 40.

Jastrzębska, A. M., P. Kurtycz, and A. R. Olszyna. 2012. Recent advances in graphene family materials toxicity investigations, *Journal of Nanoparticle Research*, **14(12)**: 1320.

Jaworski, S., E. Sawosz, M. Grodzik, A. Winnicka, M. Prasek, M. Wierzbicki, and A. Chwalibog. 2013. In vitro evaluation of the effects of graphene platelets on glioblastoma multiforme cells, *International Journal of Nanomedicine*, **8**: 413–420.

Jennifer, M. and W. Maciej. 2013. Nanoparticle technology as a double-edged sword: Cytotoxic, genotoxic and epigenetic effects on living cells, *Journal of Biomaterials and Nanobiotechnology*, **4(01)**: 53.

Jiang, J., G. Oberdörster, and P. Biswas. 2009. Characterization of size, surface charge, and agglomeration state of nanoparticle dispersions for toxicological studies, *Journal of Nanoparticle Research*, **11(1)**: 77–89.

Jiang, W., B. Y. Kim, J. T. Rutka, and W. C. Chan. 2008. Nanoparticle-mediated cellular response is size-dependent, *Nature Nanotechnology*, **3(3)**: 145.

Jiang, X., J. Dausend, M. Hafner, A. Musyanovych, C. Röcker, K. Landfester, V. Mailänder, and G. U. Nienhaus. 2010. Specific effects of surface amines on polystyrene nanoparticles in their interactions with mesenchymal stem cells, *Biomacromolecules*, **11(3)**: 748–753.

Jiao, Y., F. Qian, Y. Li, G. Wang, C. W. Saltikov, and J. A. Granick. 2011. Deciphering the electron transport pathway for graphene oxide reduction by Shewanella oneidensis MR-1, *Journal of Bacteriology*, JB. 00201-11.

Jin, C., F. Wang, Y. Tang, X. Zhang, J. Wang, and Y. Yang. 2014. Distribution of graphene oxide and TiO2-graphene oxide composite in A549 cells, *Biological Trace Element Research*, **159**(*1–3*): 393–398.

Johnston, H. J., G. R. Hutchison, F. M. Christensen, S. Peters, S. Hankin, K. Aschberger, and V. Stone. 2010. A critical review of the biological mechanisms underlying the in vivo and in vitro toxicity of carbon nanotubes: The contribution of physico-chemical characteristics, *Nanotoxicology*, **4**(**2**): 207–246.

Khim Chng, E. L., C. K. Chua, and M. Pumera. 2014. Graphene oxide nanoribbons exhibit significantly greater toxicity than graphene oxide nanoplatelets, *Nanoscale*, **6**(**18**): 10792–10797.

Khodakovskaya, M. V., K. de Silva, D. A. Nedosekin, E. Dervishi, A. S. Biris, E. V. Shashkov, E. I. Galanzha, and V. P. Zharov. 2011. Complex genetic, photothermal, and photoacoustic analysis of nanoparticle-plant interactions, *Proceedings of the National Academy of Sciences*, **108**(**3**): 1028–1033.

Khosravi-Katuli, K., E. Prato, G. Lofrano, M. Guida, G. Vale, and G. Libralato. 2017. Effects of nanoparticles in species of aquaculture interest, *Environmental Science and Pollution Research*, **24**(**21**): 17326–17346.

Kim, F., L. J. Cote, and J. Huang. 2010a. Graphene oxide: Surface activity and two-dimensional assembly, *Advanced Materials*, **22**(**17**): 1954–1958.

Kim, J., L. J. Cote, F. Kim, W. Yuan, K. R. Shull, and J. Huang. 2010b. Graphene oxide sheets at interfaces, *Journal of the American Chemical Society*, **132**(**23**): 8180–8186.

Kim, S. N., Z. Kuang, J. M. Slocik, S. E. Jones, Y. Cui, B. L. Farmer, M. C. McAlpine, and R. R. Naik. 2011. Preferential binding of peptides to graphene edges and planes, *Journal of the American Chemical Society*, **133**(**37**): 14480–14483.

Koenig, S. P., N. G. Boddeti, M. L. Dunn, and J. S. Bunch. 2011. Ultrastrong adhesion of graphene membranes, *Nature Nanotechnology*, **6**(**9**): 543.

Kostarelos, K. and K. S. Novoselov. 2014. Exploring the interface of graphene and biology, *Science*, **344**(**6181**): 261.

Kotchey, G. P., B. L. Allen, H. Vedala, N. Yanamala, A. A. Kapralov, Y. Y. Tyurina, J. Klein-Seetharaman, V. E. Kagan, and A. Star. 2011. The enzymatic oxidation of graphene oxide, *ACS Nano*, **5**(**3**): 2098–2108.

Krishnamoorthy, K., N. Umasuthan, R. Mohan, J. Lee, and S.-J. Kim. 2012. Antibacterial activity of graphene oxide nanosheets, *Science of Advanced Materials*, **4**(**11**): 1111–1117.

Kulvinder Singh, C. 2017. Graphene gets its own international standard, *Physics World*, **30**(**12**): 11.

Kumar, R., O. Baghel, S. K. Sidar, P. K. Sen, and S. K. Bohidar. 2014. Applications of nanorobotics, *International Journal of Scientific Research Engineering & Technology* (*IJSRET*), ISSN, 2278-0882.

Kurantowicz, N., B. Strojny, E. Sawosz, S. Jaworski, M. Kutwin, M. Grodzik, M. Wierzbicki, L. Lipińska, K. Mitura, and A. Chwalibog. 2015. Biodistribution of a high dose of diamond, graphite, and graphene oxide nanoparticles after multiple intraperitoneal injections in rats, *Nanoscale Research Letters*, **10**(**1**): 398.

Lalwani, G., M. D'Agati, A. M. Khan, and B. Sitharaman. 2016. Toxicology of graphene-based nanomaterials, *Advanced Drug Delivery Reviews*, **105**: 109–144.

Lammel, T. and J. M. Navas. 2014. Graphene nanoplatelets spontaneously translocate into the cytosol and physically interact with cellular organelles in the fish cell line PLHC-1, *Aquatic Toxicology*, **150**: 55–65.

Lammel, T., P. Boisseaux, M.-L. Fernández-Cruz, and J. M. Navas. 2013. Internalization and cytotoxicity of graphene oxide and carboxyl graphene nanoplatelets in the human hepatocellular carcinoma cell line Hep G2, *Particle and Fibre Toxicology*, **10**(**1**): 27.

Lanphere, J. D., Rogers, B., Luth, C., Bolster, C. H., and Walker, S. L. 2014. Stability and transport of graphene oxide nanoparticles in groundwater and surface water, *Environmental Engineering Science*, **31**(**7**): 350–359.

Lawrence, T. 2009. The nuclear factor NF-kappaB pathway in inflammation, *Cold Spring Harbor Perspectives in Biology*, **1**(**6**): a001651–a001651.

Lee, D. Y., Z. Khatun, J.-H. Lee, Y.-k. Lee, and I. In. 2011. Blood compatible graphene/heparin conjugate through noncovalent chemistry, *Biomacromolecules*, **12(2)**: 336–341.
Lee, G.-H., Y.-J. Yu, X. Cui, N. Petrone, C.-H. Lee, M. S. Choi, D.-Y. Lee, C. Lee, W. J. Yoo, and K. Watanabe. 2013. Flexible and transparent MoS2 field-effect transistors on hexagonal boron nitride-graphene heterostructures, *ACS Nano*, **7(9)**: 7931–7936.
Lee, J. K., A. Y. Jeong, J. Bae, J. H. Seok, J.-Y. Yang, H. S. Roh, J. Jeong, Y. Han, J. Jeong, and W.-S. Cho. 2017. The role of surface functionalization on the pulmonary inflammogenicity and translocation into mediastinal lymph nodes of graphene nanoplatelets in rats, *Archives of Toxicology*, **91(2)**: 667–676.
Lee, Y. A., A. Durandin, P. C. Dedon, N. E. Geacintov, and V. Shafirovich. 2008. Oxidation of guanine in G, GG, and GGG sequence contexts by aromatic pyrenyl radical cations and carbonate radical anions: Relationship between kinetics and distribution of alkali-labile lesions, *The Journal of Physical Chemistry B*, **112(6)**: 1834–1844.
Levine, B., N. Mizushima, and H. W. Virgin. 2011. Autophagy in immunity and inflammation, *Nature*, **469(7330)**: 323–335.
Li, B., J. Yang, Q. Huang, Y. Zhang, C. Peng, Y. Zhang, Y. He, J. Shi, W. Li, J. Hu, and C. Fan. 2013a. Biodistribution and pulmonary toxicity of intratracheally instilled graphene oxide in mice, *Npg Asia Materials*, **5**: e44.
Li, B., X.-Y. Zhang, J.-Z. Yang, Y.-J. Zhang, W.-X. Li, C.-H. Fan, and Q. Huang. 2014. Influence of polyethylene glycol coating on biodistribution and toxicity of nanoscale graphene oxide in mice after intravenous injection, *International Journal of Nanomedicine*, **9**: 4697.
Li, C., J. Adamcik, and R. Mezzenga. 2012a. Biodegradable nanocomposites of amyloid fibrils and graphene with shape-memory and enzyme-sensing properties, *Nature Nanotechnology*, **7**: 421.
Li, L., G. Wu, G. Yang, J. Peng, J. Zhao, and J.-J. Zhu. 2013b. Focusing on luminescent graphene quantum dots: Current status and future perspectives, *Nanoscale*, **5(10)**: 4015–4039.
Li, Y., H. Yuan, A. von dem Bussche, M. Creighton, R. H. Hurt, A. B. Kane, and H. Gao. 2013c. Graphene microsheets enter cells through spontaneous membrane penetration at edge asperities and corner sites, *Proceedings of the National Academy of Sciences*, **110(30)**: 12295.
Li, Y., Y. Liu, Y. Fu, T. Wei, L. Le Guyader, G. Gao, R.-S. Liu, Y.-Z. Chang, and C. Chen. 2012b. The triggering of apoptosis in macrophages by pristine graphene through the MAPK and TGF-beta signaling pathways, *Biomaterials*, **33(2)**: 402–411.
Liao, K.-H., Y.-S. Lin, C. W. Macosko, and C. L. Haynes. 2011. Cytotoxicity of graphene oxide and graphene in human erythrocytes and skin fibroblasts, *ACS Applied Materials & Interfaces*, **3(7)**: 2607–2615.
Lightcap, I. V., T. H. Kosel, and P. V. Kamat. 2010. Anchoring semiconductor and metal nanoparticles on a two-dimensional catalyst mat. Storing and shuttling electrons with reduced graphene oxide, *Nano Letters*, **10(2)**: 577–583.
Liu, J., L. Cui, and D. Losic. 2013a. Graphene and graphene oxide as new nanocarriers for drug delivery applications, *Acta Biomaterialia*, **9(12)**: 9243–9257.
Liu, J.-H., S.-T. Yang, H. Wang, Y. Chang, A. Cao, and Y. Liu. 2012. Effect of size and dose on the biodistribution of graphene oxide in mice, *Nanomedicine*, **7(12)**: 1801–1812.
Liu, S., T. H. Zeng, M. Hofmann, E. Burcombe, J. Wei, R. Jiang, J. Kong, and Y. Chen. 2011. Antibacterial activity of graphite, graphite oxide, graphene oxide, and reduced graphene oxide: Membrane and oxidative stress, *ACS Nano*, **5(9)**: 6971–6980.
Liu, Y., Y. Luo, J. Wu, Y. Wang, X. Yang, R. Yang, B. Wang, J. Yang, and N. Zhang. 2013b. Graphene oxide can induce in vitro and in vivo mutagenesis, *Scientific Reports*, **3**: 3469.
Longmire, M., P. L. Choyke, and H. Kobayashi. 2008. Clearance properties of nano-sized particles and molecules as imaging agents: Considerations and caveats, *Nanomedicine*, **3(5)**: 703–717.
Magdolenova, Z., A. Collins, A. Kumar, A. Dhawan, V. Stone, and M. Dusinska. 2014. Mechanisms of genotoxicity. A review of *in vitro* and *in vivo* studies with engineered nanoparticles, *Nanotoxicology*, **8(3)**: 233–278.

Ma-Hock, L., V. Strauss, S. Treumann, K. Küttler, W. Wohlleben, T. Hofmann, S. Gröters, K. Wiench, B. van Ravenzwaay, and R. Landsiedel. 2013. Comparative inhalation toxicity of multi-wall carbon nanotubes, graphene, graphite nanoplatelets and low surface carbon black, *Particle and Fibre Toxicology*, **10(1)**: 23.

Maiorano, G., S. Sabella, B. Sorce, V. Brunetti, M. A. Malvindi, R. Cingolani, and P. P. Pompa. 2010. Effects of cell culture media on the dynamic formation of protein—Nanoparticle complexes and influence on the cellular response, *ACS Nano*, **4(12)**: 7481–7491.

Majeed, W., S. Bourdo, D. M. Petibone, V. Saini, K. B. Vang, Z. A. Nima, K. M. Alghazali, E. Darrigues, A. Ghosh, F. Watanabe, D. Casciano, S. F. Ali, and A. S. Biris. 2017. The role of surface chemistry in the cytotoxicity profile of graphene, *Journal of Applied Toxicology*, **37(4)**: 462–470.

Mao, L., M. Hu, B. Pan, Y. Xie, and E. J. Petersen. 2016. Biodistribution and toxicity of radio-labeled few layer graphene in mice after intratracheal instillation, *Particle and Fibre Toxicology*, **13**: 7–7.

Markovic, Z. M., B. Z. Ristic, K. M. Arsikin, D. G. Klisic, L. M. Harhaji-Trajkovic, B. M. Todorovic-Markovic, D. P. Kepic et al. 2012. Graphene quantum dots as autophagy-inducing photodynamic agents, *Biomaterials*, **33(29)**: 7084–7092.

Matesanz, M.-C., M. Vila, M.-J. Feito, J. Linares, G. Gonçalves, M. Vallet-Regi, P.-A. A. P. Marques, and M.-T. Portolés. 2013. The effects of graphene oxide nanosheets localized on F-actin filaments on cell-cycle alterations, *Biomaterials*, **34(5)**: 1562–1569.

Matsumoto, Y., M. Koinuma, S. Ida, S. Hayami, T. Taniguchi, K. Hatakeyama, H. Tateishi, Y. Watanabe, and S. Amano. 2011. Photoreaction of graphene oxide nanosheets in water, *The Journal of Physical Chemistry C*, **115(39)**: 19280–19286.

Mendonça, M. C. P., E. S. Soares, M. B. de Jesus, H. J. Ceragioli, M. S. Ferreira, R. R. Catharino, and M. A. da Cruz-Höfling. 2015. Reduced graphene oxide induces transient blood–brain barrier opening: An *in vivo* study, *Journal of Nanobiotechnology*, **13(1)**: 78.

Mesarič, T., C. Gambardella, T. Milivojević, M. Faimali, D. Drobne, C. Falugi, D. Makovec, A. Jemec, and K. Sepčić. 2015. High surface adsorption properties of carbon-based nanomaterials are responsible for mortality, swimming inhibition, and biochemical responses in Artemia salina larvae, *Aquatic Toxicology*, **163**: 121–129.

Mesarič, T., K. Sepčič, V. Piazza, C. Gambardella, F. Garaventa, D. Drobne, and M. Faimali. 2013. Effects of nano carbon black and single-layer graphene oxide on settlement, survival and swimming behaviour of Amphibalanus amphitrite larvae, *Chemistry and Ecology*, **29(7)**: 643–652.

Meyer, J. C., A. K. Geim, M. I. Katsnelson, K. S. Novoselov, T. J. Booth, and S. Roth. 2007. The structure of suspended graphene sheets, *Nature*, **446(7131)**: 60.

Mohanty, N., M. Fahrenholtz, A. Nagaraja, D. Boyle, and V. Berry. 2011. Impermeable graphenic encasement of bacteria, *Nano Letters*, **11(3)**: 1270–1275.

Monopoli, M. P., C. Åberg, A. Salvati, and K. A. Dawson. 2012. Biomolecular coronas provide the biological identity of nanosized materials, *Nature Nanotechnology*, **7**: 779.

Montagner, A., Bosi, S., Tenori, E., Bidussi, M., Alshatwi, A. A., Tretiach, M., and Syrgiannis, Z. 2017. Ecotoxicological effects of graphene-based materials, *2D Materials*, **4(1)**: 012001.

Monteith, G. R., D. McAndrew, H. M. Faddy, and S. J. Roberts-Thomson. 2007. Calcium and cancer: Targeting Ca2+ transport, *Nature Reviews Cancer*, **7**: 519.

Montes-Navajas, P., N. G. Asenjo, R. Santamaría, R. Menéndez, A. Corma, and H. García. 2013. Surface area measurement of graphene oxide in aqueous solutions, *Langmuir*, **29(44)**: 13443–13448.

Mu, Q., G. Su, L. Li, B. O. Gilbertson, L. H. Yu, Q. Zhang, Y.-P. Sun, and B. Yan. 2012. Size-Dependent cell uptake of protein-coated graphene oxide nanosheets, *ACS Applied Materials & Interfaces*, **4(4)**: 2259–2266.

Murray, A. R., E. Kisin, S. S. Leonard, S. H. Young, C. Kommineni, V. E. Kagan, V. Castranova, and A. A. Shvedova. 2009. Oxidative stress and inflammatory response in dermal toxicity of single-walled carbon nanotubes, *Toxicology*, **257(3)**: 161–171.

Myers, C. R. and J. M. Myers. 1997. Cloning and sequence of cymA, a gene encoding a tetraheme cytochrome c required for reduction of iron(III), fumarate, and nitrate by Shewanella putrefaciens MR-1, *Journal of Bacteriology*, **179(4)**: 1143.

Nealson, K. H. and J. Scott. 2006. Ecophysiology of the Genus *Shewanella*, in *The Prokaryotes: Volume 6: Proteobacteria: Gamma Subclass*, pp. 1133–1151, M. Dworkin, S. Falkow, E. Rosenberg, K.-H. Schleifer and E. Stackebrandt, eds., Springer New York.

Nel, A. E., L. Mädler, D. Velegol, T. Xia, E. M. V. Hoek, P. Somasundaran, F. Klaessig, V. Castranova, and M. Thompson. 2009. Understanding biophysicochemical interactions at the nano–bio interface, *Nature Materials*, **8**: 543.

Nel, A., T. Xia, L. Mädler, and N. Li. 2006. Toxic potential of materials at the nanolevel, *Science*, **311(5761)**: 622–627.

Nezakati, T., B. G. Cousins, and A. M. Seifalian. 2014. Toxicology of chemically modified graphene-based materials for medical application, *Archives of Toxicology*, **88(11)**: 1987–2012.

Nguyen, H. N., S. L. Castro-Wallace, and D. F. Rodrigues. 2017. Acute toxicity of graphene nanoplatelets on biological wastewater treatment process, *Environmental Science: Nano*, **4(1)**: 160–169.

Nikalje, A. P. 2015. Nanotechnology and its applications in medicine, *Medicinal Chemistry*, **5(2)**, 081–089.

Nogueira, P. F. M., D. Nakabayashi, and V. Zucolotto, 2015, The effects of graphene oxide on green algae Raphidocelis subcapitata, *Aquatic Toxicology*, **166**: 29–35.

Nouara, A., Q. Wu, Y. Li, M. Tang, H. Wang, Y. Zhao, and D. Wang. 2013. Carboxylic acid functionalization prevents the translocation of multi-walled carbon nanotubes at predicted environmentally relevant concentrations into targeted organs of nematode Caenorhabditis elegans, *Nanoscale*, **5(13)**: 6088–6096.

Novoselov, K. S., A. K. Geim, S. Morozov, D. Jiang, Y. Zhang, S. A. Dubonos, I. Grigorieva, and A. Firsov. 2004. Electric field effect in atomically thin carbon films, *Science*, **306(5696)**: 666–669.

Novoselov, K. S., V. I. Fal'ko, L. Colombo, P. R. Gellert, M. G. Schwab, and K. Kim. 2012. A roadmap for graphene, *Nature*, **490**: 192.

Oberdörster, G., E. Oberdörster, and J. Oberdörster. 2005. Nanotoxicology: An emerging discipline evolving from studies of ultrafine particles, *Environmental Health Perspectives*, **113(7)**: 823.

Ou, L., B. Song, H. Liang, J. Liu, X. Feng, B. Deng, T. Sun, and L. Shao. 2016. Toxicity of graphene-family nanoparticles: A general review of the origins and mechanisms, *Particle and Fibre Toxicology*, **13(1)**: 57.

Ouyang, S., X. Hu, and Q. Zhou. 2015. Envelopment–internalization synergistic effects and metabolic mechanisms of graphene oxide on single-cell chlorella vulgaris are dependent on the nanomaterial particle size, *ACS Applied Materials & Interfaces*, **7(32)**: 18104–18112.

Paek, S.-M., E. Yoo, and I. Honma. 2008. Enhanced cyclic performance and lithium storage capacity of SnO2/graphene nanoporous electrodes with three-dimensionally delaminated flexible structure, *Nano Letters*, **9(1)**: 72–75.

Pan, Y., H. Bao, N. G. Sahoo, T. Wu, and L. Li. 2011. Water-soluble poly (N-isopropylacrylamide)–graphene sheets synthesized via click chemistry for drug delivery, *Advanced Functional Materials*, **21(14)**: 2754–2763.

Pan, Y., N. G. Sahoo, and L. Li. 2012. The application of graphene oxide in drug delivery, *Expert Opinion on Drug Delivery*, **9(11)**: 1365–1376.

Park, E.-J., G.-H. Lee, B. S. Han, B.-S. Lee, S. Lee, M.-H. Cho, J.-H. Kim, and D.-W. Kim. 2015. Toxic response of graphene nanoplatelets in vivo and in vitro, *Archives of Toxicology*, **89(9)**: 1557–1568.

Park, M. V. D. Z., E. A. J. Bleeker, W. Brand, F. R. Cassee, M. van Elk, I. Gosens, W. H. de Jong et al. 2017. Considerations for safe innovation: The case of graphene, *ACS Nano*, **11(10)**: 9574–9593.

Patlolla, R. and V. Vobalaboina. 2005. Pharmacokinetics and tissue distribution of etoposide delivered in parenteral emulsion, *Journal of Pharmaceutical Sciences*, **94(2)**: 437–445.

Patra, N., B. Wang, and P. Král. 2009. Nanodroplet activated and guided folding of graphene nanostructures, *Nano Letters*, **9(11)**: 3766–3771.

Paul, W. and C. P. Sharma. 2011. Blood compatibility and biomedical applications of graphene, *Trends Biomaterials & Artificial Organs*, **25(3)**: 91–94.

Peng, B., J. Tang, J. Luo, P. Wang, B. Ding, and K. C. Tam. 2018. Applications of nanotechnology in oil and gas industry: Progress and perspective, *The Canadian Journal of Chemical Engineering*, **96(1)**: 91–100.

Peng, C., W. Hu, Y. Zhou, C. Fan, and Q. Huang. 2010. Intracellular imaging with a graphene-based fluorescent probe, *Small*, **6(15)**: 1686–1692.

Peralta-Videa, J. R., L. Zhao, M. L. Lopez-Moreno, G. de la Rosa, J. Hong, and J. L. Gardea-Torresdey. 2011. Nanomaterials and the environment: A review for the biennium 2008–2010, *Journal of Hazardous Materials*, **186(1)**: 1–15.

Petros, R. A. and J. M. DeSimone. 2010. Strategies in the design of nanoparticles for therapeutic applications, *Nature Reviews Drug Discovery*, **9**: 615.

Pollard, A. J. and C. A. Clifford. 2017. Terminology: The first step towards international standardisation of graphene and related 2D materials, *Journal of Materials Science*, **52(24)**: 13685–13688.

Pretti, C., M. Oliva, R. Di Pietro, G. Monni, G. Cevasco, F. Chiellini, C. Pomelli, and C. Chiappe. 2014. Ecotoxicity of pristine graphene to marine organisms, *Ecotoxicology and Environmental Safety*, **101**: 138–145.

Pumera, M. 2014. Heteroatom modified graphenes: Electronic and electrochemical applications, *Journal of Materials Chemistry C*, **2(32)**: 6454–6461.

Qi, W., J. Bi, X. Zhang, J. Wang, J. Wang, P. Liu, Z. Li, and W. Wu. 2014. Damaging effects of multi-walled carbon nanotubes on pregnant mice with different pregnancy times, *Scientific Reports*, **4**: 4352.

Qu, G., S. Liu, S. Zhang, L. Wang, X. Wang, B. Sun, N. Yin, X. Gao, T. Xia, J.-J. Chen, and G.-B. Jiang. 2013. Graphene oxide induces toll-like receptor 4 (TLR4)-dependent necrosis in macrophages, *ACS Nano*, **7(7)**: 5732–5745.

Rafiee, J., M. A. Rafiee, Z. Z. Yu, and N. Koratkar. 2010. Superhydrophobic to superhydrophilic wetting control in graphene films, *Advanced Materials*, **22(19)**: 2151–2154.

Reina, G., J. M. González-Domínguez, A. Criado, E. Vázquez, A. Bianco, and M. Prato. 2017. Promises, facts and challenges for graphene in biomedical applications, *Chemical Society Reviews*, **46(15)**: 4400–4416.

Ren, H., C. Wang, J. Zhang, X. Zhou, D. Xu, J. Zheng, S. Guo, and J. Zhang. 2010. DNA cleavage system of nanosized graphene oxide sheets and copper ions, *ACS Nano*, **4(12)**: 7169–7174.

Reshma, S. C., S. Syama, and P. V. Mohanan. 2016. Nano-biointeractions of PEGylated and bare reduced graphene oxide on lung alveolar epithelial cells: A comparative in vitro study, *Colloids and Surfaces B: Biointerfaces*, **140**: 104–116.

Robinson, J. T., S. M. Tabakman, Y. Liang, H. Wang, H. Sanchez Casalongue, D. Vinh, and H. Dai. 2011. Ultrasmall reduced graphene oxide with high near-infrared absorbance for photothermal therapy, *Journal of the American Chemical Society*, **133(17)**: 6825–6831.

Römbke, J., S. Jänsch, and W. Didden. 2005. The use of earthworms in ecological soil classification and assessment concepts, *Ecotoxicology and Environmental Safety*, **62(2)**: 249–265.

Ruge, C. A., U. F. Schaefer, J. Herrmann, J. Kirch, O. Cañadas, M. Echaide, J. Pérez-Gil, C. Casals, R. Müller, and C.-M. Lehr. 2012. The interplay of lung surfactant proteins and lipids assimilates the macrophage clearance of nanoparticles, *PLOS One*, **7(7)**: e40775–e40775.

Ruiz, O. N., K. A. S. Fernando, B. Wang, N. A. Brown, P. G. Luo, N. D. McNamara, M. Vangsness, Y.-P. Sun, and C. E. Bunker. 2011. Graphene oxide: A nonspecific enhancer of cellular growth, *ACS Nano*, **5(10)**: 8100–8107.

Russier, J., E. Treossi, A. Scarsi, F. Perrozzi, H. Dumortier, L. Ottaviano, M. Meneghetti, V. Palermo, and A. Bianco. 2013. Evidencing the mask effect of graphene oxide: A comparative study on primary human and murine phagocytic cells, *Nanoscale*, **5(22)**: 11234–11247.

Sahu, A., W. I. Choi, and G. Tae. 2012. A stimuli-sensitive injectable graphene oxide composite hydrogel, *Chemical Communications*, **48(47)**: 5820–5822.

Salas, E. C., Z. Sun, A. Lüttge, and J. M. Tour. 2010. Reduction of graphene oxide via bacterial respiration, *ACS Nano*, **4(8)**: 4852–4856.

Salihoglu, O., S. Balci, and C. Kocabas. 2012. Plasmon-polaritons on graphene-metal surface and their use in biosensors, *Applied Physics Letters*, **100(21)**: 213110.

Sanchez, V. C., A. Jachak, R. H. Hurt, and A. B. Kane. 2011. Biological interactions of graphene-family nanomaterials: An interdisciplinary review, *Chemical Research in Toxicology*, **25(1)**: 15–34.

Sanchez-Hernandez, J. C. 2006. Earthworm biomarkers in ecological risk assessment, in *Reviews of Environmental Contamination and Toxicology: Continuation of Residue Reviews*, pp. 85–126, G. W. Ware, D. M. Whitacre, L. A. Albert, P. de Voogt, C. P. Gerba, O. Hutzinger, J. B. Knaak, F. L. Mayer, D. P. Morgan, D. L. Park, R. S. Tjeerdema, R. S. H. Yang, and F. A. Gunther, eds., Springer, New York.

Sanjuan, M. A., C. P. Dillon, S. W. G. Tait, S. Moshiach, F. Dorsey, S. Connell, M. Komatsu, K. Tanaka, J. L. Cleveland, S. Withoff, and D. R. Green. 2007. Toll-like receptor signalling in macrophages links the autophagy pathway to phagocytosis, *Nature*, **450**: 1253.

Sasidharan, A., L. Panchakarla, P. Chandran, D. Menon, S. Nair, C. Rao, and M. Koyakutty. 2011. Differential nano-bio interactions and toxicity effects of pristine versus functionalized graphene, *Nanoscale*, **3(6)**: 2461–2464.

Sasidharan, A., L. S. Panchakarla, A. R. Sadanandan, A. Ashokan, P. Chandran, C. M. Girish, D. Menon, S. V. Nair, C. Rao, and M. Koyakutty. 2012. Hemocompatibility and macrophage response of pristine and functionalized graphene, *Small*, **8(8)**: 1251–1263.

Sawangphruk, M., P. Srimuk, P. Chiochan, T. Sangsri, and P. Siwayaprahm. 2012. Synthesis and antifungal activity of reduced graphene oxide nanosheets, *Carbon*, **50(14)**: 5156–5161.

Schinwald, A., F. A. Murphy, A. Jones, W. MacNee, and K. Donaldson. 2012. Graphene-based nanoplatelets: A new risk to the respiratory system as a consequence of their unusual aerodynamic properties, *ACS Nano*, **6(1)**: 736–746.

Schinwald, A., F. Murphy, A. Askounis, V. Koutsos, K. Sefiane, K. Donaldson, and C. J. Campbell. 2014. Minimal oxidation and inflammogenicity of pristine graphene with residence in the lung, *Nanotoxicology*, **8(8)**: 824–832.

Schmid, O. and T. Stoeger. 2016. Surface area is the biologically most effective dose metric for acute nanoparticle toxicity in the lung, *Journal of Aerosol Science*, **99**: 133–143.

Seabra, A. B., A. J. Paula, R. de Lima, O. L. Alves, and N. Duran. 2014. Nanotoxicity of graphene and graphene oxide, *Chemical Research in Toxicology*, **27(2)**: 159–168.

Seo, J.-W. T., A. A. Green, A. L. Antaris, and M. C. Hersam. 2011. High-concentration aqueous dispersions of graphene using nonionic, biocompatible block copolymers, *The Journal of Physical Chemistry Letters*, **2(9)**: 1004–1008.

Shao, Y., J. Wang, H. Wu, J. Liu, I. A. Aksay, and Y. Lin. 2010. Graphene based electrochemical sensors and biosensors: A review, *Electroanalysis: An International Journal Devoted to Fundamental and Practical Aspects of Electroanalysis*, **22(10)**: 1027–1036.

Shi, X., H. Chang, S. Chen, C. Lai, A. Khademhosseini, and H. Wu. 2012. Regulating cellular behavior on few-layer reduced graphene oxide films with well-controlled reduction states, *Advanced Functional Materials*, **22(4)**: 751–759.

Shin, S. R., Y.-C. Li, H. L. Jang, P. Khoshakhlagh, M. Akbari, A. Nasajpour, Y. S. Zhang, A. Tamayol, and A. Khademhosseini. 2016. Graphene-based materials for tissue engineering, *Advanced Drug Delivery Reviews*, **105**: 255–274.

Singh, S. K., M. K. Singh, M. K. Nayak, S. Kumari, S. Shrivastava, J. J. A. Grácio, and D. Dash. 2011. Thrombus inducing property of atomically thin graphene oxide sheets, *ACS Nano*, **5(6)**: 4987–4996.

Singh, S. K., M. K. Singh, P. P. Kulkarni, V. K. Sonkar, J. J. Grácio, and D. Dash. 2012. Amine-modified graphene: Thrombo-protective safer alternative to graphene oxide for biomedical applications, *ACS Nano*, **6(3)**: 2731–2740.

Souza, J. P., J. F. Baretta, F. Santos, I. M. M. Paino, and V. Zucolotto. 2017. Toxicological effects of graphene oxide on adult zebrafish (Danio rerio), *Aquatic Toxicology*, **186**: 11–18.

Stadler, J., T. Schmid, and R. Zenobi. 2011. Nanoscale chemical imaging of single-layer graphene, *Acs Nano*, **5(10)**: 8442–8448.

Stone, V., H. Johnston, and R. P. F. Schins, 2009, Development of in vitro systems for nanotoxicology: Methodological considerations, *Critical Reviews in Toxicology*, **39(7)**: 613–626.

Su, W.-C., B. K. Ku, P. Kulkarni, and Y. S. Cheng. 2016. Deposition of graphene nanomaterial aerosols in human upper airways, *Journal of Occupational and Environmental Hygiene*, **13(1)**: 48–59.

Suárez-Iglesias, O., S. Collado, P. Oulego, and M. Díaz. 2017. Graphene-family nanomaterials in wastewater treatment plants, *Chemical Engineering Journal*, **313**: 121–135.

Sun, T. Y., N. A. Bornhöft, K. Hungerbühler, and B. Nowack. 2016. Dynamic probabilistic modeling of environmental emissions of engineered nanomaterials, *Environmental Science & Technology*, **50(9)**: 4701–4711.

Sun, X., Z. Liu, K. Welsher, J. T. Robinson, A. Goodwin, S. Zaric, and H. Dai. 2008. Nano-graphene oxide for cellular imaging and drug delivery, *Nano Research*, **1(3)**: 203–212.

Syama, S. and P. V. Mohanan. 2016. Safety and biocompatibility of graphene: A new generation nanomaterial for biomedical application, *International Journal of Biological Macromolecules*, **86**: 546–555.

Sydlik, S. A., S. Jhunjhunwala, M. J. Webber, D. G. Anderson, and R. Langer. 2015. In vivo compatibility of graphene oxide with differing oxidation states, *ACS Nano*, **9(4)**: 3866–3874.

Tang, Y., J. Tian, S. Li, C. Xue, Z. Xue, D. Yin, and S. Yu. 2015. Combined effects of graphene oxide and Cd on the photosynthetic capacity and survival of Microcystis aeruginosa, *Science of the Total Environment*, **532**: 154–161.

Tu, Y., M. Lv, P. Xiu, T. Huynh, M. Zhang, M. Castelli, Z. Liu, Q. Huang, C. Fan, and H. Fang. 2013. Destructive extraction of phospholipids from escherichia coli membranes by graphene nanosheets, *Nature Nanotechnology*, **8(8)**: 594–601.

Vallabani, N., S. Mittal, R. K. Shukla, A. K. Pandey, S. R. Dhakate, R. Pasricha, and A. Dhawan. 2011. Toxicity of graphene in normal human lung cells (BEAS-2B), *Journal of Biomedical Nanotechnology*, **7(1)**: 106–107.

van der Zande, M., R. J. Vandebriel, M. J. Groot, E. Kramer, Z. E. Herrera Rivera, K. Rasmussen, J. S. Ossenkoppele et al. 2014. Sub-chronic toxicity study in rats orally exposed to nanostructured silica, *Particle and Fibre Toxicology*, **11(1)**: 8.

Vivekchand, S., C. S. Rout, K. Subrahmanyam, A. Govindaraj, and C. Rao. 2008. Graphene-based electrochemical supercapacitors, *Journal of Chemical Sciences*, **120(1)**: 9–13.

Waiwijit, U., W. Kandhavivorn, B. Oonkhanond, T. Lomas, D. Phokaratkul, A. Wisitsoraat, and A. Tuantranont. 2014. Cytotoxicity assessment of MDA-MB-231 breast cancer cells on screen-printed graphene-carbon paste substrate, *Colloids and Surfaces B: Biointerfaces*, **113**: 190–197.

Wan, B., Z.-X. Wang, Q.-Y. Lv, P.-X. Dong, L.-X. Zhao, Y. Yang, and L.-H. Guo. 2013. Single-walled carbon nanotubes and graphene oxides induce autophagosome accumulation and lysosome impairment in primarily cultured murine peritoneal macrophages, *Toxicology Letters*, **221(2)**: 118–127.

Wang, A., K. Pu, B. Dong, Y. Liu, L. Zhang, Z. Zhang, W. Duan, and Y. Zhu. 2013a. Role of surface charge and oxidative stress in cytotoxicity and genotoxicity of graphene oxide towards human lung fibroblast cells, *Journal of Applied Toxicology*, **33(10)**: 1156–1164.

Wang, D., L. Zhu, J.-F. Chen, and L. Dai. 2015a. Can graphene quantum dots cause DNA damage in cells? *Nanoscale*, **7(21)**: 9894–9901.

Wang, Q., S. Zhao, Y. Zhao, Q. Rui, and D. Wang. 2014. Toxicity and translocation of graphene oxide in Arabidopsis plants under stress conditions, *RSC Advances*, **4(105)**: 60891–60901.

Wang, S., P. K. Ang, Z. Wang, A. L. L. Tang, J. T. Thong, and K. P. Loh. 2009. High mobility, printable, and solution-processed graphene electronics, *Nano Letters*, **10(1)**: 92–98.

Wang, Y., H. Wang, D. Liu, S. Song, X. Wang, and H. Zhang. 2013b. Graphene oxide covalently grafted upconversion nanoparticles for combined NIR mediated imaging and photothermal/photodynamic cancer therapy, *Biomaterials*, **34(31)**: 7715–7724.

Wang, Y., Z. Li, D. Hu, C.-T. Lin, J. Li, and Y. Lin. 2010. Aptamer/graphene oxide nanocomplex for in situ molecular probing in living cells, *Journal of the American Chemical Society*, **132(27)**: 9274–9276.

Wang, Z. G., R. Zhou, D. Jiang, J. E. Song, Q. Xu, J. Si, Y. P. Chen, X. Zhou, L. Gan, J. Z. Li, H. Zhang, and B. Liu. 2015b. Toxicity of graphene quantum dots in zebrafish embryo, *Biomedical and Environmental Sciences*, **28(5)**: 341–351.

Wang, Z., Y. Gao, S. Wang, H. Fang, D. Xu, and F. Zhang. 2016. Impacts of low-molecular-weight organic acids on aquatic behavior of graphene nanoplatelets and their induced algal toxicity and antioxidant capacity, *Environmental Science and Pollution Research*, **23(11)**: 10938–10945.

Warheit, D. B., R. Boatman, and S. C. Brown. 2015. Developmental toxicity studies with 6 forms of titanium dioxide test materials (3 pigment-different grade & 3 nanoscale) demonstrate an absence of effects in orally-exposed rats, *Regulatory Toxicology and Pharmacology*, **73(3)**: 887–896.

Wen, K.-P., Y.-C. Chen, C.-H. Chuang, H.-Y. Chang, C.-Y. Lee, and N.-H. Tai. 2015. Accumulation and toxicity of intravenously-injected functionalized graphene oxide in mice, *Journal of Applied Toxicology*, **35(10)**: 1211–1218.

Westervelt, R. 2008. Graphene nanoelectronics, *Science*, **320(5874)**: 324–325.

Wick, P., A. E. Louw-Gaume, M. Kucki, H. F. Krug, K. Kostarelos, B. Fadeel, K. A. Dawson et al. 2014. Classification framework for graphene-based materials, *Angewandte Chemie International Edition*, **53(30)**: 7714–7718.

Wojtoniszak, M., X. Chen, R. J. Kalenczuk, A. Wajda, J. Łapczuk, M. Kurzewski, M. Drozdzik, P. K. Chu, and E. Borowiak-Palen. 2012. Synthesis, dispersion, and cytocompatibility of graphene oxide and reduced graphene oxide, *Colloids and Surfaces B: Biointerfaces*, **89**: 79–85.

Wu, M., R. Kempaiah, P.-J. J. Huang, V. Maheshwari, and J. Liu. 2011. Adsorption and desorption of DNA on graphene oxide studied by fluorescently labeled oligonucleotides, *Langmuir*, **27(6)**: 2731–2738.

Wu, Q., L. Yin, X. Li, M. Tang, T. Zhang, and D. Wang. 2013. Contributions of altered permeability of intestinal barrier and defecation behavior to toxicity formation from graphene oxide in nematode caenorhabditis elegans, *Nanoscale*, **5(20)**: 9934–9943.

Wu, S.-Y., S. S. A. An, and J. Hulme. 2015. Current applications of graphene oxide in nanomedicine, *International Journal of Nanomedicine*, **10(Spec Iss)**: 9.

Xia, T., N. Li, and A. E. Nel. 2009. Potential health impact of nanoparticles, *Annual Review of Public Health*, **30**: 137–150.

Xie, J., Z. Ming, H. Li, H. Yang, B. Yu, R. Wu, X. Liu, Y. Bai, and S.-T. Yang. 2016. Toxicity of graphene oxide to white rot fungus phanerochaete chrysosporium, *Chemosphere*, **151**: 324–331.

Xu, M., J. Zhu, F. Wang, Y. Xiong, Y. Wu, Q. Wang, J. Weng, Z. Zhang, W. Chen, and S. Liu. 2016. Improved in vitro and in vivo biocompatibility of graphene oxide through surface modification: Poly(acrylic acid)-functionalization is superior to PEGylation, *ACS Nano*, **10(3)**: 3267–3281.

Yan, L., Y. B. Zheng, F. Zhao, S. Li, X. Gao, B. Xu, P. S. Weiss, and Y. Zhao. 2012. Chemistry and physics of a single atomic layer: Strategies and challenges for functionalization of graphene and graphene-based materials, *Chemical Society Reviews*, **41(1)**: 97–114.

Yang, K., H. Gong, X. Shi, J. Wan, Y. Zhang, and Z. Liu. 2013. *In vivo* biodistribution and toxicology of functionalized nano-graphene oxide in mice after oral and intraperitoneal administration, *Biomaterials*, **34(11)**: 2787–2795.

Yang, K., J. Wan, S. Zhang, Y. Zhang, S.-T. Lee, and Z. Liu. 2010a. *In vivo* pharmacokinetics, long-term biodistribution, and toxicology of PEGylated graphene in mice, *ACS Nano*, **5(1)**: 516–522.

Yang, K., J. Wan, S. Zhang, Y. Zhang, S.-T. Lee, and Z. Liu. 2011a. *In vivo* pharmacokinetics, long-term biodistribution, and toxicology of PEGylated graphene in mice, *ACS Nano*, **5(1)**: 516–522.

Yang, K., S. Zhang, G. Zhang, X. Sun, S.-T. Lee, and Z. Liu. 2010b. Graphene in mice: Ultrahigh in vivo tumor uptake and efficient photothermal therapy, *Nano Letters*, **10(9)**: 3318–3323.

Yang, X., J. Zhu, L. Qiu, and D. Li. 2011d. Bioinspired effective prevention of restacking in multilayered graphene films: Towards the next generation of high-performance supercapacitors, *Advanced Materials*, **23(25)**: 2833–2838.

Yang, X., L. Qiu, C. Cheng, Y. Wu, Z. F. Ma, and D. Li. 2011b. Ordered gelation of chemically converted graphene for next-generation electroconductive hydrogel films, *Angewandte Chemie*, **123(32)**: 7463–7466.

Yang, X., Y. Wang, X. Huang, Y. Ma, Y. Huang, R. Yang, H. Duan, and Y. Chen. 2011c. Multi-functionalized graphene oxide based anticancer drug-carrier with dual-targeting function and pH-sensitivity, *Journal of Materials Chemistry*, **21(10)**: 3448–3454.

Yang, Y., Z. Yu, T. Nosaka, K. Doudrick, K. Hristovski, P. Herckes, and P. Westerhoff. 2015. Interaction of carbonaceous nanomaterials with wastewater biomass, *Frontiers of Environmental Science & Engineering*, **9(5)**: 823–831.

Yin, P. T., S. Shah, M. Chhowalla, and K.-B. Lee. 2015. Design, synthesis, and characterization of graphene-nanoparticle hybrid materials for bioapplications, *Chemical Reviews*, **115(7)**: 2483–2531.

Yin, Z., J. Zhu, Q. He, X. Cao, C. Tan, H. Chen, Q. Yan, and H. Zhang. 2014. Graphene-based materials for solar cell applications, *Advanced Energy Materials*, **4(1)**: 1300574.

Yuan, J., H. Gao, and C. B. Ching. 2011. Comparative protein profile of human hepatoma HepG2 cells treated with graphene and single-walled carbon nanotubes: An iTRAQ-coupled 2D LC–MS/MS proteome analysis, *Toxicology Letters*, **207(3)**: 213–221.

Yue, H., W. Wei, Z. Yue, B. Wang, N. Luo, Y. Gao, D. Ma, G. Ma, and Z. Su. 2012. The role of the lateral dimension of graphene oxide in the regulation of cellular responses, *Biomaterials*, **33(16)**: 4013–4021.

Yue, Z.-G., W. Wei, P.-P. Lv, H. Yue, L.-Y. Wang, Z.-G. Su, and G.-H. Ma. 2011. Surface charge affects cellular uptake and intracellular trafficking of chitosan-based nanoparticles, *Biomacromolecules*, **12(7)**: 2440–2446.

Zanni, E., G. De Bellis, M. P. Bracciale, A. Broggi, M. L. Santarelli, M. S. Sarto, C. Palleschi, and D. Uccelletti. 2012. Graphite nanoplatelets and caenorhabditis elegans: Insights from an in vivo model, *Nano Letters*, **12(6)**: 2740–2744.

Zhang, P., H. Selck, S. R. Tangaa, C. Pang, and B. Zhao. 2017a. Bioaccumulation and effects of sediment-associated gold- and graphene oxide nanoparticles on Tubifex tubifex, *Journal of Environmental Sciences*, **51**: 138–145.

Zhang, S., K. Yang, L. Feng, and Z. Liu. 2011a. *In vitro* and *in vivo* behaviors of dextran functionalized graphene, *Carbon*, **49(12)**: 4040–4049.

Zhang, W., C. Wang, Z. Li, Z. Lu, Y. Li, J. J. Yin, Y. T. Zhou, X. Gao, Y. Fang, and G. Nie. 2012. Unraveling stress-induced toxicity properties of graphene oxide and the underlying mechanism, *Advanced Materials*, **24(39)**: 5391–5397.

Zhang, X., J. Yin, C. Peng, W. Hu, Z. Zhu, W. Li, C. Fan, and Q. Huang. 2011b. Distribution and biocompatibility studies of graphene oxide in mice after intravenous administration, *Carbon*, **49(3)**: 986–995.

Zhang, X., Q. Zhou, W. Zou, and X. Hu. 2017b. Molecular mechanisms of developmental toxicity induced by graphene oxide at predicted environmental concentrations, *Environmental Science & Technology*, **51(14)**: 7861–7871.

Zhang, Y., S. F. Ali, E. Dervishi, Y. Xu, Z. Li, D. Casciano, and A. S. Biris. 2010. Cytotoxicity effects of graphene and single-wall carbon nanotubes in neural phaeochromocytoma-derived PC12 Cells, *ACS Nano*, **4(6)**: 3181–3186.

Zhao, S., Q. Wang, Y. Zhao, Q. Rui, and D. Wang. 2015. Toxicity and translocation of graphene oxide in Arabidopsis thaliana, *Environmental Toxicology and Pharmacology*, **39(1)**: 145–156.

Zhao, X. 2011. Self-assembly of DNA segments on graphene and carbon nanotube arrays in aqueous solution: A molecular simulation study, *The Journal of Physical Chemistry C*, **115(14)**: 6181–6189.

Zhou, H., K. Zhao, W. Li, N. Yang, Y. Liu, C. Chen, and T. Wei. 2012a. The interactions between pristine graphene and macrophages and the production of cytokines/chemokines via TLR- and NF-κB-related signaling pathways, *Biomaterials*, **33(29)**: 6933–6942.

Zhou, X., F. Laroche, G. E. M. Lamers, V. Torraca, P. Voskamp, T. Lu, F. Chu, H. P. Spaink, J. P. Abrahams, and Z. Liu. 2012b. Ultra-small graphene oxide functionalized with polyethylenimine (PEI) for very efficient gene delivery in cell and zebrafish embryos, *Nano Research*, **5(10)**: 703–709.

Zhou, Z., J. Son, B. Harper, Z. Zhou, and S. Harper. 2015. Influence of surface chemical properties on the toxicity of engineered zinc oxide nanoparticles to embryonic zebrafish, *Beilstein Journal of Nanotechnology*, **6**: 1568–1579.

Zhu, M., G. Nie, H. Meng, T. Xia, A. Nel, and Y. Zhao. 2013. Physicochemical properties determine nanomaterial cellular uptake, transport, and fate, *Accounts of Chemical Research*, **46(3)**: 622–631.

Zhu, Y., S. Murali, M. D. Stoller, K. J. Ganesh, W. Cai, P. J. Ferreira, A. Pirkle, R. M. Wallace, K. A. Cychosz, M. Thommes, D. Su, E. A. Stach, and R. S. Ruoff. 2011. Carbon-based supercapacitors produced by activation of graphene, *Science*, **332(6037)**: 1537.

Zuo, G., X. Zhou, Q. Huang, H. Fang, and R. Zhou. 2011. Adsorption of villin headpiece onto graphene, carbon nanotube, and C60: Effect of contacting surface curvatures on binding affinity, *The Journal of Physical Chemistry C*, **115(47)**: 23323–23328.

4

Toxicology Studies of Semiconductor Nanomaterials: Environmental Applications

T. P. Nisha,
Meera Sathyan,
M. K. Kavitha,
and Honey John

4.1 Introduction

The remarkable increase in the development of nanomaterials has urged significant growth in nanotechnology and led to the significant interest in the environmental applications of nanomaterials and nanotechnology. In particular, revolutionary changes in water treatment including removal of organic pollutants, disinfection of water, water purification, etc. have been reported (Cho et al., 2005, Shannon et al., 2008, Wei et al., 1994). Similarly, the development of self-cleaning and antireflective coatings in vehicle windscreens, goggle lens, windows in armoured cars, solar panels, green-house enclosures, chemical and biological protective face masks, etc. also accelerates the environmental importance of nanomaterials. Among various kinds of nanomaterials, semiconductor nanomaterials occupy the highest position with respect to environmental applications.

Semiconductor nanomaterials possess excellent properties such as photocatalytic efficiency, adsorbing capacity, antimicrobial activities, and antibacterial activities (Gazit, 2007). Nanosemiconductor oxides are considered to be less toxic compared to those of chemical disinfectants and hence researchers are more recently focusing nanosemiconductor research on the development of materials with water purification capability. The semiconducting oxides can absorb UV light and excite electron-hole pairs. These photogenerated electrons react with molecular oxygen producing super oxide radical anions and the holes react with water leading to the generation of hydroxyl radicals. These two highly reactive radicals will decompose organic impurities leading to the enhancement of photocatalysis (Fujishima et al., 2000, Nishimoto and Bhushan, 2013). Among various photocatalysts, nano TiO_2 is the most studied nanosemiconductor oxide photocatalyst. Photocatalytic properties are utilized for the removal of pollutants from water and air. Nano TiO_2 is a good material for disinfection of bacteria and has been shown to kill both Gram-positive and Gram-negative bacteria, Hepatitis B virus, and Herpes simplex virus (Hajkova et al., 2007, Watts et al., 1995, Zan et al., 2007). Similarly, nano ZnO is also widely used in sunscreen lotions, coatings, paints, and antibacterial lotions (Franklin et al., 2007). The similarity between these two semiconductor oxides is their photocatalytic activity under UV light irradiation. The photocatalytic properties of semiconductors can be improved by creating hybrids with carbon nanotubes (CNTs) or graphene. This will form positive and negative charges on the interfaces of CNT/graphene and semiconductor oxides resulting in an area of depletion layer, which maintains charge separation in the hybrid materials and thereby increases the recombination time. The increase in recombination time enhances the photocatalytic efficiency of such hybrids (Fan et al., 2012, Jitianu et al., 2004, Ng et al., 2010, Tian et al., 2012, Woan et al., 2009).

The increased growth of nanomaterials and their applications in commercial products prompt researchers to investigate the effect of nanomaterials and their products on human health. The human body is in constant contact with the environment and the human skin acts like a barrier to foreign materials entering into the body. But due to the smaller size of nanomaterials with respect to the pores of skin, nanomaterials can enter into the body and easily translocate from these access portals into the circulatory and lymphatic systems; ultimately, the particles can enter into all body tissues and organs. This will cause adverse health effects. Even though studies on nanomaterial toxicology and the environmental impact of nanotechnology is very important, to date, only a very few investigations have been reported in the area of toxicology and the direct and indirect environmental impact of nanomaterials (Kraft et al., 2003, Warheit et al., 1995, Yang, 2002).

This chapter gives a thorough review on the environmental applications of nanosemiconductor materials with special reference to photocatalytic applications such as water purification, air purification, and green energy production. The chapter also gives a brief account on the biocompatibility of nanomaterials, biomedical applications, and the impact of nanomaterials on human health.

4.2 Photocatalytic Applications of Semiconductor Nanomaterials

In this era of rapid urbanization and industrialization, where environmental issues are a major concern, photocatalytic nanomaterials as 'green' mediators have sparked great interest in effective energy conversion and environmental remediation. Most semiconductor nanomaterials are sustainable, highly stable, environmentally benign, and easily reproducible. They are capable of utilizing the bountiful solar energy as a light source to undergo the photocatalytic mechanisms. All contemporary research works on the development and application of photocatalysts to tackle environmental issues began with the work of Fujishima and Honda (1972), when they discovered the decomposition of H_2O directly to H_2 and O_2 by visible light irradiation of a TiO_2 anode connected with Pt black cathode.

Photocatalysis begins when light is absorbed and deployed by the photocatalyst to excite its valence band (VB) electrons (e^-) to conduction band (CB), thereby leaving holes (h^+) in the VB. These photogenerated charge carriers then migrate to the surface of the catalyst and undergo redox reactions with either water or atmospheric oxygen. This is followed by the generation of various reactive oxygen species

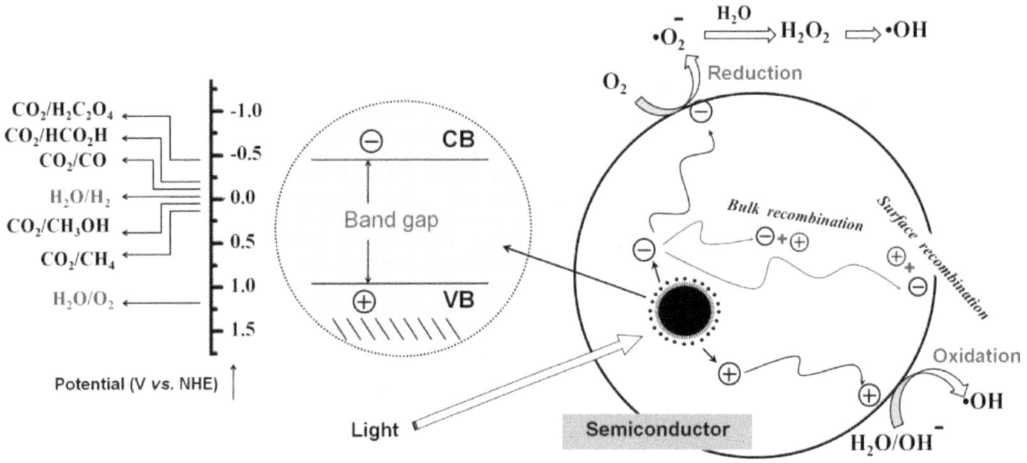

FIGURE 4.1 Various reactions occurring during the photoexcitation of a semiconductor, and the relevant redox potentials for H_2O splitting and CO_2 reduction at neutral pH.

(ROS) like hydroxyl (•OH) and superoxide ($O_2^{\bullet -}$) radicals, H_2O_2, and •OOH on the semiconductor surface, which forms the active species in photocatalytic mechanisms (Figure 4.1). The photogenerated holes and •OH radicals are widely considered as the oxidants for directly degrading the organic pollutants and microorganisms adsorbed to the photocatalyst. However, photogenerated holes and electrons can undergo recombination either in the bulk or at the surface. Thus, a semiconductor with excellent photocatalytic quantum efficiency should have competent spatial separation of photogenerated charge carriers and their transport, efficient light harvesting, proficient charge utilization, and good surface adsorption capacity.

Although a wide range of semiconductors have been developed, including d^0 metal oxides (Ti^{4+}, Zr^{4+}, Nb^{5+}, Mo^{6+}, W^{6+}), d^{10} metal oxides (Ga^{3+}, In^{3+}, Sn^{4+}, Ge^{4+}, Sb^{5+}), and f^0 metal oxides (Ce^{4+}), along with a small group of non-oxide photocatalysts such as ZnS, CdS, GaN, and InP, their efficiency is limited in the UV region. Matching the bandgap of the semiconductor and the photocatalytic mechanisms to the solar spectrum is always challenging. In the past few decades, several strategies have been employed and significant achievements have been made for the improved exploitation of solar energy, such as ion doping, surface plasmon enhancement, and dye sensitization. Additionally, some novel semiconductor photocatalysts have also been reported to be visible light responsive, such as $BiVO_4$, Cu_2O, $CuFeO_4$, graphitic-C_3N_4, and ZnTe. Size-dependent quantum confinement effect is significant for the semiconductors to exhibit high photoactivity since it directly impacts the specific reactive surface area of the catalyst. Besides, an energy shift in the LUMO of the semiconductor by an upward shift of CB to more negative potential (*vs.* NHE) with decreasing particle size reflects in the visible light response of the photocatalyst. Photocatalysts of varied morphology *viz.* nanowires, nanotubes, nanobelts, nanorods, and nanoscrolls have been developed for the efficient charge separation and transportation (An et al., 2012, Pan et al., 2012, Varghese et al., 2009, Wu et al., 2010). Significant crystallinity with crystal defects is always preferable in that it minimizes the interface recombination. Recent years have seen the upconversion of semiconductors with luminescent agents consisting of transition metals, having the capacity to absorb two or more incident low-energy photons to a single high-energy photon (Hou et al., 2012, Yin et al., 2011). Moreover, tailoring the exposed percentage of reactive crystal facets, hybridizing with graphitic monolayers and quantum dots (QDs), together with construction of homojunctions based on crystalline phase or facets and semiconductor heterojunctions (Marschall, 2014) also favoured visible-light excitation, reduced recombination, and effectual charge transport for intensified photocatalytic outcomes. Semiconductors can thus be easily tuned and applied for various environmental applications.

4.2.1 H_2 Production

Energy conversion using clean and renewable sources of energy provides viable solutions to reduce pollution and the present dependency on depleting fossil sources. Hydrogen evolution reaction (HER) by photocatalytic water splitting is considered to be an eco-friendly technology for clean hydrogen fuel production and is a path to future energy supply. H_2 production through water splitting by a semiconductor photocatalyst consists of two half reactions: the reduction of proton and the highly demanding four-electron oxidation of water. However, the organic substrate which is adsorbed onto the catalyst surface decomposes by oxidation and at the same time itself acts as a proton source. Therefore, the photoreforming process is the proton reduction coupled with the oxidative decomposition of the organic substrate (Christoforidis and Fornasiero, 2017). These reactions can be summarized as:

$$4h^+ + H_2O \rightarrow O_2 + 4H^+ \quad \text{Oxidative half reaction} \tag{4.1}$$

$$2e^- + 2H^+ \rightarrow H_2 \quad \text{Reductive half reaction} \tag{4.2}$$

$$C_xH_yO_z + (2x-z)H_2O \rightarrow (2x-z+y/2)H_2 + xCO_2 \quad \text{Photoreforming} \tag{4.3}$$

An essential requirement to facilitate these reactions is that the oxidation and reduction potentials of water should lie within the bandgap of the semiconductor photocatalyst. VB should be more positive than the oxidation potential of O_2/H_2O (1.23 V *vs.* NHE), whereas CB should be more negative than the reduction potential of H^+/H_2 (0 V *vs.* NHE) (Hisatomi et al., 2014). This forms the principle behind single-step water splitting using a single photocatalyst (Figure 4.1). Alternatively, HER can be conducted in two-step photocatalytic systems where two semiconductors with reverse redox shuttles are connected in a series. By a 'Z-scheme', water is reduced to H_2 and oxidation of the reduced mediator occurs at one photocatalyst, and simultaneously water is oxidized to O_2 and the oxidized mediator is reduced at the other photocatalyst (Sayama et al., 2002, Hisatomi et al., 2014). H_2 production is also be done by photoelectrochemical (PEC) water splitting when the semiconductor electrode is immersed into suitable electrolytes, so that electron transfer occurs between them by equilibration of Fermi level of the semiconductor with the redox potential of the electrolyte.

Over the past decades, many semiconductors have been designed and developed for H_2 production and this is still an attractive research field. Studied semiconductors include TiO_2, CdS, CdSe, Cu_2O, $ZnIn_2S_4$, $CuInS_2$, $SrTiO_3$, g-C_3N_4, and ZnO:GaN (Adeli and Taghipour, 2016, Han et al., 2017, Ye et al., 2014, Yuan et al., 2016, Yuan et al., 2018). Co-catalysts Pt, Pd, and Au (Altomare et al., 2018, Cao et al., 2018, Li et al., 2016) are used to modify semiconductors to efficiently improve visible light response and enhance the efficiency of HER. However, the practical implementation of noble-metal-containing photocatalytic systems is limited due to their cost of production. Therefore, significant interest is gaining in using metal-free semiconductors such as *g*-C_3N_4 (Shen et al., 2018, Vu et al., 2018, Xia et al., 2018).

Other potential candidates for H_2 production include materials like chalcogenides, oxysulphides, and oxynitrides, as their VB is shifted to more negative redox potentials (Hisatomi et al., 2014). However, suppressing the backward reaction is equally important to matching the bandgap of the semiconductor for water splitting. Facilitation charge separation and reaction kinetics by surface modifications, as well as constructing redox-free Z-scheme photocatalytic systems, are plausible solutions. As the target for solar-to-hydrogen conversion is set at 10% (Maeda and Domen, 2010), the photocatalyst should be active at least up to 600 nm light source with a fair quantum efficiency. The future semiconductor materials are expected to split water under red and near-infrared rays. Transportation, purification, storage, and utilizations of solar-hydrogen are the challenges met by today's researchers and still this research field has a long way to travel (Figure 4.2).

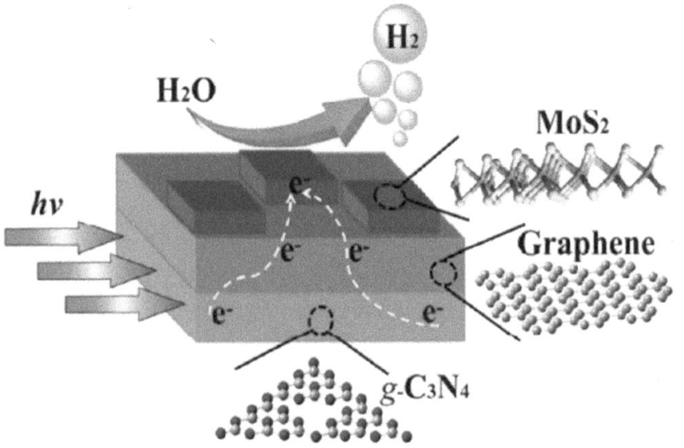

FIGURE 4.2 A ternary g-C_3N_4/graphene/MoS_2 two-dimensional nanojunction photocatalyst with 0.5% graphene and 1.2% MoS_2 achieved a high H_2 evolution rate of 317 μmol h^{-1} g^{-1}. The apparent quantum yield at 420 nm reaches 3.4% and was higher than the optimized Pt-loaded g-C_3N_4 photocatalyst. (Reprinted with permission from Yuan, Y. J. et al., *ACS Appl. Energy Mater.*, 1, 1400–1407, 2018. Copyright 2018 from American Chemical Society.)

4.2.2 Adsorption and Photoreduction of CO_2

Increased consumption of fossil fuels accompanied by anthropogenic CO_2 elimination leads to ever-increasing global warming and adverse environmental changes. An effective and promising approach to address these crises is the adsorption and solar-driven reduction of CO_2 by a photocatalyst into desirable products like methane and methanol. This strategy not only reduces the excessive emissions of greenhouse gases but at the same time replaces the traditional fossil feedstock. Only those photogenerated electrons with a more negative reduction potential than that of the expected chemical reaction alone will lose electrons and reduce CO_2 to hydrocarbons (Figure 4.1) (Lia et al., 2014). Kinetically, the generation of CH_4 and CH_3OH by multiple proton-coupled electron transfer is difficult when compared with the 2 e^- oxidation of water to H_2. Hence, this artificial photosynthesis is in acute competition with HER in the presence of water, which in turn reduces the overall efficiency of CO_2 conversion. However, various methods have been adopted to achieve effective CO_2 adsorption and its conversion to fuels and other chemicals like HCO_2H, CO, C_2H_5OH, and $H_2C_2O_4$. Much attention is deserved by the surface reactions for the CO_2 adsorption, besides the other requirements including light harvesting, charge generation, and transportation with reduced recombination. To suppress the competitive HER with increased selectivity towards carbonaceous products, an important role is played by the CO_2 adsorption and activation at the surface of the catalyst. Developing surface defects by introducing oxygen and sulphur vacancies, creating surface basic sites, and the use of noble-metal cocatalysts are the common strategies to enhance the adsorption and activation of CO_2 and thus decrease the barrier for the reduction reactions (Chang et al., 2016). However, another factor to be considered is the selectivity of products from the photoreduction of CO_2, i.e., to produce valuable hydrocarbons together with reduced cost of product separation. Additionally, the pH of the electrolyte has a significant role as it determines the solubility of CO_2 and its relative concentration with CO_3^{2-} and HCO^{3-}, thereby leading to different adsorption modes (Chang et al., 2016). Much research has been focused on the CO_2 photoreduction using TiO_2 owing to its suitable reduction potentials. TiO_2 can be used for the selective production of CH_4 by the modification of Ti^{3+} impurity level (oxygen vacancies) by fluorination. When a reduced TiO_{2-x} is fluorinated, the substitutional F replaces the exciting O vacancies and the reduction potential of the photogenerated electrons is enhanced (Xing et al., 2018). This in turn increases the selective CH_4 production yield under solar light illumination with a superior selectivity for CH_4 production.

4.2.3 Water Remediation

The advanced oxidation technology of the semiconductors is being widely employed to remove persistent organic compounds and microorganisms from water. Organic dyes as effluents from the textile and printing industries that are often difficult to be biodegraded or oxidized by chemicals can be easily adsorbed and degraded with photocatalysts. Among various hybrid materials for water treatment, semiconductors hybridized with bismuth oxyhalides (BiOX, X = Cl, Br, or I) due to their better photocatalytic activity and facile band gap tuning (Choi et al., 2016) have gained attention in recent years. The combination of TiO_2 with carbonaceous materials (like CNTs, graphene monolayers, and activated carbon) has recently been investigated for its significant environmentally friendly decomposition of pharmaceutical and personal care products (PPCPs) from water (Awfa et al., 2018), which are considered as emerging pollutants capable of causing ecological damage. Meanwhile, phenolic compounds causing diseases like cancer have been effectively removed from the water medium using a novel Ag-doped ZnO/MWCNT nanocomposite (Hosseini et al., 2018). Also, the carcinogenic and mutagenic inorganic heavy metal pollutants like hexavalent chromium (Cr [VI]) and trivalent and pentavalent arsenic (Ar [III], Ar [V]) have recently been successfully reduced with SnS_2/SnO_2 on carbon cloth (Zhang et al., 2018) (Figure 4.3).

Before commercialization of any photocatalyst for water treatment, the means to prevent secondary pollutions caused by the incomplete removal of photocatalyst from the water system should be greatly considered. In this aspect, magnetic core-shell Fe_3O_{4-}-TiO_2 nanoparticles can be completely recovered from a water system using a magnet and undergo excellent degradation of organic pollutants as high as 96% at pH 3, 1g/L photocatalyst dose, 30°C, and 90 min reaction time (Salamat et al., 2017). Additionally,

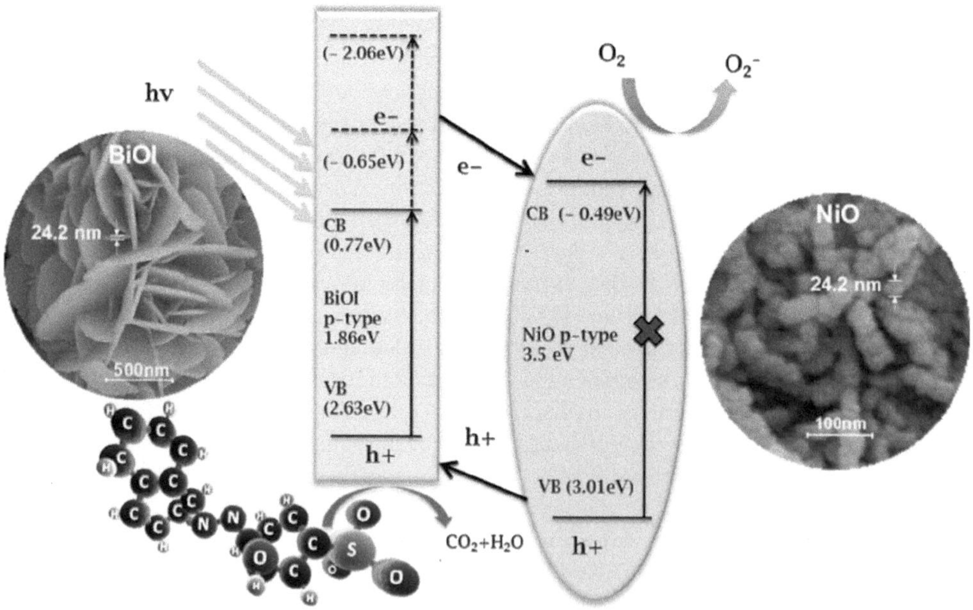

FIGURE 4.3 Reaction mechanism for the removal of acid orange from water over a novel flower-like p-p junction BiOI(x)-NiO(100−x) nanophotocatalyst with staggered band gap alignment. When exposed to visible light, only BiOI gets excited (due to smaller band gap of 1.86 eV) with formation of e^- in the CB and h^+ in the VB. As the CB potential of BiOI in the composite is more negative than that of NiO, the photogenerated e^- on the surface will migrate to the CB of NiO. Hence, holes will be in the VB of BiOI with electrons in the CB of NiO, thus alleviating the recombination of e^-–h^+ pairs. (Reprinted from Yosefi, L. and Haghighi, M., *Appl. Catal. B Environ.*, 220, 367–378, 2018. Copyright 2017 permission from Elsevier.)

Fe_3O_4 has the capability of adsorbing heavy metal ions in the water. Likewise, photocatalysts grown on flexible supporters like aerogels (Sacco et al., 2018, Maleki and Hüsing, 2018) and carbon cloth (Shi et al., 2017, Zhang et al., 2018) can address well the loss of the photocatalyst during repeated cycles of reaction as well as the concomitant secondary pollution.

Antibactericidal effects of nanomaterials can be significantly used for the disinfection of drinking water since the work of Matsunaga et al. (1985). The strategy is particularly important in the biomedical field tackling pathogenic microorganisms causing diseases like cholera and diarrhoea, and substitutes the present carcinogenic-byproduct-producing disinfectants like chlorine, chloramines, and ozone. The same criteria are utilized for the non-thermal inactivation of microorganisms in foods (Zhu et al., 2018). Ireland et al. (1993) established that OH radicals formed during UV excitation of a TiO_2 suspension in water are responsible for inactivation of *Escherichia coli* in water. Since then, the microbicidal activity of TiO_2 has been investigated for the killing of several other microorganisms including *Staphylococcus aureus*, *Cryptosporidium parvum*, *Klebsiella pneumoniae*, and *Bacillus subtilis* (Cho et al., 2011). A recent study reported multifunctional TiO_2 nanofibres photosensitized with Ag_2S disinfected both Gram-positive and Gram-negative bacteria under simulated solar light illumination, with an additional behaviour of Cr (VI) photoreduction (Ghafoor et al., 2018).

Semiconductor materials can be additionally used as a surface-enhanced Raman scattering (SERS) substrate, which is a powerful tool that can be used for environmental monitoring, biomedical research, and chemical analysis due to its ultrasensitivity, non-destructing, and fingerprint effect. On a SERS substrate, the Raman signal of an analyte is amplified to 10^8 times or more, and hence can be used for the detection of trace amounts of herbicides like paraquat in water (Halvorson and Vikesland et al., 2010). Originally, the SERS activity was recognized only for noble metal nanoparticles and transition metals, but now some semiconductors also display notable SERS activity, *viz.* NiO, Ag_2O, Cu_2O, ZnO, TiO_2, α-Fe_2O_3, and InAs QDs (Qi et al., 2014). Defect engineering by inducing oxygen vacancies in α-MoO_3 transferred the weak SERS active substrate into a SERS-active substrate with an enhancement factor as high as 1.8×10^7 and a detection limit of 10^{-8} M for Rhodamine 6G, which is highest among the so-far reported semiconductors (Wu et al., 2017).

4.2.4 Air Purification

Photocatalytic oxidation of semiconductors can be employed to destruct a broad range of contaminants like volatile organic compounds (VOCs) and inorganic pollutants into harmless products at low or room temperature without significant energy input. TiO_2 has been extensively studied for the removal of VOCs like toluene, benzene, acetaldehyde, pyridine, formaldehyde, ethanol, and acetone, and inorganic chemicals like NO_x, SO_x, CO, H_2S, ozone, and others. Tailored porous structure, BET surface area, and shape control with exposed high energy facets are the major factors that contribute to the efficiency in air purification. Recently, hierarchical TiO_2 nanofibers decorated with high-energy TiO_2 nanocrystals showed enhanced oxidation of NO and acetone (Duan et al., 2018). The nanofibers were fabricated by electrospinning tetrabutyltitanate solution containing cubic {001} faceted $TiOF_2$ nanocrystals followed by calcination. A highest photooxidation rate of 97.2 $ppmh^{-1}$ is due to a z-scheme charge transfer pathway in the so-formed homojunction. An elevated e^-—h^+ separation efficiency found at the p-n heterojunction $Bi_2O_2CO_3/ZnFe_2O_4$ interface by the influence of an internal electric field is identified as a visible responsive photocatalyst with high NO conversion rate and low NO_2 selectivity (Huang et al., 2018). Lately, macroporous/mesoporous TiO_2 foams prepared by a simple mechanical stirring with the aid of short-chain amphiphilic molecules demonstrated the photoassisted degradation of NO and can be directly used for air purification devices (Jami et al., 2018).

4.2.5 Self-cleaning/Anti-reflective Surfaces

Superhydrophilic self-cleaning surfaces can be fabricated using semiconductor photocatalysts and thus bio-mimicking the variegated forms of nature like lotus leaf, wings of butterfly, and legs of water strider. Photocatalysts like TiO_2 and ZnO have an additional property that they show extreme wettability on photoirradiation. The mechanism of superhydrophilicity begins when photoinduced electrons reduce the metal centres (for example, Ti^{4+} gets reduced to Ti^{3+}) and the photogenerated holes formed in the VB oxidize O^{2-} anions of the semiconductor, causing the oxygen atoms to be ejected from the semiconductor surface and leaving vacancies. The hydroxyl anions formed during the photocatalytic mechanism or the water molecule itself get adsorbed to these oxygen vacancies leading to reconstruction of Ti-OH bonds (Banerjee et al., 2015), thereby increasing the affinity of the surface to water. The photogenerated holes are more vital than electrons for the superhydrophilic behaviour of the surface and hence their diffusion to the surface of the photocatalyst is crucial (Emeline et al., 2013). Concurrently, the photogenerated electrons will decompose the organic matter adsorbed on the photocatalyst surface. Both the photodecomposition and photoinduced superhydrophilicity (where the water sheets over the surface and water contact angle approaches zero) render the surface clean.

Self-cleaning coatings with anti-reflective property are of great importance for optoelectronic devices like solar panels, green houses, and glare-reduced displays in eyeglasses and automobile glasses. The semiconductor photocatalyst usually with a high refractive index is made of composites with those having low refractive index (e.g., SiO_2, MgF_2, Al_2O_3 with n = 1.5, 1.3, and 1.7, respectively). A double layer of Al_2O_3/TiO_2 coated over a Si wafer exhibited self-cleaning anti-reflective property with a 4.74% reflectance in the wavelength region 400–1000 nm, and the fabricated solar cell attained conversion efficiency of ~13.95% (Jung et al., 2018). A hierarchical macro-mesoporous SiO_2 thin film onto which TiO_2 is dip-coated showed a superior broadband anti-reflection with an average reflectance of 3.45% in the wavelength range from 350 nm to 1200 nm. The composite porous film showed a water contact angle of 2.4° showing good super hydrophilicity and retained with an increase to 7° even after 50 cycles of the sandpaper abrasion test (Jin et al., 2017).

A clear practical significance in clothing is shown by self-cleaning fabrics. When a semiconductor is covalently grafted to the cotton/polyester fibre, the self-cleaning activity can be addressed especially after fabric laundering. Introducing C = C bonds on the surface of TiO_2 by an esterification reaction with maleic anhydride and finally co-grafting onto the cotton fabric by inducing γ-ray irradiation along with 2-hydroxyethyl acrylate showed photocatalyzed self-cleaning activity after repeated cycles of the laundering durability test (Yu et al., 2013). Again, introducing graphene into the formula can increase the bonding uniformity between the coating and the fabric with reduced recombination rate.

4.3 Toxicology Studies of Photocatalytic Materials

Photocatalytic nanomaterials show broad material diversity and are utilized in diverse applications. During their production, use and disposal, they may be discharged into the environment causing a risk to ecosystems, including humans. The knowledge about their harmful effects is very limited and there are no specific regulations implemented for their usage. Therefore, it is inevitable to investigate the toxic effects of photocatalytic materials on the environment and living organisms. Due to the smaller size and the larger surface area of nano particles, they easily get adhered to cells and tissues and enter easily even to sensitive organs across various cellular barriers. The immune system of the body is mostly unable to detect and eliminate nanoparticles due to their smaller size, which finally results in DNA mutations and mitochondrial damage and, finally, cell death. Hence, the size of nanoparticles can be considered inversely proportional to their potential toxicity. Nanoparticles less than 10 nm behave as gas to the human body, easily being absorbed by the cells and even nucleus, severely effecting their normal biological functioning. Nanoparticles having a longer lifetime as compared to larger particles of the same chemical composition pose a greater health risk and cause accumulation in the liver (Vishwakarma et al., 2010). In addition to the size and surface area of nanoparticles, other physicochemical properties that contribute to their toxicity are shape, surface functionalisation, chemical stability or reactivity, polarity, agglomeration phase, and composition. In case of photocatalytic semiconductor nanoparticles, light-induced toxicity is another fearful concern due to the generation of ROS.

Nanoparticles mainly enter and interact with the human body cells in three ways: directly through skin cells, inhaling to lung cells, and orally to epithelial cells of the digestive tract. Of these, the latter two are major concern. Nanoparticles can enter a cell either by diffusion through cell membrane, adhesion, or endocytosis (Klaine et al., 2008).

After penetrating into the cells, nanoparticles can interact with biomolecules like proteins, nucleic acids, and lipids and can affect their biological activity. Their interaction with bacterial cells depends on interfacial force between them. It is observed that nanoparticles with fewer interfacial energy barriers can have strong interaction with cells and induce cytotoxicity. A quantitative correlation between interfacial interaction between nanoparticles and cell membrane can be established by the Derjaguin-Landau-Verwey-Overbeek (DLVO) theory (Li et al., 2012a) and it has been observed that acute toxicity of seven different metal oxide nanoparticles towards paramecia are in the order of Al_2O_3 < TiO_2 < CeO_2 < ZnO < SiO_2 < CuO <Fe_2O_3 (Figure 4.4).

Inflammation of the cell by the effect of nanoparticles has been studied by measuring inflammatory biomarkers such as IL-6, IL-8, and tumour necrosis factor using ELISA; cell viability assays such as MTT, MTS, and WST-1; integrity of the cell membrane using lactase dehydrogenase (LDH) assay; and various *in-vitro* toxicity models to understand the extent of toxicity by nanoparticles in animal cells. Nanoparticles interact with these assay materials and systems, and hence the data acquired by such strategies are often contradictory and incoherent (Bahadar et al., 2016). Therefore, it is quite difficult to judge whether an economically valuable nanoparticle is more toxic to the environment and list them accordingly. However, the cytotoxic effects of a photocatalyst on living organisms can occur broadly in two ways: due to direct contact or uptake of nanoparticles into the living cells and photo-induced toxicity. Phototoxic effects of nanoparticles by biological uptake occur either due to the generation of ROS or the photocorrosion of nanoparticles.

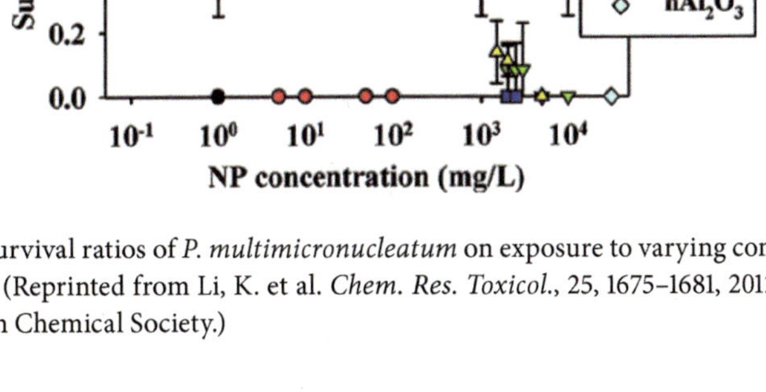

FIGURE 4.4 Mean survival ratios of *P. multimicronucleatum* on exposure to varying concentrations of different nanoparticles for 48 h. (Reprinted from Li, K. et al. *Chem. Res. Toxicol.*, 25, 1675–1681, 2012a. Copyright 2012 permission from American Chemical Society.)

4.3.1 Photo-Induced Toxicity: Reactive Oxygen Species

Phototoxicity refers to photo-induced toxic effects of materials having photocatalytic activity. In semiconductor photocatalysts, ROS are formed when photon with appropriate energy is absorbed. The photogenerated charge carriers reacts with H_2O and O_2 to yield ROS such as $^{\bullet}OH$, $O_2^{\bullet -}$ and singlet oxygen ($^1O_2$). In cells, ROS in lower concentrations are formed as byproducts during the cellular oxidative metabolisms including the synthesis of ATP, and play relevant physiological roles in inducing mitogenic responses and in cellular signalling. But their greater concentrations in the presence of semiconductors cause adverse effects on the normal biochemical environment of cells.

Oxidative stress is the major cause of toxicity due to the interaction of ROS with cellular biomolecules. Nanoparticles due to their large surface to volume ratio provide a larger interface for the redox reactions and hence their oxidation potential is higher when compared to their bigger bulk particles. Hence, they have a higher impact on the ecosystems due to the higher amount of ROS generated. This highly reactive ROS can interact with biomolecules causing lipid damage in the cell membrane, oxidation of amino acids, and genotoxicity. ROS oxidize the double bonds in fatty acids of phospholipids in the membrane, a process called lipid peroxidation, leading to increased membrane permeability and fluidity. This causes damage to the integrity of cell membranes, hindering the nutrient uptake of cells, and also makes them susceptible to osmotic stress. Simultaneously, the oxidized fatty acids can generate more ROS as byproducts, thereby creating more imbalances to the cells. Another significant cellular target of ROS is DNA. Highly reactive $^{\bullet}OH$ radicals can quickly damage DNA in the vicinity, causing base and sugar lesions, crosslinks of DNA with proteins, breaks in single and double strands, and abasic site formation (Valko et al., 2006). Consequently, ROS contributes to oxidative stress in biological systems and is associated with many degenerative diseases like sclerosis, cardiovascular disease, arthritis, inflammation, Parkinson's disease, Alzheimer's disease, diabetes, and cancer. Cells have several defence mechanisms to deactivate ROS including removal of repaired biomolecules, either enzymatic or non-enzymatic. Superoxide dismutases (SODs), catalases, and peroxidases are prominent antioxidant enzymes that deactivate the produced ROS as shown in equations 4 and 5. Disruption of these functions, however, leads to the accumulation of ROS and causes further oxidative stress or cellular damage.

$$2H_2O_2 \xrightarrow{\text{Catalase}} 2H_2O + O_2 \tag{4.4}$$

$$2O_2^{\bullet -} + 2H^+ \xrightarrow{\text{Superoxide dimutase}} H_2O_2 \tag{4.5}$$

Several engineered photocatalyst nanoparticles such as TiO_2, ZnO, SiO_2, CuO, graphene, and CNTs cause oxidative stress by the generation of ROS. ROS generation mainly depends on type and concentration of nanoparticles as well as light irradiation intensity wavelength and duration. The concentration and type of ROS generated is strongly related to the particle size and band edge structure of the photocatalyst (Jiang et al., 2009, Karlsson et al., 2009). Oxidative stress induced by ROS is quantitatively related to the antibacterial activity of the nanoparticles. This can be examined by the viability of *E. coli* cells in aqueous suspension of nanoparticles (Carré et al., 2014, Leung et al., 2016, Li et al., 2012b). Carré et al. (2014) demonstrated two-dimensional electrophoresis proteomic analysis as the first study to gain insight into the antimicrobial effect of a TiO_2 photocatalyst is to identify potential protein targets modified during the cytotoxic treatment in dark and in the presence of UV-A irradiation. This proteomics method can provide quantitative evidence for toxicity by detecting changes in protein expression as a reaction of exposure of bacteria to the nanomaterial. In the dark, TiO_2 shows no cytotoxic effect while UV-A photocatalytic treatment results in lipid peroxidation from the reaction with the $O_2^{\bullet -}$ superoxide radical ROS. Leung et al. studied the toxicity of TiO_2 and ZnO nanoparticles using *E. coli* as a model organism and examined cell count, ROS generation, and proteomics investigations. Toxicity of these nanoparticles can be mainly due to the interaction with liposaccharide molecules in the cell membrane and ROS-induced oxidative stress under illumination. Both TiO_2 and ZnO are good photocatalysts, which can inactivate microorganisms. It is found that ZnO shows less anti-bacterial activity compared to TiO_2. However, ZnO-treated bacterial cells show significant up-regulation of ROS-related proteins (glutathione reductase and glutathione dehydrogenase). For TiO_2, ROS-related protein bacterial response is not significant, but the outer membrane protein expression is observed (Leung et al., 2016).

The toxicity of a semiconductor nanoparticle can be investigated by several *in vitro* assays probing its oxidative reactivity, i.e., the quantitative capacity to produce ROS. Various fluorescent probes have been developed to detect ROS in biological and non-biological environments. Higher sensitivity, ease of data collection, and high resolution in microscopic imaging methods make them excellent sensors of ROS. Most widely used fluorescent probes are dihydroethidium and 1,3-diphenlylisobenzofuran for detecting $O_2^{\bullet -}$ radicals; 2,7-dichlorodihydrofluorescein (DCFH), homovanillic acid, and Scopoletin for H_2O_2; 9,10-dimethylanthracene and DMAX for $^1O_2$; and 1,3-cyclohexanedione and sodium terephthalate for $^{\bullet}OH$ radicals (Gomes et al., 2005). The DCFH assay is among the most widely used, as it is more non-specific to ROS and can be discriminatorily oxidized by many functional groups of ROS. It is usually employed *in-vitro* in the form of its acetate form (DCFH-DA), which easily permeates through the cell membrane and is hydrolysed enzymatically to DCHF by the action of intracellular esterases (Figure 4.5).

The nanoparticles show enhanced ROS generation compared to their bulk counterpart due to their large surface area, which provides more reactive sites for light absorption (Verma et al., 2018). Li et al. (2012b) demonstrated the ROS generation kinetics in seven different metal oxide nanoparticles and their bulk counter parts under UV irradiation. The quantitative measurement of ROS generation was performed using indicators. XTT (2,3-bis(2-methoxy-4-nitro-5-sulfophehyl)-2H-tetrazolium-5-carboxanilide) can be used as an indicator for superoxide radical. p-Chlorobenzoic acid and furfuryl alcohol can be used as indicators for hydroxyl and singlet oxygen, respectively (Li et al., 2012a, 2012b). ZnO nanoparticles can induce cellular toxicity by ROS generation. TiO_2 and ZnO generate three types of ROS,

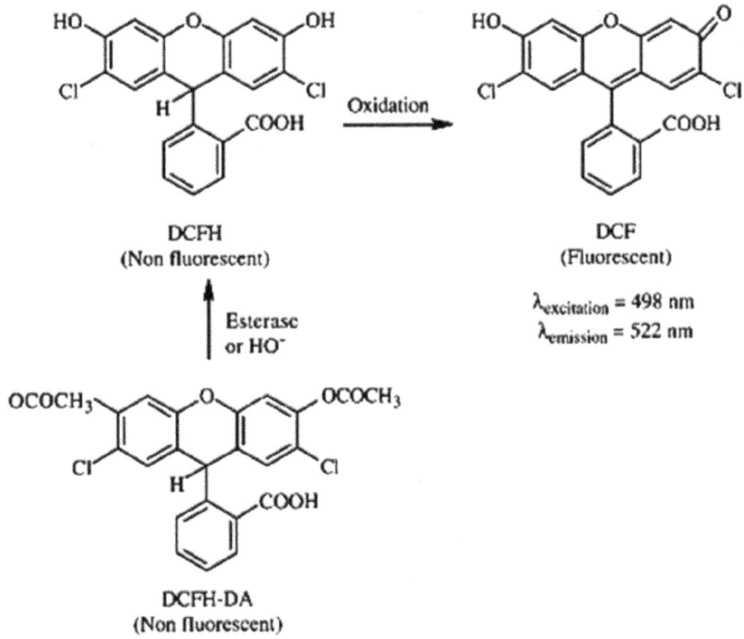

FIGURE 4.5 De-esterification mechanism of DCFH-DA to DCFH by the cellular enzyme esterase and the oxidation of DCFH to DCF by ROS. (Reprinted from Gomes, A. et al., *J. Biochem. Biophys. Methods*, 65, 45–80, 2005. Copyright 2005 permission from Elsevier.)

namely superoxide radical, hydroxyl radical, and singlet oxygen. Fe_2O_3 nanoparticles produce significant amounts of superoxide radical and minimal amounts of hydroxyl radical. SiO_2 and Al_2O_3 nanoparticles generate singlet oxygen while CeO_2 contributes superoxide radical. Also, ZnO and CuO exhibit toxicity due to the release of toxic ions during irradiation; the metal ion release from particles can be investigated using inductively coupled plasma-mass spectrometry (ICP-MS). Photochromic tungsten oxide nanoparticles show UV-induced toxicity against bacterial and mammalian cells. These nanoparticles show both time- and dose-dependent cytotoxicity due to the decreased dehydrogenase activity (Popov et al., 2018).

4.3.2 Photocorrosion

It is always agreed that the main cause for the cell death by the photocatalytic semiconductor nanomaterials is due to ROS generation. However, Dasari et al. in 2013 found that CuO nanoparticles with no enhanced ROS generation showed GSH and LDH depletion in *E. coli.* This light-induced hypertoxicity is due to the dissolution of the semiconductor into free metal ions. The surfaces of most semiconductors are susceptible to photocorrosion because their decomposition potential lies in their own band gaps. The process is strongly sensitive to surface states and the mobility of photogenerated electrons (Aziziyan et al., 2016). When mobility of the electrons is higher, the photogenerated charge carriers move to the surface and undergo photocatalytic mechanisms. When this mobility is lower, a portion of charge carriers take part in the oxidation of the metal centres completing the photocatalytic process, thereby corroding the semiconductor. Comparing TiO_2 and ZnO, the former exhibits excellent chemical stability while the latter is stable only in a narrow pH range.

Most studies have suggested that the major contributor to ZnO toxicity is release and dissolution of ionic zinc (Zn^{2+}) (Miao et al., 2010, Miller et al., 2010). Dissolution takes place both in the presence of light and under acidic conditions (Equations 4.6 and 4.7), and due to its enhanced dissolution it is a major hazard to aquatic organisms.

$$ZnO + 2h^+ \longrightarrow Zn^{2+} + \frac{1}{2}O_2 \qquad \text{Photocorrosion} \tag{4.6}$$

$$ZnO + 2H^+ \longrightarrow Zn^{2+} + H_2O \qquad \text{Acidic condition} \tag{4.7}$$

During photocorrosion, the photogenerated holes attack Zn-O bond thereby releasing Zn^{2+} ions. Similar for the photocorrosion of CuO and CdSe, holes oxidize the metal centres to Cu^{2+} and Cd^{2+} metal ions, respectively, with release of O_2 or H_2O_2 for CuO and elemental Se for CdSe. Additionally, by a process called the Fenton reaction, these released ions can further produce ROS (Formanowicz et al., 2018). The Fenton process plays an important role in transforming poor reactive radicals to high reactive ones like hydroxyl radicals. For example.

$$Cu^{2+} + H_2O_2 \longrightarrow Cu^{3+} + HO^- + HO^\bullet \tag{4.8}$$

Hence, photocorrosion ultimately leads to metal-induced generation of ROS resulting in cell proliferation, DNA damage, lipid peroxidation, and impairment of protein synthesis.

4.4 Semiconductor Nanomaterials and Biocompatibility Analysis

Large-scale experimental and theoretical research is being conducted on semiconductor nanomaterials because of their size-dependent optical and electronic properties, which are due to quantum phenomena (Woggon, 1997). Usually these materials are referred to as quantum dots (QDs) and their properties have been intensively investigated in the past few years. But biocompatibility of these semiconductor materials is a matter of discussion as these materials are widely used in biological and biomimetic systems for

FIGURE 4.6 Proposed mechanism of Cd release from the QD surface via either solvent-mediated or UV-catalysed surface oxidation. (Reprinted from Austin, M. et al., *Nano. Lett.*, 4, 11–18, 2004. Copyright 2004 permission from American Chemical Society.)

the investigation of complex cell processes (Medintz et al., 2005). Biocompatibility of a material depends on its non-maleficence towards living tissues. It can be better understood by analysing its interaction with living tissues. In order to make QDs biologically applicable, the cytocompatibility analysis has to be done before applying them in a biological system. For example, CdSe QDs are widely used as an alternative for organic dyes for biological labelling but their potential toxicity is a concern. Austin et al. (2004) studied the hepatic injury caused by CdSe QDs, as liver is the primary site of attack by heavy metals. They selected hepatocytes from rats as a representative model for studying cytotoxicity and found that CdSe QDs were toxic under certain conditions especially when any surface oxidation leading to the formation of reduced Cd occurs. But use of surface coatings can significantly reduce these problems (Austin et al. 2004). The cytotoxicity and biocompatibility of many semiconductor nanomaterials has been studied but the underlying mechanisms of their interactions with biological systems are not yet fully understood (Figure 4.6).

4.4.1 Making of Biocompatible Semiconductor Nanomaterials

Semiconductor nanomaterials can be synthesized by means of physical, chemical, and biological methods. Physical methods such as physical vapour deposition (Jagannadham et al., 2010) or laser irradiation technique (Wu et al., 2015) can lead to defective surfaces and poor quantum yield. Chemical methods can produce semiconductor nanoparticles with uniform size and better quantum yield (Wang et al., 2012); however, vigorous reaction conditions and the contamination of the products with reagents, leading to toxicity, are associated problems. The leverage of bio-synthesis methods is associated with the production of nano semiconductors with good biostability and biocompatibility. The principle features associated with biosynthesis are biodetoxification by *in situ* reduction of toxic metal ions through binding them with biomolecules such as amino acids or enzymes, regulation of crystal growth, and size distribution by means of genetic engineering, abate reaction conditions like moderate temperature, and mild nontoxic agents (Cui et al., 2009).

4.4.2 Synthesis in Biological and Biomimetic Systems

The synthesis can be done inside the living organisms so that bio-detoxification and bio-mineralization takes place along with the advantage of having enzymes which catalyse the synthesis reaction. So far, many nanosemiconductors have been prepared in different living organisms. Cadmium sulphide was the first nanosemiconductor to be prepared in a biological system by Dameron et al. (1989). The yeast cells were cultured with cadmium salts where the cadmium ions were chelated with amino acid and reacted with intracellular sulphide to produce CdS crystallites. Similarly, PbS and ZnS (Mala et al., 2014) were also prepared. *E. coli* have been used as a biofactory to synthesise different semiconductor nanomaterials including CdZn, CdSe, CdTe, and SeZn with controlled size by varying the metal ion concentration. The semiconductor QDs showed enhanced fluorescence, which was further used for imaging human fibroblast cells (Park et al., 2010, Sweeney et al., 2004).

Many other living organisms which show strong reducing capacity towards semiconductor precursors are also being used to produce diverse nanosemiconductors. Xiong et al. adopted space-time coupling for transforming *Staphylococcus aureus* bacterial cells to cellular beacons for biotargeting viruses, bacteria, and tumour cells. *S. aureus* cells were co-incubated with Se and Cd precursors for a specified time to intracellularly produce $CdS_{0.5}Se_{0.5}$. These cells were further coupled with monoclonal antibodies (mAbs) and immunomagnetic beads resulting in the formation of a nano bioprobe (Xiong et al., 2014). Metal detoxification ability of the earthworm *Lumbricus rubellus* had been exploited to produce CdTe QDs by exposing these worms to soil spiked and equilibrated with $CdCl_2$ and $NaTeO_3$ salts for 11 days. These QDs were successfully employed for live-cell imaging (Sturzenbaum et al., 2012). The synthesis can also be widened to biomimetic systems that utilize bio molecules like DNA, RNA, enzymes, and proteins as templates for the growth of semiconductor nanomaterials. In addition to these natural systems, artificial membranes such as biopolymers, N-isopropylamide, or gelatine-based hydrogel encapsulated with living cell extract have also been used as templates (Lee et al., 2012) (Figure 4.7).

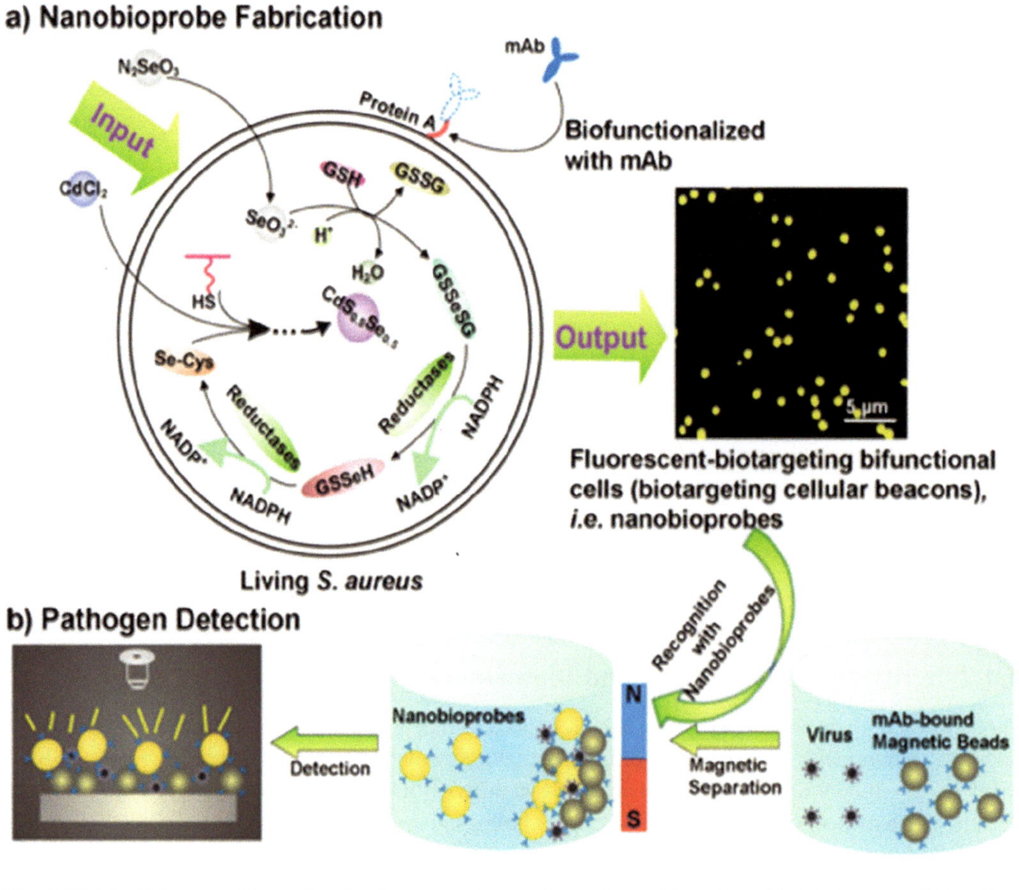

FIGURE 4.7 Schematic illustration for the generation and use of nanobioprobes. (a) Fabrication of bioprobes. Space-time coupling strategy producing $CdS_{0.5}Se_{0.5}$ QDs. These cells conjugate with mAbs to form biotargeting cellular beacons. (b) Using fluorescent-biotargeting bifunctional cells as bioprobes for pathogen detection. (Reprinted from Xiong, L.H. et al., *ACS Nano*, 8, 5116–5124, 2014. Copyright 2014 permission from American Chemical Society.)

4.4.3 Biocompatibility Testing

Broadly, the biocompatibility tests can be categorized into *in vivo* and *in vitro*. It is generally easier for the scientists to do cytotoxicity *in vivo* analysis as it is done in its natural environment, but it sometimes has to meet the exacting standards concerned with the analysis performed in living systems. In contrast, *in vitro* analysis is performed outside the living system in a highly controlled artificial environment without harming the living organism but sporadically may not be able to exactly predict the conditions inside a living system.

Biocompatibility is concerned with biosafety as well as the biofunctionality of the material used. Biosafety includes cytotoxicity analysis while biofunctionality deals with the capability of the material to perform the biological functions like cell adhesion, cell spreading, and cell proliferation as well as biosynthetic functions. Commonly, the quantitative examination of cytotoxic effects includes MTT assay, i.e., the response of mitochondrial enzymes to a tetrazolium salt, 3-(4,5dimethylthiazol-2yl)-2, 5-diphenyl tetrazolium bromide. The enzymes are capable of reducing MTT to insoluble purple-coloured formazan whose concentration is directly proportional to the number of viable cells and can be quantified by spectrophotometer (Dekker et al., 1994). Another conventional method to check cell viability is to examine the state of plasmalemma, i.e., the permeability of plasma membrane which acts as a wall between the cell and its surroundings. Any harmful effects on the cell could create alternations in structural and functional integrity of the plasma membrane. A graceful examination for checking cell integrity was described by Rotman and Papermaster (1966). The test involves the incubation of cells with fluorescien-diacetate (FDA), a non-fluorescent fatty acid ester of fluorescein and ethidium bromide (ED). FDA can be taken up intracellularly and is converted to fluorescein, which gives green fluorescence. EB cannot pass through intact cell membranes but can pass through a damaged one and can give orange to red fluorescence. Another substitute for FDA is acridine orange (AO) (Kirkpatrick and Dekker 1992). Both the MTT assay and the membrane integrity test are regarded as reliable quantitative methods to check the cell viability towards various materials.

Haemolysis assays can be carried out to determine the hemocompatibility (compatibility with red blood cells) of the semiconductor QDs. Generally, the erythrocytes and toxin are added to a buffer and incubated for 20 min to 2 h at a suitable *in vivo* temperature (usually normal human body temperature). Unlysed cells and cell membranes are then removed by centrifugation. Haemolysis is then determined by spectrochemically analysing the supernatant and comparing it with the same cell concentration in buffer without toxin (Gail et al., 1994). Liu and Yu investigated the hemocompatibility of glutathione-capped CdTe/ZnS QDs by incubating them in phosphate buffer. It was found that on prolonged incubation, the haemolysis rates increased from 0.2% to 10.7% for GSH-CdTe/ZnS QDs however better than the uncapped ones (Liu and Yu, 2010).

4.4.4 Improving Bio-compatibility of Semiconductor Nanomaterials Through Chemical Modifications

The incompatibility of semiconductor nanomaterials with bio-background can be revamped by chemical modification of these materials (Zhou et al., 2015). There are several main strategies.

4.4.4.1 Hydrophilicity

Most of the semiconductor nanomaterials are hydrophobic and one of the strategies to improve biocompatibility is to mould these materials to hydrophilic from their hydrophobic nature. Most chemical synthesis methods use the encapsulation of nanomaterials with hydrophobic stabilizers. Replacing these stabilizers or surfactants with water-soluble ones could make them suitable for biological applications. These capping agents can bestow the nanomaterials with uniform size, stability, and excellent properties. Ligands containing groups like amino, carboxyl, or PEG are ideal for hydrophilic environments, but the widely used ligands contain thiol groups due to their higher efficiency. The use of capping agents to make biocompatible nanosemiconductors began in 1998 by Chan and Nie. They coupled luminescent ZnS and CdSe conjugate with biomolecules for sensitive detection of proteins, nucleic acids, and viruses. To attach biomolecules with ZnS/CdSe semiconductor, they reacted it with mercaptoacetic acid whose polar carboxylic acid group made it soluble in water (Chan and Nie, 1998). Presently many thiol-containing ligands including cysteine (Tohgha et al., 2013) and artificially thiolated agents like cyclodextrin (Palaniappan et al., 2006), silane (Gerion et al., 2001), proteins (Willard et al., 2001), and nucleic acid (Mitchell et al., 1999) have been used to synthesis hydrophilic and biocompatible semiconductor nanomaterials (Figure 4.8).

Hydrophobicity of semiconductor nanomaterials can also be improved by encapsulation with amphiphobic molecules like phospholipids or liposomes. The hydrophilic nature of these materials makes them water soluble whereas the hydrophobic domain helps the encapsulation of semiconductor nanomaterials. The hydrophobic fatty acid group and hydrophilic phosphate group of phospholipids help them to form lipid bilayers and can efficiently enclose the semiconductor QDs. Phospholipid micelles like poly ethylene glycol phosphatidyl ethanol amine (PEG-PE), or phosphatidyl choline (PC) can improve the biocompatibility (Dubertret et al., 2002). Liposomes, another class of material used for enveloping semiconductor QDs, are composed of an internal hydrophilic chamber and hydrophobic phosphor lipid bilayer. These nanomaterials have been used as carriers of pharmaceutical drugs, making them ready to use for biological applications (Al-Jamal et al., 2008).

Silica is another class of material that is used for improving hydrophilicity of nanosemiconductors. Silica shells can prevent leakage of toxic ions into biological systems as well as provide the possibilities of binding the semiconductor with many functional groups including thiol, amine, phosphate, carboxylate, or PEG present on biomolecules without loss of photoluminescence and surface characteristics (Acebron et al., 2015). The encapsulated semiconductor nanomaterials show good photoluminescence, appreciable quantum yield, and good stability, making these nanomaterials suitable for bioimaging applications. Thus, hydrophobic nanosemiconductors which are not suitable to be used in biological systems can be functionalized with biocompatible ligands or can be encapsulated with multifunctional shells which make them hydrophilic. These hydrophilic QDs can be attached with biomolecules including nucleic acids, antibodies, proteins, and peptides by chemical reactions or absorption which further makes them biocompatible.

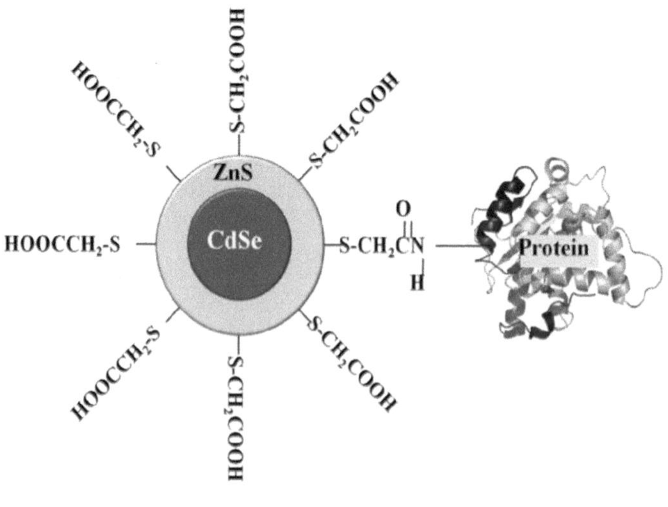

FIGURE 4.8 Diagrammatic sketch of a ZnS-capped CdSe QD covalently bonded to a protein by mercaptoacetic acid. (Adapted from Chan, W.C.W. and Nie, S.M., *Science*, 281, 2016–2018, 1998.)

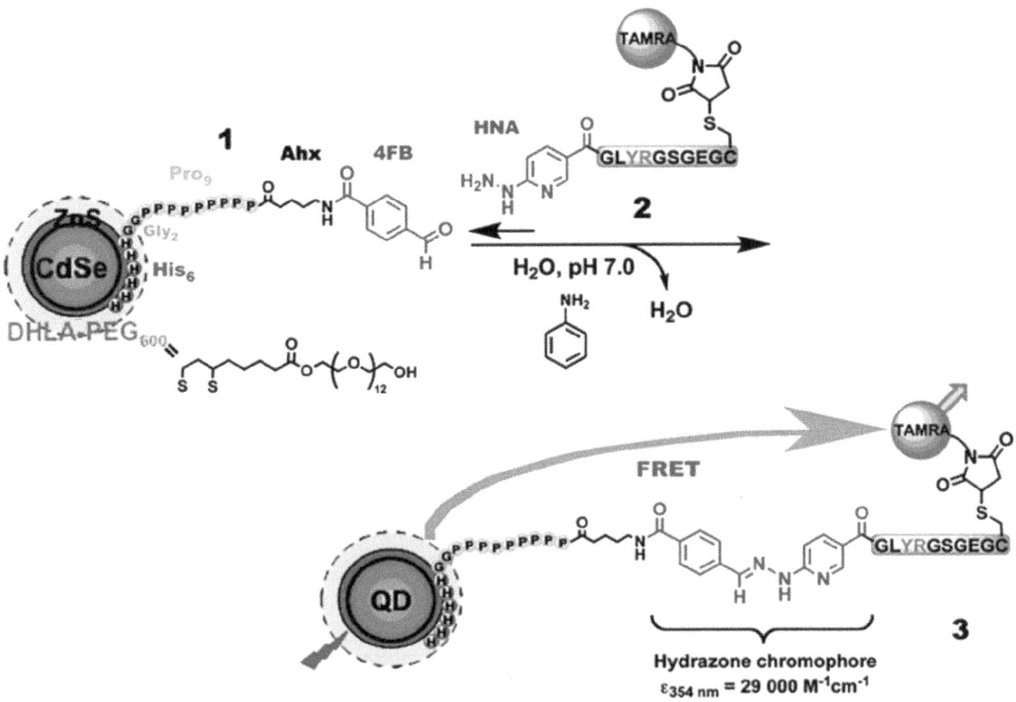

FIGURE 4.9 General covalent conjugation strategy followed for the ligation of peptides to PEGylated QDs. Peptide 2 is ligated to QD/peptide 1 to yield the QD/peptide 3 product. The highlighted Y and R residues in peptide 3 are subsequently cleaved by addition of the proteases chymotrypsin or trypsin. (Reprinted from Blanco-Canosa J.B. et al., *J. Am. Chem. Soc.*, 132, 10027–10033, 2010. Copyright 2010 permission from American Chemical Society.)

4.4.4.2 Covalent Binding

Covalent binding of biomolecules with nanosemiconductors can be made possible by using coupling agents like carbodiimide reagents (Chan et al., 1998). Most biomolecules contain amino and carboxyl groups. The reaction between these groups to form amide linkage can be catalysed by carbodiimide reagents. As an example, InP/ZnS semiconductor QDs were activated with dicyclohexylcarbodiimide (DCC) to yield a carboxylated intermediate which can react with the amino group of folic acid. These folate InS/ZnS conjugates were used for folate receptor-mediated delivery. Similar to carbodiimide conjugation, other strategies that are capable of covalent binding biomolecules to QDs include amine sulfhydryl reaction (Hildebrandt, 2011) and aniline-catalysed hydrazone (Blanco-Canosa et al., 2010) reactions (Figure 4.9).

4.4.4.3 Ionic Interaction

Another strategy for improving biocompatibility by chemical modification of semiconductor can be done by coupling biomolecules onto the surface of nanosemiconductor with electrostatic interaction. In their work, Mattoussi and co-workers failed to covalently attach recombinant maltose binding proteins (MBP) to dihydrolipoic acid-coated CdSe-ZnS nanohybrid (Mattoussi et al., 2000). So, they made use of ionic interaction between negatively charged lipoic acid-capped CdSe-ZnS nano semiconductors. They prepared CdSe-ZnS/MBP QDs that retained biological activity and fluorescence. In another similar work, hyaluronic acid (HA) (Bhang et al., 2009), a major component of the extracellular matrix, was electrostatically conjugated with a ternary core shell nanosemiconductor CdSe/CdS/ZnS QDs. A surface ligand with dithiol and amine was used for creating positive charge on CdSe/CdS/ZnS, which was electrostatically coupled with negatively charged HA. These hybrids showed potential activity towards cancer imaging. Even though electrostatic interactions are better than covalent conjugation in particle agglomerations, the electrostatic assembly approach can sometimes lead to unfavourable effects on function of attached element.

4.4.4.4 Biomolecule Encapsulation

Biomolecules, in addition to serving as template for the synthesis of biocompatible QDs, can also be used as functional ligands for the modification purposes. Virus-based cages have been extensively used to modifying nanosemiconductor materials. The caspid proteins of viruses are capable of forming nanocages which can enclose the nanomaterials, providing them with biocompatibility. Gao and coworkers studied the assembly of simian virus caspid proteins to nanocages in the presence of QDs and also the encapsulation of QDs by the nanocages. By controlling the self-assembly of caspid proteins, the size of nanocages can be tuned (Gao et al., 2013). The interaction between the virus protein and the semiconductor QDs can be electrostatic, van der Waals, or hydrogen bond. All these interactions may lead to self-organization of these caspid proteins into cage-like structures that may be used for tracking virus infections.

4.4.5 Biomedical Applications

The distinct optical features of semiconductor nanomaterials, especially QDs, combined with their biocompatible nature, make them extensively useful in the biomedical field. Their idiosyncratic properties include tuneable bandgap, good luminescence, stability, narrow spectral bandwidth, and large surface area; all these properties have been exploited in targeted cellular imaging, sensing, drug delivery, and cancer therapy.

By combining semiconductor QDs with bioligands, targeted cellular labelling and imaging of various biological objects such as proteins, nucleic acids, stem cells, and tumour cells have been made possible in the last few decades. The unique optical properties make them useful for multicolour and multimodel imaging. In multimodel imaging, the imaging property of QDs is merged with other imaging techniques like magnetic resonance imaging (MRI).

Semiconductor QD-based biosensors target the specific biomolecules with the help of sensing ligands and the number of biomolecules is recognized from the change of luminescence. Besides the QD-based conventional biosensors, QD-based energy transfer sensors like FRET have also been developed. In QD-based FRET biosensors, QDs act as the donor and the fluorescence dye acts as acceptor and has been widely used for the detection of amino acids, insulin, intracellular pH, proteolytic activity assay, and monitoring DNA cleavage (Algar et al., 2014, Geissler et al., 2014). Similar to QD – FRET system, QD – BRET (bioluminescence resonance energy transfer) (Alam et al., 2014) and QD – CRET (chemiluminescence energy transfer) (Chen et al., 2014a) systems are also available (Figure 4.10).

Other than imaging and sensing applications, one prominent area of application of QDs involves the construction of drug delivery (Yang et al., 2012, Chen et al., 2014b) systems for cancer and tumour therapy. They can simultaneously act as drug carriers as well as imaging agents to track the drug delivery. These QD can be functionalized by anticancer drugs and cancer-targeting groups like folic acid or aptamers. The targeting moiety leads the system to an affected area and when the target is reached, the QD drug system dissociates to release the drug. The ability of these QDs such as loading capacity, stability, target selectivity, and controlled drug release make them excellent in this field.

A possible method of making biocompatible QDs for drug delivery is by integrating them with silica. Zhu et al. employed ZnO QDs to seal the nanopores of mesoporous silica nanoparticles (MSNs) in order to control doxorubicin release. On administration to HeLa cells, the ZnO QD may serve as a lid for releasing the loaded drug upon acid dissolution and also has a synergistic antitumor effect on cancer cells (Muhammad et al. 2011) (Figure 4.11).

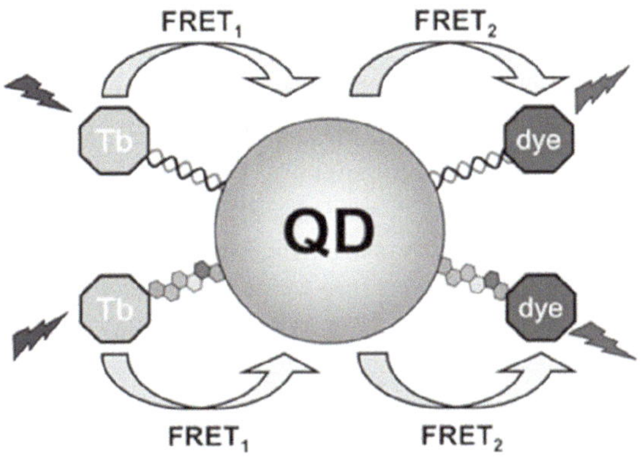

FIGURE 4.10 Terbium complexes and dyes conjugated with peptides or oligonucleotides are assembled on the same QD. Pulsed UV excitation excites terbium and QD and prompts FRET from QD to dye (FRET2). After several microseconds, there is still much terbium in the excited state, whereas the QD has decayed to its ground state, and efficient time-delayed FRET from terbium to QD (FRET1) becomes possible. The newly excited QD can then sensitize the dyes again by time-gated FRET2. (Reprinted from Geissler, D. et al., *Inorg. Chem.*, 53, 1824–1838, 2014. Copyright 2014 permission from American Chemical Society.)

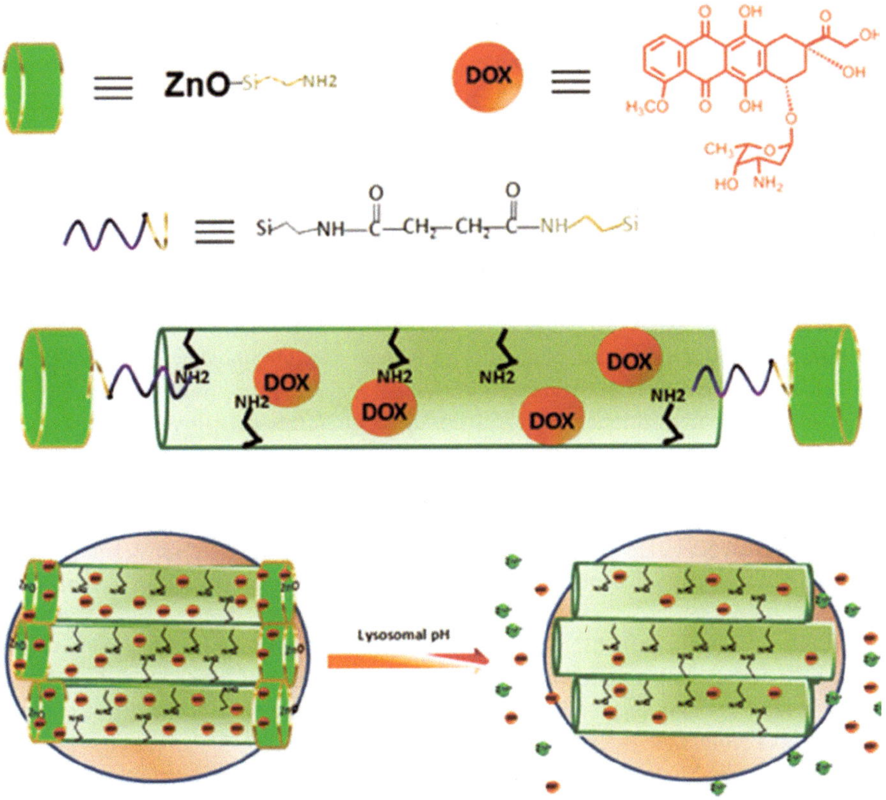

FIGURE 4.11 Schematic illustration of the preparation of ZnO@MSNs-DOX for the controlled release of anticancer drug (DOX) via acidic dissolution of ZnO QDs in the intracellular compartments of cancer cells. (Reprinted from Muhammad, F. et al., *J. Am. Chem. Soc.*, 133, 8778–8781, 2011. Copyright 2011 permission from American Chemical Society.)

Another versatile nanocarrier that can be employed for QD-based drug delivery involves polymeric micelles made from biodegradable polymers like chitosan, polyesters, and dextran. Yuan et al. combined blue-light-emitting ZnO QDs with biodegradable chitosan (N-acetylglucosamine) for tumour-targeted drug delivery. ZnO QDs were synthesized by a chemical hydrolysis method, encapsulated with chitosan, which enhanced hydrophilicity, making them more stable, and were then loaded with the anticancer drug doxorubicin. The DOX-loaded ZnO-QD–chitosan–folate showed a controlled drug release (Yuan et al., 2010). Stimuli-responsive drug release is possible by making use of polymers like poly N-isopropylacrylamide (temperature responsive) or poly (N-(N′,N′-diisopropylaminoethyl)-aspartamide) (pH responsive) (Akimoto et al., 2014, Wang et al., 2012)

4.5 Conclusions and Future Outlook

The rapid progress in manipulation of nanomaterials led to the tremendous increase in the use of nanomaterials in environmental products. Even though the nanomaterials find a place in the upcoming water purification and biomedical applications, several challenges exist for the effective utilization of antimicrobial, antibacterial, and biocompatibility properties of nanomaterials in the products. Scientists and engineers are now more concerned about the risk associated with the implementation of such products. A proactive research is required for the assessment of risk associated with the products based on nanomaterials. The problems associated with the nanomaterial research are multidisciplinary and an accurate knowledge on the impact of the release of these nanomaterials on the environment is required. It is also challenging to characterise the environmental behaviour of nanomaterials and ensure a sustainable nanotechnology for the future.

References

Acebron, M., Galisteo-Lopez, J. F. et al. 2015. Protective ligand shells for luminescent SiO_2_coated alloyed semiconductor nanocrystals. *ACS Appl. Mater. Interfaces*, 7, 6935–6945.

Adeli, B., Taghipour, F. 2016. Facile synthesis of highly efficient nano-structured gallium zinc oxynitride solid solution photocatalyst for visible-light overall water splitting. *Appl. Catal. A Gen.*, 521, 250–258.

Akimoto, J., Nakayama, M., Okano, T. 2014. Temperature-responsive polymeric micelles for optimizing drug targeting to solid tumors. *J. Control. Release*, 193, 2–8.

Alam, R., Karam, L. M., Doane, T. L. 2014. Near infrared bioluminescence resonance energy transfer from firefly luciferase-quantum dot bionanoconjugates. *Nanotechnology*, 25, 495–606.

Algar, W. R., Kim, H., Medintz, I. L., Hildebrandt, N. 2014. Emerging non-traditional Forster resonance energy transfer configurations with semiconductor quantum dots: Investigations and applications. *Coord. Chem. Rev.*, 263, 65–85.

Al-Jamal, W. T., Al-Jamal, K. T., Tian, B. et al. 2008. Lipid-quantum dot bilayer vesicles enhance tumor cell uptake and retention in vitro and in vivo. *ACS Nano*, 2, 408–418.

Altomare, M., Nguyen, N. T., Hejazi, S., Schmuki, P. 2018. A cocatalytic electron-transfer cascade site-selectively placed on TiO_2 nanotubes yields enhanced photocatalytic H_2 evolution. *Adv. Funct. Mater.*, 28 (2), 1704259.

An, X., Yu, J. C., Wang, Y. et al. 2012. WO_3 nanorods/graphene nanocomposites for high-efficiency visible-light driven photocatalysis and NO_2 gas sensing. *J. Mater. Chem.*, 22, 8525–8531.

Austin, M., Warren, C., Chan, W., and Sangeeta, N. B. 2004. Probing the cytotoxicity of semiconductor quantum dots. *Nano Lett.*, 4, 11–18.

Awfa, D., Ateia, M., Fujii, M., Johnson, M. S., Yoshimura, C. 2018. Photodegradation of pharmaceuticals and personal care products in water treatment using carbonaceous-TiO_2 composites: A critical review of recent literature. *Water Res.*, 142, 26–45.

Aziziyan, M. R., Hassen, W. M., Morris, D., Dubowski, J. J. 2016. Photonic biosensor based on photocorrosion of GaAs/AlGaAs quantum heterostructures for detection of *Legionella pneumophila*. *Biointerphases*, 11(1), 019301.

Bahadar, H., Maqbool, F., Niaz, K., Abdollahi, M. 2016. Toxicity of nanoparticles and an overview of current experimental models. *Iran Biomed. J.*, 20, 1–11.

Banerjee, S., Dionysiou, D. D., Pillai, S. C. 2015. Self-cleaning applications of TiO_2 by photo-induced hydrophilicity and photocatalysis. *Appl. Catal. B Environ.*, 176–177, 396–428.

Bhang, S. H., Won, N., Lee, T.-J. et al. 2009. Hyaluronic acid-quantum dot conjugates for in vivo lymphatic vessel imaging. *ACS Nano*, 3, 1389–1398.

Blanco-Canosa, J. B., Medintz, I. L., Farrell, D., Mattoussi, H., Dawson, P. E. 2010. Rapid covalent ligation of fluorescent peptides to water solubilized quantum dots. *J. Am. Chem. Soc.*, 132, 10027–10033.

Cao, S., Li, H., Zhu, B. Yu, J. 2018. Dependence of exposed facets on Pd on photocatalytic H_2-production activity. *ACS Sustainable Chem. Eng.*, 6, 6478–6487.

Carré, G., Hamon, E., Ennahar, S. et al. 2014. TiO_2 photocatalysis damages lipids and proteins in escherichia coli. *Appl. Environ. Microbiol.*, 80, 2573–2581.

Chan, W. C. W., Nie, S. M. 1998. Quantum dot bioconjugates for ultrasensitive nonisotopic detection. *Science*, 281, 2016–2018.

Chang, X., Wang T., Gong, J. 2016. CO_2 photo-reduction: Insights into CO_2 activation and reaction on surfaces of photocatalysts. *Energy Environ. Sci.*, 9, 2177–2196.

Chen, H., Lin, L., Li, H., Lin, J.-M. 2014a. Quantum dots-enhanced chemiluminescence: Mechanism and application. *Coord. Chem. Rev.*, 263, 86–100.

Chen, X., Tang, Y., Cai, B., Fan, H. 2014b. 'One-Pot' synthesis of multifunctional GSH-CdTe quantum dots for targeted drug delivery. *Nanotechnology*, 25, 235101.

Cho, M., Chung, H., Choi, W., Yoon, J. 2005. Different inactivation behavior of MS-2 phage and escherichia coli in TiO_2 photocatalytic disinfection. *Appl. Environ. Microbiol.*, 71(1), 270–275.

Cho, M., Cates, E. L., Kim, J.-H. 2011. Inactivation and surface interactions of MS-2 bacteriophage in a TiO_2 photoelectrocatalytic reactor. *Water Res.*, 45 (5), 2104–2110.

Choi, Y. I., Jeon, K. H., Kim, H. S. et al. 2016. TiO_2/BiOX (X = Cl, Br, I) hybrid microspheres for artificial waste water and real sample treatment under visible light irradiation. *Sep. Purif. Technol.*, 160, 28–42.

Christoforidis, K. C., Fornasiero, P. 2017. Photocatalytic hydrogen production: A rift into the future energy supply. *Chem. Cat. Chem.*, 9, 1523–1544.

Cui, R., Liu, H. H., Xie, H. et al. 2009. Living yeast cells as a controllable biosynthesizer for fluorescent quantum dots. *Adv. Funct. Mater.*, 19, 2359–2364.

Dameron, C. T., Reese, R. N., Mehra, R. K. et al. 1989. Biosynthesis of cadmium sulphide quantum semiconductor crystallites. *Nature*, 338, 596–597.

Dekker, A., Panfil, C., Valdor, M., Pennartz, G., Richter, H., Mittermayer, C. H. and Kirkpatrick, C. J. 1994. Quantitative methods for in vitro cytotoxicity testing of biomaterials. *Cells Mater.*, 4, 101–112.

Duan, Y., Liang, L., LV, K., Li, Q., Li, M. 2018. TiO_2 faceted nanocrystals on the nanofibers: Homojunction TiO_2 based Z-scheme photocatalyst for air purification. *Appl. Surf. Sci.*, 456, 817–826.

Dubertret, B., Skourides, P., Norris, D. J., Noireaux, V., Brivanlou, A. H., Libchaber, A. 2002. In vivo imaging of quantum dots encapsulated in phospholipid micelles. *Science*, 298, 1759–1762.

Emeline, A. V., Rudakova, A. V., Sakai, M., Murakami, T., Fujishima, A. 2013. Factors affecting UV-induced superhydrophilic conversion of a TiO_2 surface. *J. Phys. Chem. C*, 177 (23), 12086–12092.

Fan, H., Zhao, X., Yang, J. et al. 2012. ZnO-graphene composite for photocatalytic degradation of methylene blue dye. *Catal. Commun.*, 29, 29–34.

Formanowicz, D., Radom, M., Rybarczyk, A., Formanowicz, P. 2018. The role of Fenton reaction in ROS-induced toxicity underlying atherosclerosis—modeled and analyzed using a petri net-based approach. *Biosystems*, 165, 71–87.

Franklin, N. M., Rogers, N. J., Apte, S. C. et al. 2007. Comparative toxicity of nanoparticulate ZnO, bulk ZnO, and $ZnCl_2$ to a freshwater microalga (*Pseudokirchneriellasubcapitata*): The importance of particle solubility. *Environ. Sci. Technol.*, 41(24), 8484–8490.

Fujishima, A., Honda, K. 1972. Electrochemical photolysis of water at a semiconductor electrode. *Nature*, 238 (5358), 37–38.

Fujishima, A., Rao, T. N., Tryk, D. A. 2000. Titanium dioxide photocatalysis. *J. Photochem. Photobiolo. C: Photochem. Reviews*, 1, 1–21.

Gail, E. R., and Rodney, A. W., 1994. Assays of hemolytic toxins. *Method. Enzymol.*, 235, 657–667.

Gao, D., Zhang, Z.-P., Li, F. 2013. Quantum dot-induced viral capsid assembling in dissociation buffer. *Int. J. Nanomed.*, 8, 2119–2128.

Gazit, E. 2007. Self-assembled peptide nanostructures: the design of molecular building blocks and their technological utilization. *Chem. Soc. Rev.*, 36(8), 1263–1269.

Geissler, D., Linden, S., Liermann, K., Wegner, K. D., Charbonniere, L. J., Hildebrandt, N. 2014. Lanthanides and quantum dots as Forster resonance energy transfer agents for diagnostics and cellular imaging. *Inorg. Chem.*, 53, 1824–1838.

Gerion, D., Pinaud, F., Williams, S. C. 2001. Synthesis and properties of biocompatible water-soluble silica-coated CdSe/ZnS semiconductor quantum dots. *J. Phys. Chem. B*, 105, 8861–8871.

Ghafoor, S., Hussain, S. Z., Waseem, S., Arshad, S. N. 2018. Photo-reduction of heavy metal ions and photo-disinfection of pathogenic bacteria under simulated solar light using photosensitized TiO_2 nanofibers. *RSC Adv.*, 8(36), 20354–20362.

Gomes, A., Fernandes, E., Lima, J. L. F. C. 2005. Fluorescence probes used for detection of reactive oxygen species. *J. Biochem. Biophys. Methods*, 65, 45–80.

Hajkova, P., Spatenka, P., Horsky, J., Horska, I., Kolouch, A. 2007. Photocatalytic effect of TiO_2 films on viruses and bacteria. *Plasma Process. Polym.*, 4, S397–S401.

Halvorson R. A., Vikesland, P. J. 2010. Surface-enhanced Raman spectroscopy (SERS) for environmental analyses. *Environ. Sci. Technol.*, 44, 7749–7755.

Han, K., Kreuger, T., Mei, B., Mul, G. 2017. Transient behavior of Ni@NiOx functionalized $SrTiO_3$ in overall water splitting. *ACS Catal.*, 7, 1610–1614.

Hildebrandt, N. 2011. Biofunctional quantum dots: Controlled conjugation for multiplexed biosensors. *ACS Nano*, 5, 5286–5290.

Hisatomi, T., Kubota J., Domen, K. 2014. Recent advances in semiconductors for photocatalytic and photoelectrochemical water splitting. *Chem. Soc. Rev.*, 43, 7520–7535.

Hosseini, F., Kasaeian, A., Pourfayaz, F., Sheikhpour, M., Wen, D. 2018. Novel ZnO-Ag/MWCNT nanocomposite for the photocatalytic degradation of phenol. *Mater. Sci. Semicond. Process*, 83, 175–185.

Hou, D. X., Feng, L., Zhang, J. B. et al. 2012. Preparation, characterization and performance of a novel visible light responsive spherical activated carbon-supported and Er^{3+}:$YFeO_3$-doped TiO_2 photocatalyst. *J. Hazard. Mater.*, 199, 301–308.

Huang, Y., Zhu, D., Zhang, Q. et al. 2018. Synthesis of a $Bi_2O_2CO_3/ZnFe_2O_4$ heterojunction with enhanced photocatalytic activity for visible light irradiation-induced NO removal. *Appl. Catal. B Environ.*, 234, 70–78.

Ireland, J. C., Klostermann, P., Rice, E. W., Clark, R. M. 1993. Inactivation of *Escherichia coli* by titanium dioxide photocatalytic oxidation. *Appl. Environ. Microbiol.*, 59(5), 1668–1670.

Jagannadham, K., Howe, J., Allard, L. F. 2010. Laser physical vapor deposition of nanocrystalline dots using nanopore filters. *Appl. Phys. A Mater. Sci. Process*, 98, 285–292.

Jami, M., Dillert, R., Suo, Y., Bahnemann, D. W., Wark, M. 2018. Photoactivity of titanium dioxide foams. *Int. J. Photoenergy*, 5057814, 1–9.

Jiang, W., Mashayekhi, H. and Xing, B. 2009. Bacterial toxicity comparison between nano- and micro-scaled oxide particles. *Environ. Pollut.*, 157, 1619–1625.

Jin, B., He, J., Yao, L., Zhang, Y., Li, J. 2017. Rational design and construction of well-organized macro-mesoporous SiO_2/TiO_2 nanostructure toward robust high-performance self-cleaning antireflective thin films. *ACS Appl. Mater. Interfaces*, 9 (20), 17466–17475.

Jitianu, T., Cacciaguerra, R., Benoit, S. et al. 2004. Synthesis and characterization of carbon nanotubes-TiO_2 nanocomposites. *Carbon*, 42, 1147–1151.

Jung, J., Jannat, A., Akhtar, M. S., Yang, O.-B. 2018. Sol-gel deposited double layer TiO_2 and Al_2O_3 antireflection coating for silicon solar cell. *J. Nanosci. Nanotechnol.*, 18 (2), 1274–1278.

Karlsson, H. L., Gustafsson, J., Cronholm, P., Möller, L. 2009. Size-dependent toxicity of metal oxide particles—A comparison between nano- and micrometer size. *Toxicol. Lett.*, 188, 112–118.

Kirkpatrick, C. J., Dekker, A. 1992. Quantitative evaluation of cell interaction with biomaterials in vitro, *Adv. Biomater.*, 10, 31–41.

Klaine, S. J., Alvarez, P. J. J., Batley, G. E. et al. 2008. Nanomaterials in the environment: Behavior, fate, bioavailability, and effects. *Environ. Toxicol. Chem.*, 27, 1825–1851.

Kraft, C. N., Diedrich, O., Burian, B., Schmitt, O., Wimmer, M. A. 2003. Microvascular response of striated muscle to metal debris: A comparative in vivo study with titanium and stainless steel. *J. Bone Joint Surg.* 85B, 133–141.

Lee, K. G., Hong, J., Wang, K. W. et al. 2012. In vitro biosynthesis of metal nanoparticles in microdroplets. *ACS Nano*, 6, 6998–7008.

Leung, Y. H., Xu, X., Ma, A. P. Y. et al. 2016. Toxicity of ZnO and TiO_2 to *Escherichia coli* cells. *Sci. Rep.*, 6, 35243.

Li, K., Chen, Y., Zhang, W. et al. 2012a. Surface interactions affect the toxicity of engineered metal oxide nanoparticles toward paramecium. *Chem. Res. Toxicol.*, 25, 1675–1681.

Li, Y., Zhang, W., Niu, J., Chen, Y. 2012b. Mechanism of photogenerated reactive oxygen species and correlation with the antibacterial properties of engineered metal-oxide nanoparticles. *ACS Nano*, 6, 5164–5173.

Li, X., Bi, W., Zhang, L. et al. 2016. Single-atom Pt as co-catalyst for enhanced photocatalytic H_2 evolution. *Adv. Mater.*, 28 (12), 2427–2431.

Lia, K., Ana, X., Parka, K. H., Khraishehb, M. 2014. A critical review of CO_2 photoconversion: Catalysts and reactors. *Catal. Today*, 224, 3–12.

Liu, Y. F., Yu, J. S., 2010. In situ synthesis of highly luminescent glutathione-capped CdTe/ZnS quantum dots with biocompatibility. *J. Colloid Interface Sci.*, 351, 1–9.

Maeda, K., Domen, K. 2010. Photocatalytic water splitting: Recent progress and future challenges. *J. Phys. Chem. Lett.*, 1, 2655–2661.

Mala, J. G. S., Rose, C. et al. 2014. Facile production of ZnS quantum dot nanoparticles by saccharomyces cerevisiae MTCC 2918. *J. Biotechnol.*, 170, 73–78.

Maleki, H., Hüsing, N. 2018. Current status, opportunities and challenges in catalytic and photocatalytic applications of aerogels: Environmental protection aspects. *Appl. Catal. B Environ.*, 221, 530–555.

Marschall, R. 2014. Semiconductor composites: Strategies for enhancing charge carrier separation to improve photocatalytic activity. *Adv. Funct. Mater.*, 24(17), 2421–2440.

Matsunaga, T., Tomoda, R., Nakajima, T., Wake, H. 1985. Photoelectrochemical sterilization of microbial cells by semiconductor powders. *FEMS Microbiol. Lett.*, 29(1–2), 211–214.

Mattoussi, H., Mauro, J. M., Goldman, E. R. et al. 2000. Self-Assembly of CdSe-ZnS quantum dot bioconjugates using an engineered recombinant protein. *J. Am. Chem. Soc.*, 122, 12142–12150.

Medintz, I. L., Uyeda, H. T., Goldman, E. R., Mattoussi, H. 2005. Quantum dot bioconjugates for imaging, labelling and sensing. *Nat. Mater.*, 4, 435–446.

Miao, A-J., Luo, Z., Quigg, A. et al. 2010. Zinc oxide-engineered nanoparticles: Dissolution and toxicity to marine phytoplankton. *Environ. Toxic. Chem.*, 29, 2814–2822.

Miller, R. J., Lenihan, H. S., Keller, A. A. et al. 2010. Impact of metal oxide nanoparticles on marine phytoplankton. *Environ. Sci. Technol.*, 44(19), 7329–7334.

Mitchell, G. P., Mirkin, C. A., Letsinger, R. L. 1999. Programmed assembly of DNA functionalized quantum dots. *J. Am. Chem. Soc.*, 121, 8122–8123.

Muhammad, F., Guo, M., Qi, W. et al. 2011. pH-Triggered controlled drug release from mesoporous silica nanoparticles via intracellular dissolution of ZnO nanolids. *J. Am. Chem. Soc.*, 133, 8778–8781.

Ng, Y. H., Iwase, A., Kudo, A., Amal, R. 2010. Reducing graphene oxide on a visible-light $BiVO_4$ photocatalyst for an enhanced photoelectrochemical water splitting. *J. Phys. Chem. Lett.*, 1, 2607–2612.

Nishimoto, S., Bhushan, B. 2013. Bioinspired self-cleaning surfaces with superhydrophobicity, superoleophobicity, and superhydrophilicity. *RSC Adv.*, 3, 671–690.
Palaniappan, K., Xue, C. H., Arumugam, G., Hackney, S. A., Liu, J. 2006. Water-soluble, cyclodextrin-modified CdSe-CdS core-shell structured quantum dots. *Chem. Mater.*, 18, 1275–1280.
Pan, X., Zhao, Y., Liu, S. et al. 2012. Comparing graphene-TiO_2 nanowire and graphene-TiO_2 nanoparticle composite photocatalysts. *ACS Appl. Mater. Interfaces*, 4(8), 3944–3950.
Park, T. J., Lee, S. Y., Heo, N. S., Seo, T. S. 2010. In vivo synthesis of diverse metal nanoparticles by recombinant *Escherichia coli*. *Angew. Chem., Int. Ed. Engl.*, 49, 7019–7024.
Popov, A. L., Zholobak, N. M., Balko, O. I. et al. 2018. Photo-induced toxicity of tungsten oxide photochromic nanoparticles. *J. Photochem. Photobiol. B*, 178, 395–403.
Qi, D., Lu, L., Wang, L., Zhang, J. 2014. Improved SERS sensitivity on plasmon-free TiO_2 photonic microarray by enhancing light-matter coupling. *J. Am. Chem. Soc.*, 136, 9886–9889.
Rotman, B., Papermaster, B. W. 1966. Membrane properties of living mammalian cells as studied by enzymatic hydrolysis of fluorogenic esters. *Proc. Natl. Acad. Sci. U. S. A.*, 55(1), 134–141.
Sacco, O., Vaiano, V., Daniel, C., Navarra, W., Venditto, V. 2018. Removal of phenol in aqueous media by N-doped TiO_2 based photocatalytic aerogels, *Mater. Sci. Semicon. Proc.*, 80, 104–110.
Salamat, S., Younesi, H., Bahramifar, N. 2017. Synthesis of magnetic core-shell Fe_3O_4@TiO_2 nanoparticles from electric arc furnace dust for photocatalytic degradation of steel mill wastewater. *RSC Adv.*, 7 (31), 19391–19405.
Sayama, K., Mukasa, K., Abe, R., Abe, Y., Arakawa, H. 2002. A new photocatalytic water splitting system under visible light irradiation mimicking a Z-scheme mechanism in photosynthesis. *J. Photochem. Photobiol. A*, 148(1–3), 71–77.
Shannon, M., Bohn, P. W., Elimelech, M. et al. 2008. Science and technology for water purification in the coming decades. *Nature*, 452, 301–310.
Shen, R., Xie, J., Guo, P. et al. 2018. Bridging the g-C_3N_4 nanosheets and robust CuS cocatalysts by metallic acetylene black interface mediators for active and durable photocatalytic H_2 production. *ACS Appl. Energy Mater.*, 1, 2232–2241.
Shi, H., Zhang, S., Zhu, X. et al. 2017. Uniform gold-nanoparticle-decorated {001}-faceted anatase TiO_2 nanosheets for enhanced solar-light photocatalytic reactions. *ACS Appl. Mater. Interfaces*, 9, 36907–36916.
Sturzenbaum, S. R., Hoeckner, M. et al. 2013. Biosynthesis of luminescent quantum dots in an earthworm. *Nat. Nanotechnol.*, 8, 57–60.
Sweeney, R. Y., Mao, C. B., Gao, X. X. et al. 2004. Bacterial biosynthesis of cadmium sulfide nanocrystals. *Chem. Biol.*, 11, 1553–1559.
Tian, L., Ye, L., Liu J., Zan, L. 2012. Solvothermal synthesis of CNTs-WO_3 hybrid nanostructures with high photocatalytic activity under visbile light. *Catal. Commun.*, 17, 99–103.
Tohgha, U., Varga, K., Balaz, M. 2013. Achiral CdSe quantum dots exhibit optical activity in the visible region upon post-synthetic ligand exchange with D- or L-cysteine. *Chem. Commun.*, 49, 1844–1846.
Valko, M., Rhodes, C. J., Moncol, J., Izakovic, M., Mazur, M. 2006. Free radicals, metals and antioxidants in oxidative stress-induced cancer. *Chem. Biol. Interact.*, 160, 1–40.
Varghese, O. M., Paulose, M., Grimes, C. A. 2009. Long vertically aligned titania nanotubes on transparent conducting oxide for highly efficient solar cells, *Nat. Nanotechnol.*, 4, 592–597.
Verma, S. K., Jha, E., Panda, P. K. et al. 2018. Mechanistic insight into size-dependent enhanced cytotoxicity of industrial antibacterial titanium oxide nanoparticles on colon cells because of reactive oxygen species quenching and neutral lipid alteration. *ACS Omega*, 3, 1244–1262.
Vishwakarma, V., Samal, S. S., Manoharan, N. 2010. Safety and risk associated with nanoparticles: A review. *JMMCE*, 9, 455.
Vu, M-H., Sakar, M., Nguyen C-C., Do, T-O. 2018. Chemically bonded Ni cocatalyst onto the S doped g-C_3N_4 nanosheets and their synergistic enhancement in H_2 production under sunlight irradiation. *ACS Sustainable Chem. Eng*, 6, 4194–4420.

Wang, W., Cheng, D., Gong, F., Miao, X., Shuai, X. 2012. Design of multifunctional micelle for tumor-targeted intracellular drug release and fluorescent imaging. *Adv. Mater.*, 24, 115–120.

Wang, Q., Fang, T., Liu, P., Deng, B., Min, X., Li, X. 2012. Direct synthesis of high-quality water-soluble CdTe: Zn^{2+} quantum dots. *Inorg. Chem.*, 51, 9208–9213.

Warheit, D. B., McHugh, T. A., Hartsky, M. A. 1995. Differential pulmonary responses in rats inhaling crystalline, colloidal or amorphous silica dusts. *Scand. J. Work Environ. Health*, 21, 19–21.

Watts, R. J., Kong, S., Orr, M. P., Miller, G. C., Henry, B. E. 1995. Photocatalytic inactivation of coliform bacteria and viruses in secondary wastewater effluent. *Water Res.*, 29, 95–100.

Wei, C., Lin, W. Y., Zainal, Z. et al. 1994. Bactericidal activity of TiO_2 photocatalyst in aqueous media: Toward a solar-assisted water disinfection system. *Environ. Sci. Technol.*, 28(5), 934–938.

Willard, D. M., Carillo, L. L., Jung, J., Van Orden, A. 2001. CdSe-ZnS quantum dots as resonance energy transfer donors in a model protein-protein binding assay. *Nano Lett.*, 1, 469–474.

Woan, K., Pyrgiotakis G., Sigmund, W. 2009, Photocatalytic carbon-nanotube-TiO_2 composites. *Adv. Mater.*, 21, 2233–2239.

Woggon, U. 1997. Optical properties of semiconductor quantum dots. *Tr. Mod. Phys.*, 36. Springer.

Wu, N., Wang, J., Tafen, D. N. et al. 2010. Shape-enhanced photocatalytic activity of single-crystalline anatase TiO_2 (101) nanobelts. *J. Am. Chem. Soc.*, 132 (19), 6679–6685.

Wu, W. T., Liu, H., Dong, C. et al. 2015. Gain high-quality colloidal quantum dots directly from natural minerals. *Langmuir*, 31, 2251–2255.

Wu, H., Wang, H., Li, G. 2017. Metal oxide semiconductor SERS-active substrates by defect engineering. *Analyst*, 142, 326–335.

Xia, P., Liu, M., Cheng, B., Yu, J., Zhang, L. 2018. Dopamine modified g-C_3N_4 and its enhanced visible-light photocatalytic H_2-production activity. *ACS Sustainable Chem. Eng.*, 6, 8945–8953.

Xing, M., Zhou, Y., Dong, C. et al. 2018. Modulation of the reduction potential of TiO_{2-x} by fluorination for efficient and selective CH_4 generation from CO_2 photoreduction, *Nano Lett.*, 18, 3384–3390.

Xiong, L. H., Cui, R., Zhang, Z. L. et al., 2014. Uniform fluorescent nanobioprobes for pathogen detection. *ACS Nano*, 8, 5116–5124.

Yang, A. 2002. In vitro cytotoxicity testing with fluorescence-based assays in cultured human lung and dermal cells. *Cell Biol. Toxicol.*, 18, 97–108.

Yang, H., Xiong, H., Yu, S. 2012. Quantum dots-based drug delivery system. *Prog. Chem.*, 24 (11), 2234–2246.

Ye, L., Fu, J. L., Xu, Z., Yuan, R. S., Li, Z. H. 2014. Facile one-pot solvothermal method to synthesize sheet-on-sheet reduced graphene oxide (RGO)/$ZnIn_2S_4$ nanocomposites with superior photocatalytic performance. *ACS Appl. Mater. Interfaces*, 6, 3483–3490.

Yin, L. N., Gao, J. G., Wang, J. et al. 2011. Enhancement of sonocatalytic performance of TiO_2 by coating Er^{3+}:$YAlO_3$ in azo dye degradation. *Sep. Purif. Technol.*, 81, 94–100.

Yosefi, L., Haghighi, M. 2018. Fabrication of nanostructured flowerlike p-BiOI/p-NiO heterostructure and its efficient photocatalytic performance in water treatment under visible-light irradiation. *Appl. Catal. B Environ.*, 220, 367–378.

Yu, M., Wang, Z., Liu, H. et al. 2013. Laundering durability of photo-catalyzed self-cleaning cotton fabric with TiO_2 nanoparticles covalently immobilized. *ACS Appl. Mater. Interfaces*, 5 (9), 3697–3703.

Yuan, Q., Hein, S., Misra, R. D. K. 2010. New generation of chitosan-encapsulated ZnO quantum dots loaded with drug: Synthesis, characterization and in vitro drug delivery response. *Acta Biomater.*, 6, 2732–2739.

Yuan, Y. J., Chen, D. Q., Huang, Y. W. et al. 2016. MoS_2 nanosheet-modified $CuInS_2$ photocatalyst for visible-light-driven hydrogen production from water, *Chem. Sus. Chem.*, 9 (9), 1003–1009.

Yuan, Y-J., Yang, Y., Li, Z. et al. 2018. Promoting charge separation in g-C_3N_4/Graphene/MoS_2 photocatalysts by two-dimensional nanojunction for enhanced photocatalytic H_2 production. *ACS Appl. Energy Mater.*, 1, 1400–1407.

Zan, L., Fa, W., Peng, T. P., Gong, Z. K. 2007. Photocatalysis effect of nanometer TiO_2 and TiO_2-coated ceramic plate on hepatitis B virus. *J. Photochem. Photobiol. B. Biol.*, 86(2), 165–169.

Zhang, G., Chen, D., Li, N. et al. 2018. SnS_2/SnO_2 heterostructured nanosheet arrays grown on carbon cloth for efficient photocatalytic reduction of Cr (VI). *J. Colloid Interface Sci.*, 514, 306–315.

Zhou, J., Yang, Y., Zhang, C.-Y. 2015. Toward biocompatible semiconductor quantum dots: From biosynthesis and bioconjugation to biomedical application. *Chem. Rev.*, 115, 11669–11717.

Zhu, Z., Cai, H., Sun, D.-W. 2018. Titanium dioxide (TiO_2) photocatalysis technology for nonthermal inactivation of microorganisms in foods. *Trends Food Sci. Technol.*, 75, 23–35.

5

Toxicity Analysis of Ag and Au Nanoparticles

Sami Rtimi

5.1 Introduction to Metal Nanoparticles

Engineered nanomaterials (ENMs) are becoming progressively part of our everyday lives. The novel properties afforded by ENMs lead to their utilization in new products and production processes. Nanoparticles (also called nanocrystals or nano-powders) are defined as materials that are intentionally engineered or manufactured to have one or more dimensions, which are typically less than 100 nanometres (nm) [1].

Nowadays, nanoparticles are potentially used in different biological and medical applications, such as disinfectants [2], contrast agents for medical imaging [3], or cell delivery of therapeutic drugs [4,5]. Some of these applications require that drug-loaded nanoparticles be taken up by the target cells [6]. Very often, inorganic nanoparticles show different physico-chemical properties compared to their bulk behaviour. These properties in combination with the dimensions (size and shape) of the particles allow their use as novel diagnostic tools. In addition to gold NPs, quantum dots and iron oxides are the most interesting nanoparticles that are still under investigation and development [7]. Super-paramagnetic iron oxide nanoparticles are showing a novel combination of properties when compared to other nanoparticles, which is related to their magnetization. This enables them to be directed to a defined location [3,7].

Coated super paramagnetic iron oxide nanoparticles are multifunctional particles with a significant potential for medical application including molecular imaging, protein separation, cancer diagnosis and treatment, and tissue soldering [8]. This potential can only be realized if the particles are the correct size, have coatings, and more importantly have a sufficient colloidal stability in physiological liquids [9,10]. Therefore, a short discussion of different aspects regarding the reproducible manufacturing of complex coated particles is urgently needed. Because *in-vivo* diagnosis plays an important role, some recent results regarding the behaviour of nanoparticles in contact with bio-fluids and cells showed

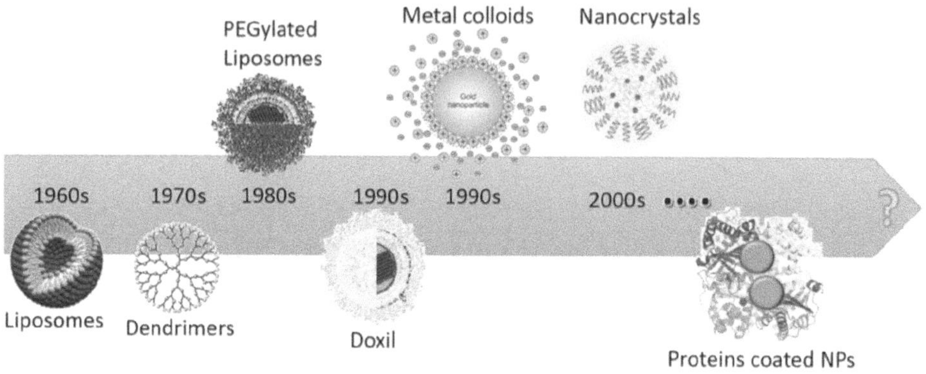

FIGURE 5.1 Historical evolution of the nano-bio-interface: combining bio-entities with ENMs.

distinct (sometimes contradictory) behaviour [11]. Figure 5.1 shows the historical evolution of the use of nano-entities for drug delivery exploring combinatorial nano-bio interfaces. ENMs appeared at the end of the nineties as promising drug delivery and contrast agents.

The primary means for evaluating the safety of ENMs are currently the standard *in vitro* methodologies based on model cells. A vast variety of particles exist and that requires urgent assessment of the used methodologies. However, cell-based models can give an indication of the behaviour of the NMs when in contact with cells but are limited by the fact that they are not representative of tissues/organs/organisms presenting complex physiological environments. The behaviour of NPs and consequent bio-responses are dependent upon the surrounding cells, liquids, proteins, and fatty acids. Therefore, standard *in vitro* models will likely produce inaccurate NP behavioural analyses and erroneous results concerning their safety.

Metallic NPs such as Au, Ag, Pt, and Pd exhibit high mechanical strength, a definite electrical conduction, and photonic/optical properties. These NPs show different dimensions going from zero-dimensional, to 1D (nanorods, nanowires), to 2D (thin layers, nanosheets), up to 3D structures (porous metallic foam, nanospheres). In recent years, biosynthesized metallic NPs (BSM-NPs) for *in vitro* and *in vivo* diagnostics have attracted a lot of attention. BSM-NPs are good candidates for the fabrication of biosensors and/or implants requiring high corrosion resistance. BSM-NPs can be prepared to encapsulate or immobilize enzymes, antibodies, DNA, or cells without folding their inner proteins structure for *in vivo* sensing. These nanostructures can be used as transducer materials for biosensors. Furthermore, they can be employed as bone plates, knee or hip joint replacements, screws, or dental implants. In addition, due to their electrical conductivity and high resistance corrosion, they can be employed as electrode tips for pacemakers.

Owing to their plasmonic property, Au nanocrystals have been shown to be useful in *in vivo* cancer imaging. With respect to this application, Au-nanoclusters stabilized with trypsin were shown to detect heparin or to target Hela-cells after folic acid surface modification as shown in Figure 5.2 [12]. The fluorescence probe used for Au nanocrystals can provide selective affinity toward folate receptor (FR)-positive Hela tumour. An *in vivo* study reported the detection of glutathione disulphide and oxidative stress using covalently immobilized glutathione reductase and nicotinamide adenine dinucleotide phosphate on Au-NPs deposited on poly[2,20:50,200-terthiophene-30-(p-benzoic acid)].

Recently, it was reported that alloy NPs have attracted attention due to their promising properties in comparison with their monometallic counterparts. Biosynthesized Au–Ag alloy NPs were reported as an electrochemical vanillin sensor showing high sensitivity [13].

Biosynthesis of nanostructured Au-NPs assisted by poly(vinylpyrrolidone)-functionalized graphene oxide (GO) was reported as an excellent surface-enhanced Raman scattering probe for ultrasensitive

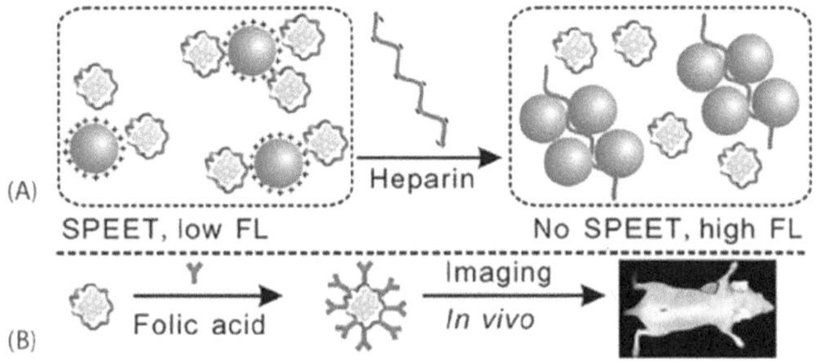

FIGURE 5.2 (A) Schematic illustration for selective detection of heparin based on surface plasmon enhanced energy transfer between cyst-AuNPs and try-AuNCs. (B) Schematic illustration for the application of folic acid modified try-AuNCs for in vivo cancer imaging. (From Malhotra, B.D. and Azahar Ali, M.D. Chapter 7: Nanostructured biomaterials for in vivo biosensors, *Nanomaterials for Biosensors*, 183–219, 2018.)

quantification of cellular components of cancer cells situated in the cytoplasm, nucleoplasm, and nucleolus [14]. This reflects the biocompatibility of these alloys and their insignificant toxicity allowing imaging the cells during their usual metabolism.

The potential use of other types of inorganic and metallic NPs for diagnosis and/or disinfection applications are currently under investigation [15]. Surface plasmon, photoluminescence, and catalytic properties can be combined in the future to design novel functional materials for advanced medical applications. Recently Pillai and his co-authors reported on the chemical preparation of size- and shape-defined Ag-NPs and investigated their impact on skin (HaCat) and lung (A549) cell lines [16]. Findings show that lung cells are more sensitive than skin cells to the toxicity induced by Ag-NPs.

5.2 NPs for Bacterial Inactivation and Infections Prevention

The control of pathogenic microbes has been a major challenge across the world to prevent infections especially nosocomial infections. Pathogenic microbes such as viruses, bacteria, or fungi are the major cause of these infections. These infections are spread through contaminated surfaces, food, water, and air. According to the World Health Organization (WHO), each year infectious diseases kill millions of people and the recent outbreaks have placed significant burden on global economies [17].

There is an urgent need to develop a testing system that will enable the identification and quantification of potential NP side effects for human health. This will allow the bridging between fundamental theories and safer applications of the manufactured NPs for human use. These methods should be precise and reproducible to enable drawing mechanisms of the interactions between the NPs and biological entities/cells.

Due to their high tensile strength, high conductivity (thermal and electrical), low weight, and controllable optical properties, NPs are increasingly used in commercial products. Unfortunately, these properties can unexpectedly interact with the environment or the human body. Figure 5.3 shows the possible interactions between NPs and human cells.

During the last decade, Teeguarden et al. reported on the principles of multidimensional aspects of *in vitro* nanomaterial dosimetry in biological medium [18]. From a general perspective, *in vitro* dosimetry of chemicals presents some limitations that can be summarised in two main dimensions, namely amount and time. Once in solution, a chemical presents some physicochemical properties (hydrophilicity or lipophilicity, solubility, etc.) that allow it to be unaffected by the surrounding solution/entity. These unchanged properties are mainly due to the absence of macroscale physical characteristics of this chemical (e.g., shape and surface chemistry). Contrary to NPs, the absence of macroscale dimensions affects the distribution and delivery of a chemical in solution (e.g., biological liquids). *In vitro* dosimetry for NPs is a problem of multiple dimensions. In addition to the amount and time of exposure, particle

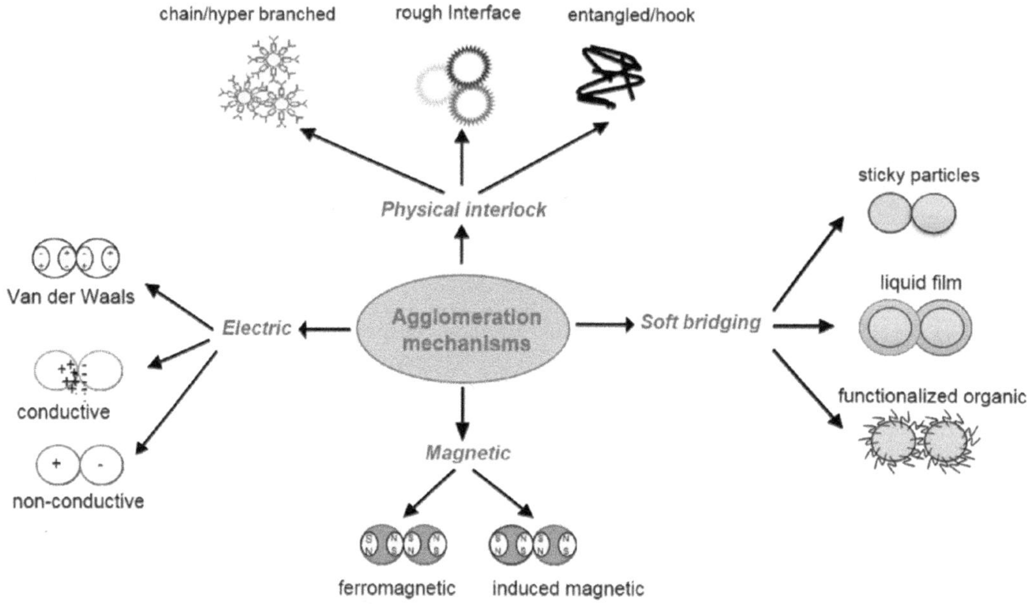

FIGURE 5.3 Agglomeration of NPs in a solution based on their charge, shape, and medium bridging.

physical characteristics are also very important, e.g., surface chemistry, shape, agglomeration, and size among many others. Of these aspects, the physicochemical characteristics draw the most attention [19]. Without a clear understanding of the physical and physicochemical characteristics of the nanomaterial, it is incorrect to draw conclusions regarding the toxicity of nanoparticles. Oberdorster et al. assessed the related issues of material characterization and dosimetry *in vitro* [20], presenting a characterization prioritization scheme focusing on the most important material properties.

While physicochemical characterization is critically important, it should not be equated to dosimetry. The other aspects of dosimetry *in vitro*, time and amount, have not been the subject of as thorough an analysis [18]. Larger, denser particles are delivered to cells *in vitro* more rapidly and more completely over shorter durations than smaller, less dense particles. These differences in transport would be expected to affect the magnitude and timing of cellular responses. Thus, there is a need to extend the paradigm for *in vitro* particle and nanomaterial dosimetry to these other elements and develop appropriate dose metrics for nanomaterials that are consistent with their unique characteristics and behaviours [18]. Figure 5.4 shows NP transport and distribution as a function of their size/density.

Figure 5.5 illustrates the influence of the transport-rate of NPs on the dose reaching the cells *in vitro* as a function of time overlaid with the time scale for common measures of biological response.

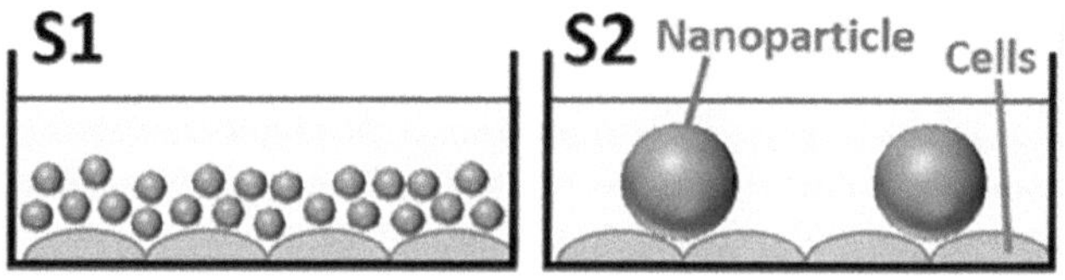

FIGURE 5.4 A scheme of the same mass of nanoparticles (with different diameters, d, d1 < d2) distributed on the surface of cells that are at the bottom of a well. (From Mionić Ebersold, M. et al., *Analyst*, 143, 837–842, 2018.)

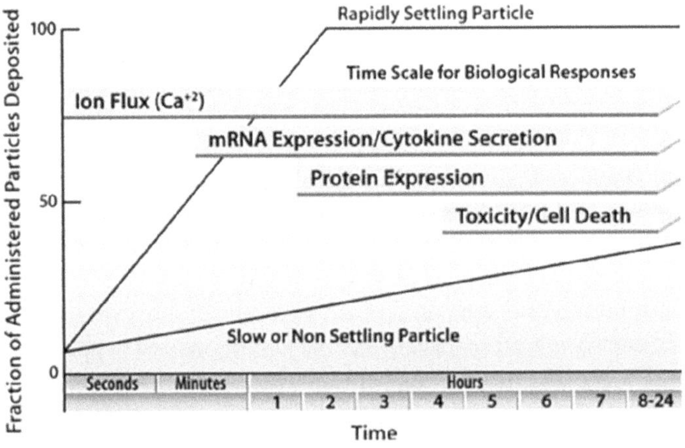

FIGURE 5.5 Influence of particle transport rate on cellular dose *in vitro* as a function of time for common measures of biological response. (From Teeguarden, J.G. et al., *Toxicol. Sci.*, 95, 300–312, 2007.)

Teeguarden et al. reported that both the rate and the time of transport of NPs vary with settling and diffusion rate and can differ significantly for test materials within the period of typical *in vitro* experiments. Au-NPs with size average of 100 nm were reported to settle in less than 90 minutes after traversing a media of 2 mm deep [18].

Although there are many studies on the bio-interaction of NPs with the surrounding cells in the literature, there are few testing methods enabling result comparison. Many efforts have started recently attempting to design a testing strategy that can act as standard tools allowing industries, regulatory bodies/agencies, and academia to characterize engineered nanoparticles *in vitro* in a robust, reliable, and comparable way.

Many people think that these nanoparticles are always toxic. This is why it is very important to develop methods to get a prediction of the behaviour and the toxicity of the NPs surrounding us in everyday life. It will then be much easier to bring these products and nanoparticles with new properties to the market without uncontrollable side effects. This approach should involve correlating the physico-chemical properties of the NPs to their biological effects. Current research in the study of the properties, synthesis, applications, and toxicity of model NPs, namely Au- and Ag-NPs, includes chemical syntheses and biomedical diagnosis.

Zhou et al. [22] developed a triangular Ag-NPs biosensor for the detection of p53 in head and neck squamous cell carcinoma. Dual-imaging/therapy-immuno-targeted nano-shells of Ag-NPs detecting cancer cells were also reported. These nanostructures absorb light and destroy the cancer cells utilising a photo-thermal mechanism. This latter mechanism is produced by the excitation of the Ag-NMs by the photons leading to the production of localized heat resulting in death of the cells.

Thin films of Ag, Au, TiO_2-Ag, and TiO_2-Au prepared on different substrates (glass, plates, textiles, and polymers) is an emerging area of interest with respect to antibacterial surfaces [1,2]. Commercial products based on the traditional sol-gel method are known for some decades. Ag, Au, TiO_2, ZnO, ZrO_2, and many other thin films were prepared on heat-resistant substrates using the sol-gel method [3–5]. Nevertheless, the thickness of these coatings is not reproducible, the coatings were not mechanically stable, and they have exhibited low adhesion to the substrate. Colloid depositions on substrates require a temperature of a few hundred degrees for an adequate adherence to the selected substrate. Many research groups reported the preparation of Ag thin films by chemical vapour deposition or physical vapour deposition. These last two methods lead to controllable thin film growth, high reproducibility, and mechanical stability [23,24]. The renewed interest in silver-based antibacterial solutions is due to the decline of antibiotic efficacy together with the emergence of stable coated surfaces killing bacteria by contact. Contact killing was defined by Espirito Santo et al. (2011) as damages on the cell-wall membrane occurring after the contact of the microbe with a surface, plus an increase of the intracellular ion content [25]. These films/coatings have been shown to kill bacteria in the minute range without loss of the content (TiO_2, Ag, Au) [26]. Ballo et al. reported sputtered Ag and Cu on flexible substrates showing quasi-instantaneous pathogens inactivation [27]. The prepared thin films were reported to kill *Staphylococcus aureus*. The latter microorganism represents the second pathogen causing hospital-acquired infections (HAIs) as reported by the Centers for Disease Control and Prevention, 2011–2014 [28].

The Ag-Cu coatings prepared by Ballo et al. were tested in contact with fresh human skin and they exhibit high biocompatibility. Histopathological analysis of human skin exposed to Ag and Cu-NPs (for 24 hours) revealed no morphological changes compared to skin exposed to phosphate buffer solution as shown in Figure 5.6 [27]. Thus, *in vitro*, the Ag (and Cu) nanoparticles did not cause harmful effects on the fresh human skin.

Antibiotic-impregnated catheters have been reported for some time to reduce device acquired infections. Vancomycin, cefazolin, and minocycline/rifampin-impregnated catheters were prepared and commercialized. These antibacterial catheters showed moderate effectiveness against some microorganisms but two limitations were raised: (i) the fast leaching of the active compounds rendering the surface inactive and (ii) the development of bacterial resistance to the used antibiotics with respect to their therapeutic spectra.

Tuberculosis is an infectious disease caused by *Mycobacterium tuberculosis*. This disease has attracted attention of the WHO as the leading cause of mortality worldwide from infections. During the development of anti-tuberculosis drugs, Bacillus Calmette-Guérin (BCG) is generally used as a surrogate.

Gold and silver NPs were investigated on BCG and *Escherichia coli*. Zhou et al. reported the effect of Au- and Ag-NPs as potential anti-TB compounds [29]. They studied the effect of particle size and shape using transmission electron microscopy. Experimentally, they added Au- or Ag-NPs to Luria-Bertani medium for four hours to reach predesigned concentrations. *E. coli* was exposed to Au-citrate (0.1, 1, 5, or 10 μg/mL), Au-NPs complexed with poly-allylamine hydrochloride, and Ag NPs (1 or 10 μg/mL). The authors showed that bacteria uptake single NPs (with a broad size range) leading to their cytoplasm rearrangement. They proposed a mechanism of uptake for the different NPs that were prepared as shown in Figure 5.7.

Different antibacterial mechanisms were reported for Au-PAH and Au-citrate NPs. Au-PAH composite did not further aggregate in culture media (See Figure 5.7B). PAH facilitated the delivery of a large number of Au NPs that strongly bond to PAH on the bacterial cell surface [29]. Positively charged PAH tend to interact with negatively charged bacterial membranes leading to Au-NPs accumulation on the membrane interface. These Au-NPs will then penetrate the bacterial cells reaching the cytoplasm. To differentiate between the bacterial inactivation caused by the Ag or Au-NPs or the PAH, it

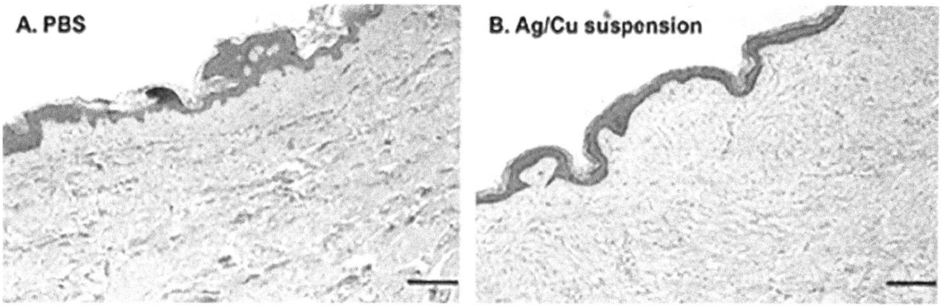

FIGURE 5.6 Cross-section of human skin exposed to PBS (A) or to Ag and Cu nanoparticles (B). No irritation signs were observed. Black bars, 200 μm. (From Ballo, M.K.S. et al., *Antimicrob. Agents Chemother.*, 60, 5349–5356, 2016.)

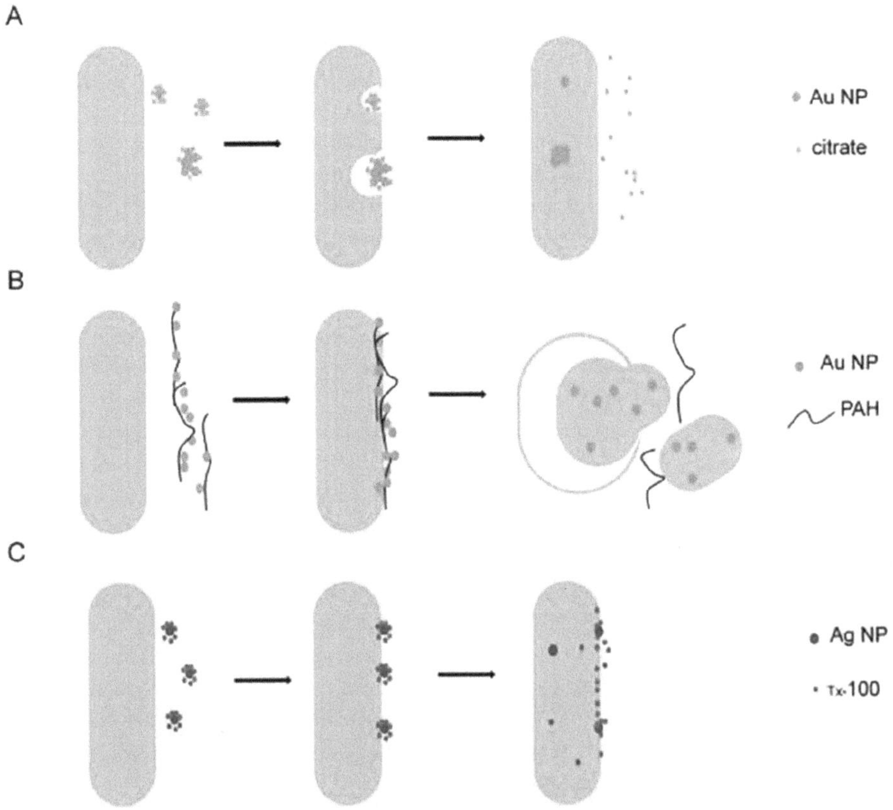

FIGURE 5.7 The mechanism of interaction between NPs and bacteria. (A) uptake of citrate Au-NPs. (B) uptake of Au NPs facilitated by PAH leading to bacterial cells lysis. (C) Most of Ag-NPs stayed trapped in cell walls. (From Zhou, Y. et al., *J. Nanobiotechnology*, 10, 19, 2012.)

is worth mentioning that the minimum inhibitory concentration of PAH is well above the NP dose. Goodmann et al. investigated the differential toxicity of Au-NPs against different cell types using 2 nm core particles. They showed that anionic Au-particles are nontoxic, while cationic particles are moderately toxic [30].

NPs-based antibacterial tests against BCG showed the reduction of the fluorescence signal reflecting the reduction of viable BCG numbers. This assay provides a fast way to monitor the growth of slow-growing mycobacteria. This opens the doors for a new manufacturing approach of antibacterial drugs [29].

5.3 NPs for Medical and Environmental Applications

By exploiting the unique properties of NPs, modern nano-medicine addresses some of the most important and current challenges in diagnosis and treatment. This field comprises the design of NPs capable of targeting certain cells to deliver drugs, genetic material, or nanofibers used for tissue regeneration/engineering or nano-sensors. Noteworthy progress has been fulfilled in the field of engineered NPs during the last years by designing NPs with a variety of shapes and sizes. This allows researchers to take advantage of the physicochemical properties of NPs for a targeted bio-response or environmental use. However, once in contact with living cells or microorganisms, the NP surface is unavoidably confronted with biological fluids (i.e., small biomolecules, such as proteins, lipids, antibodies, ions, vitamins). This interaction has critical consequences on the activity of NPs. The so-called hard corona cover is formed by the tightly bound immobile protein layers on NPs. On the top of this hard corona, a weakly associated mobile layer can be formed and can possibly recruit more proteins forming the soft corona.

For many decades, nano-silver has been used as an antibacterial material for water/wastewater treatment. Ag-nanoparticles cannot be compared to conventional chemicals or bulk materials when elucidating environmental applications (i.e., bacterial inactivation or organic pollutant degradation). Products containing silver NPs have been commercially available for decades and have been used in different applications [31]. The exact toxicity mechanism of NPs against the large variety of bacterial strains in an environmental sample is not completely revealed. Some NPs have the ability to attach to the waterborne bacterial cell-wall membrane leading to the disruption of their membrane. In general, nano-toxicity of Ag, Cu, and Au particles is triggered by an oxidative stress through the formation of free radicals referred to as reactive oxygen species (ROS).

Ag-healing pads releasing Ag-ions have been also studied. Due to the increased resistance of bacteria to antibiotics in recent years, there has been a renewed interest in silver coatings for biomedical applications [32]. The bacterial inactivation by Ag has been reported to proceed by the release of silver-ions. These ions are able to penetrate the bacterial cytoplasm through the cell-wall [32]. Recently, Rtimi and co-authors reported on the different effects of the bacterial inactivation mediating Ag-coatings. They attributed the bacterial inactivation to two main mechanisms: (i) the release of Ag-ions inactivating the bacteria by interfering with the cell machinery after penetrating inside the cell through the porins and (ii) the Ag-NPs in contact with the bacterial outer-wall inducing the contact-killing [33]. Ag-NPs/ions have been reported to inactivate bacteria and other pathogens through the oligodynamic effect involving the diffusion of 10^{-6}–10^{-9} mol/L (ppb-ppm) amounts in contact with the outer cell envelope [34]. Furthermore, Ag-based NPs or coatings can disinfect pathogens through the destruction of cell wall by the photo-generated ROS when Ag-oxides are formed [34,35].

The Ag toxicity towards bacteria, viruses, and fungi is significantly higher compared to the toxicity reported for mammalian cells [36]. Silver ions interfere with the electron transport chain of the cell wall porins diffusing to the interior of the bacteria. This is facilitated because of the high affinity of Ag to sulfhydryl groups located at the cell wall. Ag-ions were also reported to block the respiratory chain of microorganisms. Bacterial resistance towards silver ions has been reported [37]. The concentrations required for bactericidal activity are in the ppb range (10^{-9} mol/L). Metallic silver particles react with moisture, releasing highly oxidative Ag-ions. These Ag-ions complex with the microorganism genome limiting their replication in the environment. Recent reviews report that the size, shape, and concentration of Ag-NPs control their antimicrobial kinetics and efficiency for indoor and outdoor environmental remediation.

Treatment of thermal burns by Ag-NPs to preclude infections during the healing process has been known for over 100 years [38]. Due to their small size, the NPs and their associated ions can easily penetrate a biofilm and abolish it. Current research focuses on the hindrance of the biofilm formation based on the fast microorganism inactivation kinetics of Cu or Ag-NPs. In addition to their small dimensions (normally less than 100 nm) and in contrast to antibiotics, metal-NPs present a larger surface area-to-mass ratio [33,34,38]. Toxic biofilms spread highly toxic pathogens in indoor environments like hospital facilities and public places. Biofilms provide protection to the forming bacterial strains making them more resistant to antibiotics that are unable to penetrate the biofilm. The extracellular polymeric network structure strengthens the adhesion (and the communication) between bacteria and the biofilm and the substrate [39]. This allows the film matrix to remain stable for a long-time on biotic or abiotic substrates [39]. The hydrophilic-hydrophobic balance between the surface of polymer films and the bacteria envelope has a great influence on the release of Ag-NPs-ions deposited on a polymer network [38].

Today, further studies are required to better understand the basic principles on the action of Ag-NPs. Only a few environmental application mechanisms utilising light have been reported until now and

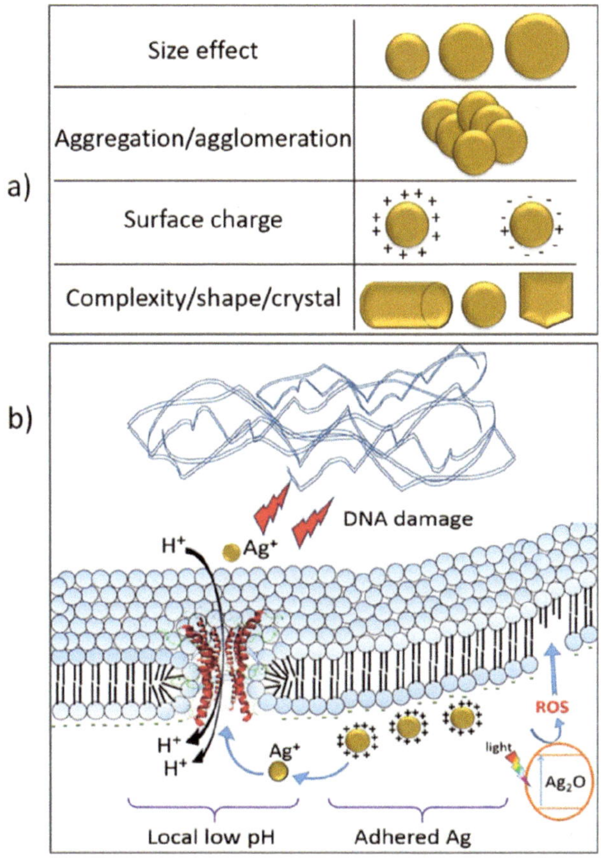

FIGURE 5.8 (a) Factors affecting the bacterial inactivation by Ag-NPs such as the NPs-size, agglomerations, charge, and shape, (b) scheme of bacterial inactivation by Ag-films involving ion penetration, surface contact killing, and oxidative killing by ROS. (From Rtimi, S. et al., *Appl. Catal. B Environ.*, 240, 291–318, 2019.)

more details are needed on this subject. Figures 5.8 and 5.9 show the effect of shape, size, and surface charge of Ag-NPs on bacterial inactivation.

Park et al. [40] developed a magnetic hybrid colloid (MHC) decorated with Ag-NPs for an antiviral application against bacteriophage ϕX174, murine norovirus (MNV), and adenovirus serotype 2 (AdV2). The viruses were exposed to the composite (containing Ag-NPs) for 1, 3, and 6 hours. The toxicity of the composite against the targeted viruses was analysed by plaque assay and real-time TaqMan PCR. The authors showed that Ag-NP of 30 nm (fixed on the MHC) displayed the highest efficacy for virus inactivation. The ϕX174 and MNV were reduced by more than 2 log10 after exposure to 4.6×10^9 Ag30-MHCs/mL for 1 hour. Shimabuku et al. [41] prepared Ag-NPs functionalized with granular activated carbon (GAC) for T4 bacteriophage inactivation. Ag-GAC showed a 3-log reduction of the targeted virus with few ppb Ag-ions released, far below the recommended limits by regulatory bodies.

Coupling Ag-NPs with antibiotics (such as amoxicillin, clindamycin, erythromycin, penicillin, and vancomycin) was shown to increase the antibacterial activities. However, prolonged exposure of bacteria to Ag-NPs may lead to the development of bacterial resistance. For example, *E. coli* K12 MG1655 strain has established resistance toward Ag-NPs, although the bacterium does not express resistance genes towards Ag.

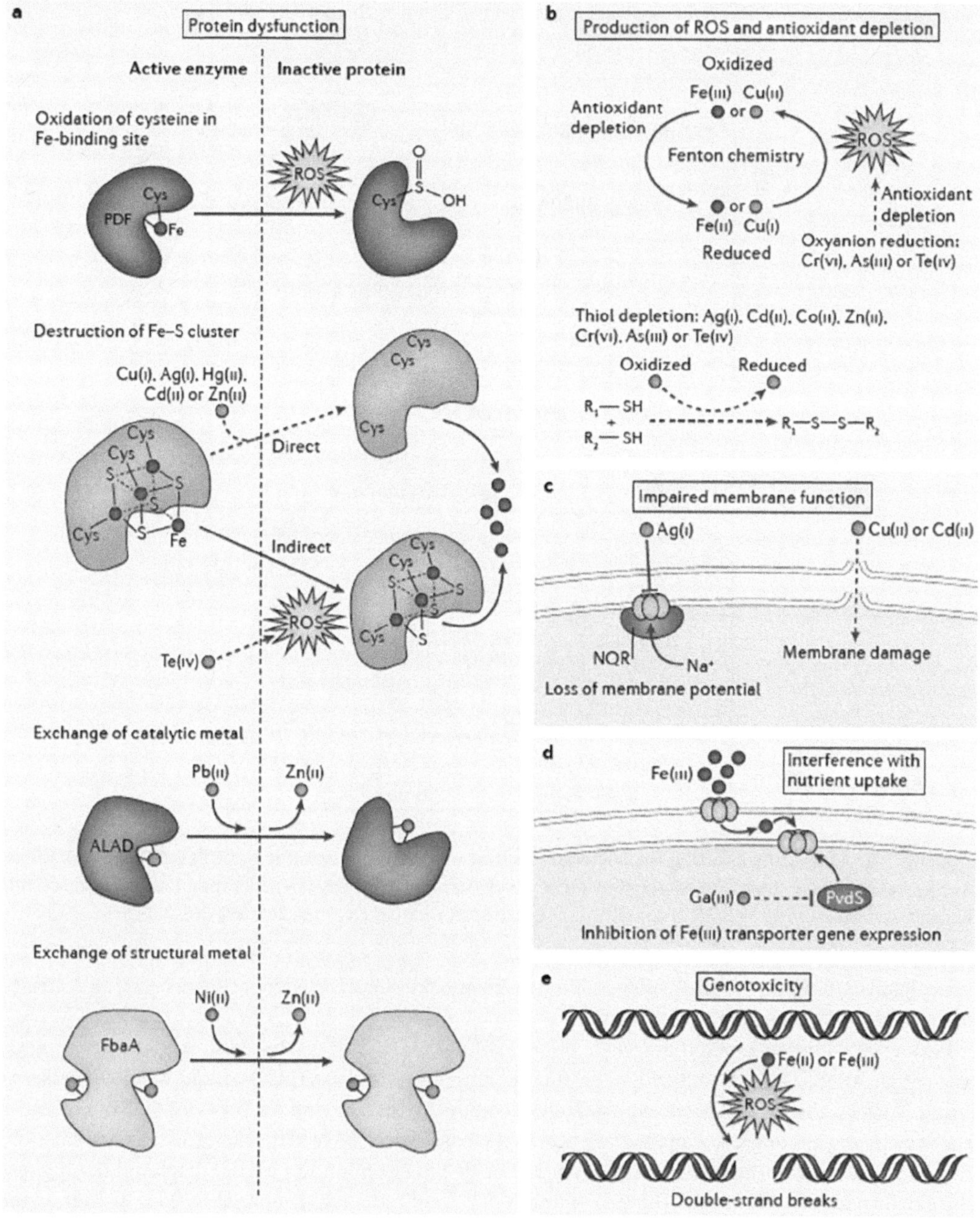

FIGURE 5.9 Antibacterial mechanisms of metal toxicity: (a) Metals leading to protein dysfunction, (b) production of ROS, (c) impairing the membrane function, (d) interference with nutrient assimilation, (e) genotoxicity. Solid arrows represent elucidated pathways and dashed arrows represent routes where the biochemistry is still unclear. ALAD, δ–aminolevulinic acid dehydratase; FbaA, fructose-1,6–bisphosphate aldolase; NQR, NADH: quinone oxidoreductase; PDF, peptide deformylase; PvdS, a σ-factor (σ24) from *Pseudomonas aeruginosa*. (From Lemire, J. A. et al., *Nat. Rev. Microbiol.*, 11, 371–384, 2013.)

5.4 Nanoparticles in Biological Media: Dose/Effect and Corona Formation

There is a famous adage indicating the basic principle of toxicology as concluded by Paracelsus (a Swiss alchemist pioneer of medical revolution) '*sola dosis facit venenum*' or 'The dose makes the poison'. This means that a substance can produce a harmful effect related to its toxic properties only if it reaches a susceptible biological system in a high enough dose: a small amount has no significant effect, and a large amount is fatal. The dose–response relationship describes the effects caused by differing doses of a stressor (Au or Ag-NPs) after a certain exposure time on an organism.

Once in contact with liquids (either biological/physiological or not), NPs can easily agglomerate. This agglomeration can be due to differing mechanisms. Recently, Mionić Ebersold et al. [21] investigated the competitive proteins formation on the NP surface and linked them to the NP size/surface area.

The NP flow rate in a solution (physiological liquids or incubation media) is a parameter that has often been neglected when studying the transport processes of NPs. NP transport experiments have been performed under static conditions in a Petri dish. Upright and inverted cell cultures have been studied by Cho et al., reporting that cellular uptake of Au-NPs depends on the sedimentation and diffusion velocities of the NPs [43]. Other researchers explored the gravitational settling of NPs [44] and the time linearity of the intracellular concentration of NPs. They showed also that the uptake of NPs is essentially irreversible [44]. These studies were performed under static conditions and the NPs were tested in a Petri dish following the NP settlement onto a cell culture. Quartz crystal microbalance with dissipation monitoring (QCM-D) was used in a flow chamber to follow the particles deposition/sedimentation under flow conditions [45]. Although the flow in a QCM-D chamber does not precisely depict how particles/NPs will move over a real human system, it brings researchers one step further in mimicking a real *in vivo* experiment in which the NPs are exposed to flow rates. To get a glimpse into understanding how the flow rate affects the NPs' deposition rate, researchers studied the NPs' behaviour under no-flow conditions and under different flow rates (from 50 to 250 μL/min).

The obtained cell viabilities were different with the same concentration and were lower for smaller NPs compared to larger ones with the same concentration. This is typically interpreted as: the smaller the NPs are, the higher their cytotoxicity is. In this interpretation, the different surface areas, cross-section areas, number, and colloidal behaviours (thus different diffusion and sedimentation behaviour, and thus different mass reaching the cells) were much neglected. This can profoundly influence our understanding of the possible nano-bio-interactions. For *in vitro* testing, the authors introduced also the number of NP layers deposited on the tested cells. The expression of the cell viability was recently shown to lead to different interpretations of the observed decrease of the cell's viability correlated with the neglected properties.

It was also reported that cell viability could be influenced by multiple factors, such as the size of the NPs and the covering of cells with multiple layers of NPs. This latter factor limits the delivery of nutrients or drugs to the targeted cells. The size distribution and the stability of NPs are directly related to their preparation protocol. The stability of colloidal mixtures containing particles with a large disparity in size has been experimentally studied based on the measurement of turbidity. The main goal was to permit the investigations of comparable particle concentrations of two or more colloidal populations. Binary mixtures containing a poly(vinyl acetate) (PVAc) latex and a Ludox AS-40 silica sol were investigated. The silica particles were much smaller than the latex ones. The experimental stability factors were compared with the theoretical values computed on the basis of the Kihira–Ryde–Matijevic model [46] for interaction between spherical particles with unevenly distributed surface charges. The experimental results showed that even when both sols are negatively charged the small silica particles are adsorbed onto the latex surface. The hetero-aggregates, which are composed of PVAc cores surrounded with silica particles, can be modelled as PVAc particles having 'modified' surface characteristics (i.e., average Stern potential and varying extents of the surface charge segregation). These aspects influence to a large extent the possible interactions of these NPs with cells and thus induce toxicity as shown previously in Figures 5.8 and 5.9.

5.5 Cyto-Toxicity and Biocompatibility Analysis of Metal Nanoparticles

In vitro screening assays are important for rapid, cost-effective, and high-throughput toxicological screening and characterization of NPs. These assays complement and/or supplement the time-consuming and costly *in vivo* tests. Testing NPs using human cell cultures has the potential to exclude the need for interspecies extrapolations and interpretations. This increases the testing efficiencies and the conclusive results and reduces the use of animal models.

Metallic nanoparticles such as Au, Ag, or TiO_2 are produced at large scales and are included in many commercial products. Most metallic NPs were reported to induce some toxicity in humans. This raises concerns about these manufactured nanomaterials/NPs. Many studies have been investigating this subject; nevertheless, almost all these studies have been conducted *in vitro* on cancer or transformed cell lines. A comparative evaluation of some metallic NPs on untransformed (normal) fibroblasts (GM07492) detected cyto- and geno-toxic responses after exposure to NPs. Pillai et al. [16] studied the cyto-toxicity of Ag-NPs on lung cells (A549). Microscopic observation of untreated and treated A549 cells incubated with Ag-NPs for 24 hours (Figure 5.10) shows extensive damage of the A549 monolayer cells revealing a high cyto-toxicity of these NPs actually used in many products of daily use. These NPs can be easily inhaled and thus cause the damages as observed *in vitro*.

When a NP reaches the blood stream, the liver will be the primary organ involved in the metabolism and detoxification of this xenobiotic. The liver filters the blood carrying the toxicant before being circulated to other parts of the body. The high rate of blood reaching the liver leads to the delivery of high concentrations of NPs/xenobiotics to this organ. NPs can be inhaled, absorbed through the skin, ingested, or administered by intravenous injections and medical devices. Once they reach the blood circulation, they are rapidly translocated to the liver. Some studies have suggested that NPs are entrapped by the reticuloendothelial system, suggesting liver and spleen as the main target organs. Therefore, nanomaterials might be potential hepatotoxicants, and, therefore, hepatotoxicity testing is an important testing strategy for safety assessment of nanomaterials [47]. Au-NP induced hepatotoxicity was investigated by Hwang et al. on rodents with healthy or damaged livers [47]. For 4 weeks, mice were fed a methionine- and choline-deficient (MCD) diet to induce a liver injury. The biodistribution of PEGylated Au-NPs of 15 nm were studied by transmission electron microscopy. Using an automatic

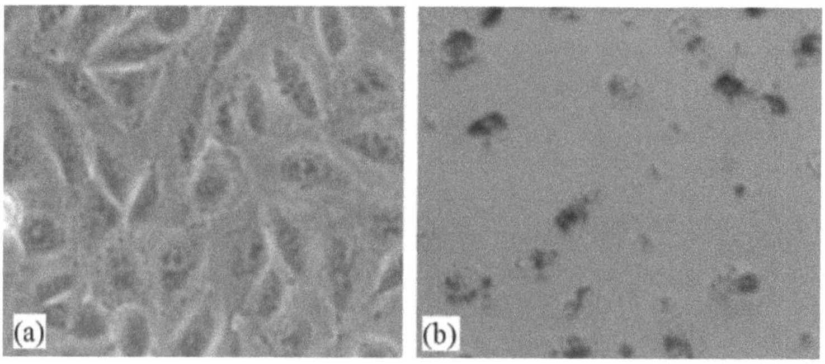

FIGURE 5.10 Images of (a) untreated control lung cells and (b) lung cells incubated 24 hours with 20 ppm Ag-NPs. (From Garvey, M. et al., *J. Toxicol. Pharmacol.*, 1, 16, 2017.)

chemical analyzer, alanine aminotransferase (ALT) and aspartate aminotransferase (AST) were analyzed. Levels of ALT and AST in the serum are common markers for hepatic toxicity. Sheth et al. showed that the levels of ALT and AST proteins increase rapidly in sera when the liver is damaged by any cause, including hepatitis or hepatic cirrhosis [48]. Mice on a MCD diet expressed severe inflammation, hepatic cell damage, and high levels of ROS production leading to cell/tissue apoptosis, which was observed in the livers after Au-NP-injection. Mice fed with normal diet did not show liver injuries. These studies showed that Au-NPs exhibit high toxicity accelerating stress-induced apoptosis. These conclusions point out the importance of considering health conditions, including liver damage, in medical applications of Au-NPs.

Orally administered Au-NPs in mice were captured by the gastrointestinal tract, translocated by the blood, and reached sensitive organs such as the liver, kidney, lungs, heart, spleen, and brain [49]. Many studies reporting on the uptake of NPs showed that it occurs in the Peyer's patch regions of the small intestine, with little translocation occurring through non-lymphoid gut tissue [49–51]. Nicklin et al. reported on the uptake of micrometer-sized Au-NPs by persorption through breaks in the tips of villi [52]. Gastrointestinal uptake of Au-NPs by persorption through gaps created by extruding enterocytes was reported by Hillyer et al. [49]. Moreover, the uptake of different-sized NPs showed that this mechanism is also particle-size dependent: smaller particles are persorbed more readily.

Studies in mice suggest that most NMs/xenobiotics accumulate in the liver following oral, inhalation, and intravenous exposures as reported by [53]. Recent study by van der Zande et al. orally exposed rats to Ag-NPs 15–20 nm for 28 days [54]. The same authors observed the presence of Ag-NPs in all the examined organs with high levels in the liver and spleen. These high Ag-NPs concentrations were mainly attributed to the amount of the released Ag-ions (from the NPs suspension) able to pass the intestines of the exposed rats.

Besides the gastrointestinal tract and the skin, the main gateway of NP entry is the respiration mediated by the 150 m^2 alveolar surface area. Inhaled airborne NPs reach the alveoli, the peripheral parts of the lungs. Once in this location, NPs are only a few hundreds of nanometres away from the blood stream by jumping the air-blood tissue barrier to reach all organs and tissues. It still remains a big question where, why, and how these nanoparticles leave the blood stream and enter the tissue of organs [9,55,56].

Few studies conducted as part of InLiveTox and ENPRA European projects tried to compare *in vitro* and *in vivo* liver models, i.e., hepatocyte cell line, primary human hepatocytes, and liver tissues. The studies suggest that, even though they present some limitations, *in vitro* tests can be very valuable in predicting the possible liver response *in vivo*. Semmler-Behnke et al. exposed female Wistar rats to 1.4 and 18 nm Au-NPs [57]. The rats translocated Au-NPs from the lungs to secondary organs (up to 8% of total dose). One of these secondary organs was the liver.

Pulmonary aerosol can reach the alveoli mainly through buccal or nasal cavities. During their transit, these inhaled materials can deposit in different parts of the respiratory system based on their size, polarity, and surface chemistry. Figure 5.11 shows the possible deposition zones of inhaled particles based on their sizes. Inhaled nanoparticle deposition in early stages of the respiratory tract based on their surface chemistry has not been investigated in detail yet.

NPs have the ability to passively enter cells by non-endocytic mechanisms. They end up in the intracellular medium without bounding to the cell membrane. Moreover, NPs have been microscopically imaged in intracellular compartments that were not expected, such as in the nucleus or the mitochondria. NP interactions with the cytoskeleton can drastically affect the cell motility and/or the organization of the organelles or the cell signalling (by interacting with the surrounding environment). Cells containing NPs try to get rid of it by destroying or excreting it. Mostly, the cell produces ROS to do so. This, by turn, leads to an oxidative stress ultimately followed by an inflammatory response. In some cases, before the ROS production, cells may produce pro-inflammatory products that may persist for long periods. An aggregation of NPs presents higher surface per unit volume compared to micron-sized particles. This may influence the amount of ROS generated by the cells and their inflammation.

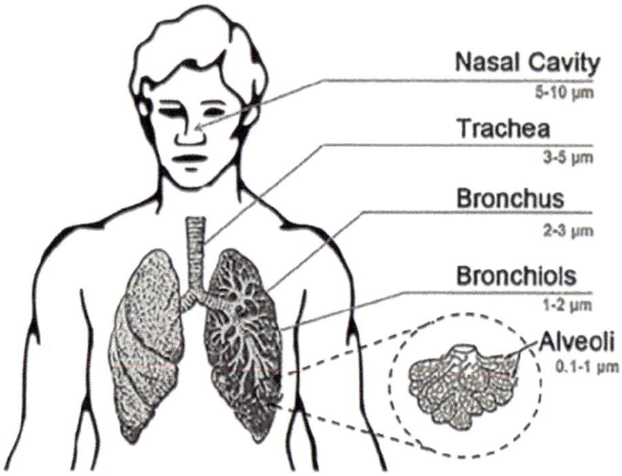

FIGURE 5.11 Regional deposition of NPs within the lungs, depending on the size of the particles and the overall inhalation process. (From Puisney, C. H. et al., *Cellular and Molecular Toxicology of Nanoparticles*, 1048, 21–36, 2018.)

5.6 Outlook

Today, there is an urgent need to develop better and more accurate tools to evaluate the safety of engineered NMs. The actual understanding of the effects of NMs exposure is very limited. The information collected from current literature indicates that nanomaterials found in human environment may have potential toxicological effects. Figure 5.12 shows the source of the anthropogenic nanoparticles surrounding us in everyday life. However, the current data on toxicological effects of NMs is non-harmonized and used differently *in vivo* and *in vitro* test models, different NPs, different characterization methods, and different experimental conditions.

The test methods need to be standardized. Therefore, in the absence of standardized validated methods any specific regulatory testing requirements for NMs are currently premature.

For the future, many questions are still open and urgently need to be addressed. These questions were recently reported by P. Gehr [11]: (1) What happens with time after the exposure of cells with nanoparticles, cells as the basis of biological systems? (2) Internal blood-tissue barriers: How do nanoparticles 'know' where, i.e., in which organ, they have to leave the blood circulation? (3) How can they do this? (4) By what mechanism is a possible effect caused?

FIGURE 5.12 Anthropogenic nanoparticles (incidental and engineered NPs) surround us in everyday life. (From http://sustainable-nano.com/2013/03/25/nanoparticles-are-all-around-us/.)

5.7 Conclusions

ENMs are becoming progressively part of our everyday lives. Au and Ag-NPs gained a lot of attention during the last number of decades due to their physico-chemical properties and the induced bioresponses when administered/inoculated to cells/tissues/organs/organisms. These NPs showed many promising results *in vitro*, but their *in vivo* counterparts were not very encouraging. This was attributed to the NP surface charge, pH, dissolution/oxidation, and interaction with the surrounding physiological liquid rich in proteins. These proteins can aggregate at the surface of the uptaken NPs forming the corona. Many research groups studied these coronas in order to understand the mechanism and the order of their formation, but results are very divergent and non-conclusive. Others studied the tendency of these NPs to aggregate and investigate the forces driving their sedimentation and stickiness to the cells *in vitro*. These *in vitro* studies illustrating NP dynamics cannot reflect the real behaviour in *in vivo* systems driven by physiological differences and/or stabilities. Today, an urgent and more accurate testing approach taking into account the physiological parameters of the targeted organism is needed.

References

1. A. Groso, A. Petri-Fink, B. Rothen-Rutishauser, H. Hofmann and T.H. Meyer, Engineered nanomaterials: Toward effective safety management in research laboratories, *J Nanobiotechnol.* (2016) 14:21.
2. S. Rtimi and J. Kiwi, Update on sputtered films with Cu in ppm concentrations showing drastic acceleration in bacterial inactivation, *Catal Today.* (2018) doi:10.1016/j.cattod.2018.06.016.
3. K. Mahmoudi, A. Bouras, D. Bozec, R. Ivkov, C. Hadjipanayis, Magnetic hyperthermia therapy for the treatment of glioblastoma: A review of the therapy's history, efficacy and application in humans, *Int J Hyperth.* 34 (2018) 1316–1328.
4. L.C. Kennedy, L.R. Bickford, N.A. Lewinski, A.J. Coughlin, Y. Hu, E.S. Day, J.L. West, R.A. Drezek, A new era for cancer treatment: Gold-nanoparticle-mediated thermal therapies, *Small* 7 (2011) 169–183.
5. P. Pengo, M. Şologan, L. Pasquato, F. Guida, S. Pacor, A. Tossi, F. Stellacci, D. Marson, S. Boccardo, S. Pricl, P. Posocco, Gold nanoparticles with patterned surface monolayers for nanomedicine: Current perspectives, *Eur Biophys J.* 46 (2017) 749–771.
6. N. Bohmer, A. Rippl, S. May, A. Walter, M. B. Heo, M. Kwak, M. Roesslein, N. W. Song, P. Wick, C. Hirsch, Interference of engineered nanomaterials in flow cytometry: A case study, *Colloids Surf B Biointerfaces.* 172 (2018) 635–645.
7. D. Bonvin, J.A.M. Bastiaansen, M. Stuber, H. Hofmann, M.M. Ebersold, Chelating agents as coating molecules for iron oxide nanoparticles, *RSC Adv.* 7 (2017) 55598–55609.
8. S. Sruthi, L. Maurizi, T. Nury, F. Sallem, J. Boudon, J.M. Riedinger, N. Millot, F. Bouyer, G. Lizard, Cellular interactions of functionalized superparamagnetic iron oxide nanoparticles on oligodendrocytes without detrimental side effects: Cell death induction, oxidative stress and inflammation, *Colloids Surf B Biointerfaces.* 170 (2018) 454–462.
9. C.H. Puisney, A. Baeza-Squiban, S. Boland, Mechanisms of uptake and translocation of nanomaterials in the lung, in Q. Saquib, M. Faisal, A. A. Al-Khedhairy, A. A. Alatar (Eds.), *Cellular and Molecular Toxicology of Nanoparticles* 1048 (2018) pp. 21–36, Cham, Switzerland, Springer International Publishing.
10. V. Uricanu, J.R. Eastman, and B. Vincent, Stability in colloidal mixtures containing particles with a large disparity in size, *J Colloid Interface Sci.* 233 (2001) 1–11.
11. P. Gehr, Interaction of nanoparticles with biological systems, *Colloids Surf B Biointerfaces.* 172 (2018) 395–399.
12. B.D. Malhotra, M.D. Azahar Ali, Chapter 7: Nanostructured biomaterials for in vivo biosensors, in *Nanomaterials for Biosensors: Fundamentals and Applications—A volume in Micro and Nano Technologies.* (2018) 183–219, Elsevier.
13. D. Zheng, CH. Hu, T. Gan, X. Dang, S. Hu, Preparation and application of a novel vanillin sensor based on biosynthesis of Au–Ag alloy nanoparticles, *Sensors Actuat B Chem.* 148 (2010) 247–252.

14. J-M. Liu, J-T. Chen, and X-P. Yan, near infrared fluorescent trypsin stabilized gold nanoclusters as surface plasmon enhanced energy transfer biosensor and in vivo cancer imaging bioprobe, *Anal Chem.* 85 (2013) 3238–3245.
15. P. Cai, X. Zhang, M. Wang, Y-L. Wu, and X. Chen, Combinatorial nano–bio interfaces, *ACS Nano.* 12 (2018) 5078–5084.
16. M. Garvey, S.C. Padmanabhan, S.C. Pillai, M. Cruz-Romero, J.P.; Kerry, M.A. Morris, In vitro cytotoxicity of water soluble silver (Ag) nanoparticles on HaCat and A549 cell lines, *J Toxicol Pharmacol.* 1 (2017) 16.
17. N.I. Nii-Trebi, Emerging and neglected infectious diseases: Insights, advances, and challenges, *Biomed Res Int.* 2017 (2017) 5245021.
18. J.G. Teeguarden, P.M. Hinderliter, G. Orr, B.D. Thrall, and J.G. Pounds, Partico-kinetics in vitro: Dosimetry considerations for in vitro nanoparticle toxicity assessments, *Toxicol Sci.* 95 (2007) 300–312.
19. G. Oberdörster, T.A.J. Kuhlbusch, In vivo effects: Methodologies and biokinetics of inhaled nanomaterials, *NanoImpact.* 10 (2018) 38–60.
20. G. Oberdörster, Nanotoxicology: In vitro-in vivo dosimetry, *Environ Health Perspect.* 120 (2012) A13.
21. M. Mionić Ebersold, D. Bonvin and H. Hofmann, Neglected nano-effects of nanoparticles in the interpretation of their toxicity, *Analyst* 143 (2018) 837–842.
22. Zhou, W., Y. Ma, H. Yang, Y. Ding, and X. Luo. A label-free biosensor based on silver nanoparticles array for clinical detection of serum p53 in head and neck squamous cell carcinoma, *Int J Nanomedicine.* 6 (2011) 381–386.
23. S. Rtimi and J. Kiwi, Bactericide effects of transparent polyethylene photocatalytic films coated by oxides under visible light, *Appl Catal B-Environ.* 213 (2017) 62–73.
24. S. Rtimi, Indoor light enhanced photocatalytic ultra-thin films on flexible non-heat resistant substrates reducing bacterial infection risks, *Catalysts* 7 (2017) 57.
25. C. Espírito Santo, N. Taudte, D.H. Nies, G. Grass, Contribution of copper ion resistance to survival of Escherichia coli on metallic copper surfaces, *Appl Environ Microbiol.* 74 (2008) 977–986.
26. S. Rtimi, S. Giannakis and C. Pulgarin, Self-sterilizing sputtered films for applications in hospital facilities, *Molecules* 22 (2017) 1074.
27. M.K.S. Ballo, S. Rtimi, C. Pulgarin, N. Hopf, A. Berthet, J. Kiwi, P.H. Moreillon, J.M. Entenza, A. Bizzini, In vitro and in vivo effectiveness of an innovative silver-copper nanoparticle coating of catheters to prevent methicillin-resistant staphylococcus aureus infection, *Antimicrob Agents Chemother.* 60 (2016) 5349–5356.
28. L.M. Weiner, A. K. Webb, B. Limbago, M. A. Dudeck, J. Patel, A. J. Kallen, J. R. Edwards and D. M. Sievert, Antimicrobial-resistant pathogens associated with healthcare-associated infections: Summary of data reported to the National Healthcare Safety Network at the Centers for Disease Control and Prevention, 2011–2014. *Infect Control Hosp Epidemiol.* 37 (2016) 1288–1301.
29. Y. Zhou, Y. Kong, S. Kundu, J.D. Cirillo, and H. Liang, Antibacterial activities of gold and silver nanoparticles against *Escherichia coli* and bacillus Calmette-Guérin, *J Nanobiotechnology.* 10 (2012) 19.
30. C.M. Goodman, C.D. McCusker, T. Yilmaz, and V.M. Rotello, Toxicity of gold nanoparticles functionalized with cationic and anionic side chains, *Bioconjugate Chem.* 15 (2004) 897–900.
31. T.C. An, H.J. Zhao, P.K. Wong, *Advances in Photocatalysis Disinfection*, Springer, Berlin, Germany, 2016.
32. P. Lejeune, Contamination of abiotic surfaces: What a colonizing bacterium sees and how to blur it, *Trends Microbiol.* 11 (2003) 179–184.
33. S. Rtimi, V. Nadtochenko, I. Khmel, M. Bensimon and J. Kiwi, First unambiguous evidence for distinct ionic and surface-contact effects during photocatalytic bacterial inactivation on Cu–Ag films: Kinetics, mechanism and energetics, *Mater Today Chem.* 6 (2017) 62–74.

34. S. Rtimi, M. Pascu, R. Sanjines, C. Pulgarin, M. Ben-Simon, A. Houas, J.-C. Lavanchy, J. Kiwi, ZrNO-Ag co-sputtered surfaces leading to *E. coli* inactivation under actinic light: Evidence for the oligodynamic effect, *Appl Catal B Environ.* 138–139 (2013) 113–121.
35. D. Xia, T. An, G. Li, W. Wang, H. Zhao, P.K. Wong, Synergistic photocatalytic inactivation mechanisms of bacteria by graphene sheets grafted plasmonic AgAgX (X = Cl, Br, I) composite photocatalyst under visible light irradiation, *Water Res.* 99 (2016) 149–161.
36. W. Wang, G. Li, D. Xia, T. An, H. Zhao, P.K. Wong, Photocatalytic nanomaterials for solar-driven bacterial inactivation: Recent progress and challenges, *Environ Sci Nano.* 4 (2017) 782–799.
37. H. Shi, G. Li, H. Sun, T. An, P. Wong, Visible-light-driven photocatalytic inactivation of *E. coli* by Ag/AgX-CNTs (X = Cl, Br, I) plasmonic photocatalysts: Bacterial performance and deactivation mechanism, *Appl Catal B.* 158 (2014) 301–307.
38. S. Rtimi, D.D. Dionysiou, S.C. Pillai, J. Kiwi, Advances in bacterial inactivation by Ag, Cu, Cu-Ag coated surfaces & medical devices, *Appl Catal B Environ.* 240 (2019) 291–318.
39. F. Lamari, I. Chakroun, and S. Rtimi, Assessment of the correlation among antibiotic resistance, adherence to abiotic and biotic surfaces, invasion and cytotoxicity of *P. aeruginosa* strains isolated from diseased cultured fish products, *Colloids Surf B Biointerfaces.* 158 (2017) 229–236.
40. S.J. Park, H. H. Park, S. Y. Kim, S. J. Kim, K. Woo, and G.P. Ko, Antiviral properties of silver nanoparticles on a magnetic hybrid colloid, *Appl Environ Microbiol.* 80 (2014) 2343–2350.
41. Q.L. Shimabuku, F.S. Arakawa, M. Fernandes Silva, P. Ferri Coldebella, T. Ueda-Nakamura, M.R. Fagundes-Klen, R. Bergamasco, Water treatment with exceptional virus inactivation using activated carbon modified with silver (Ag) and copper oxide (CuO) nanoparticles, *Environ Technol.* 38 (2017) 2058–2069.
42. J.A. Lemire, J.J. Harrison, R.J. Turner, Antimicrobial activity of metals: Mechanisms, molecular targets and applications, *Nat Rev Microbiol.* 11 (2013) 371–384.
43. E.C. Cho, Q. Zhang, and Y. Xia, The effect of sedimentation and diffusion on cellular uptake of gold nanoparticles, *Nat Nanotechnol.* 6 (2011) 385–391.
44. C.M. Alexander, J.C. Dabrowiak, J. Goodisman, Gravitational sedimentation of gold nanoparticles, *J Colloid Interface Sci.* 396 (2013) 53–62.
45. P.M. Hinderliter, K.R. Minard, G. Orr, W.B. Chrisler, B.D. Thrall, J.G. Pounds, J.G. Teeguarden, ISDD: A computational model of particle sedimentation, diffusion and target cell dosimetry for in vitro toxicity studies, *Part Fibre Toxicol.* 7 (2010) 36–56.
46. H. Kihira, N. Ryde and E. Matijević, Kinetics of heterocoagulation. Part. 2—The effect of the discreteness of surface charge, *J Chem Soc Faraday Trans.* 88 (1992) 2379–2386.
47. J.H. Hwang, S.J. Kim, Y.H. Kim, J.R. Noh, G.T. Gang, B.H. Chung, N.W. Song, C.H. Lee. Susceptibility to gold nanoparticle-induced hepatotoxicity is enhanced in a mouse model of nonalcoholic steatohepatitis, *Toxicology* 294 (2012) 27–35.
48. S.G. Sheth, S.L. Flamm, F.D. Gordon, S. Chopra, AST/ALT ratio predicts cirrhosis in patients with chronic hepatitis C virus infection. *Am J Gastroenterol.* 93 (1998) 44–48.
49. J.F. Hillyer, R.M. Albrecht, Gastrointestinal persorption and tissue distribution of differently sized colloidal gold nanoparticles. *J Pharm Sci.* 90 (2001) 1927–1936.
50. K.E. Carr, R.A. Hazzard, S. Reid, G.M. Hodges, The effect of size on uptake of orally administered latex micro-particles in the small intestine and transport to mesenteric lymph nodes. *Pharm Res.* 13 (1996) 1205–1209.
51. M.P. Desai, V. Labhasetwar, G.L. Amidon, R.J. Levy, Gastrointestinal uptake of biodegradable microparticles: Effect of particle size. *Pharm Res.* 13 (1996) 1838–1845.
52. S. Nicklin, K. Miller, Effect of orally administered food-grade carrageenans on antibody mediated and cell-mediated immunity in the inbred rat. *Food Chem Toxicol.* 22 (1984) 615–621.
53. A. Kermanizadeh, BK. Gaiser, H. Hohnston, Toxicological effect of engineered nanomaterials on the liver. *Br J Pharmacol.* 171 (2014) 3980–3987.

54. M. van der Zande, R.J. Vandebriel, E. Van Doren, Distribution, elimination, and toxicity of silver nanoparticles and silver ions in rats after 28-day oral exposure, *ACS Nano.* 6 (2012) 7427–7442.
55. M. Semmler, J. Seitz, F. Erbe, P. Mayer, J. Heyder, G. Oberdörster and W.G. Kreyling. Long-term clearance kinetics of inhaled ultrafine insoluble iridium particles from the rat lung, including transient translocation into secondary organs. *Inhal Toxicol.* 16 (2004) 453–459.
56. A.J. Thorley, P. Ruenraroengsak, T.E. Potter, T.D. Tetley, Critical determinants of uptake and translocation of nanoparticles by the human pulmonary alveolar epithelium, *ACS Nano.* 8 (2014) 11778–11789.
57. M. Semmler-Behnke, W.G. Kreyling, J. Lipka, S. Fertsch, A. Wenk, S. Takenaka, G. Schmid, W. Brandau, Biodistribution of 1.4- and 18-nm gold particles in rats. *Small* 12 (2008) 2108–2111.
58. S. Lohse, Nanoparticles are all around us, Sustainable Nano (a blog by the Center for Sustainable Nanotechnology), 25 March 2013 (updated on Sept 10, 2015). http://sustainable-nano.com/2013/03/25/nanoparticles-are-all-around-us/.

6

Designing Smart Nanotherapeutics

A. Joseph Nathanael,
Tae Hwan Oh, and
Vignesh Kumaravel

6.1 Introduction

The necessities of medical needs are enormously growing worldwide. Numerous carcinogenic and biohazardous diseases are the consequences of change in life style, catastrophic transformation of the environment, and easy flow of chemical substances in day-to-day life. Hence, the design of smart therapeutics is of utmost importance. In recent years, nanotherapeutics have received significant attention for drug delivery, diagnostics, and imaging (Couvreur 2013, Elsabahy and Wooley 2012, Prasad et al. 2018, Singh and Lillard Jr 2009, Wong and Choi 2015). The most significant fact is the size of nanomaterials is more similar to the biological macromolecules that simplify the utilization of nanomaterials for bio-medical applications both *in vivo* and *in vitro*. Targeted and controlled drug delivery is achieved by stimuli-responsive nanomaterials (Prasad et al. 2018). Nanotherapeutics respond to external stimuli such as temperature, light, magnetic field, electric field, and ultrasound. The advantages of nanotherapeutics are low cytotoxicity, easy accumulation at the target sites, minimum side effects, high bioavailability, and circulation time (Blanco et al. 2015). The efficiency and toxicity are governed by characteristics of nanotherapeutics.

The biodistribution of nanotherapeutics is influenced by size, shape, surface charge/nature/topology, and rigidity (Blanco et al. 2015). The biological functions such as circulation half-life and extravasation are controlled by the size of nanotherapeutics (Blanco et al. 2015). Haemo-rheological dynamics, *in vivo* fate, and cellular uptake are affected by the morphology. The circulation half-life and the accumulation of nanotherapeutics at specific sites could be altered by the surface charge of nanoparticles. Hydrophobic/hydrophilic behaviour of nanomaterials plays an important role in protein adsorption and cellular uptake. For example, nanomaterials with more hydrophobicity than the cell membrane could be easily taken up by the cells (Blanco et al. 2015). The surface topology (smooth or rough surface) influences the binding of nanomaterials at the target sites. Particle rigidity also plays an important role in the biodistribution. For example, the elimination/accumulation time of synthetic blood cells is controlled by their rigidity (Blanco et al. 2015).

6.2 Emergence of Nanotherapeutics in Diagnosis and Treatment of Cancer

Conventional cancer therapies (chemo and radiation therapies) have inherent limitations and side effects. The success rate of the chemotherapy highly depends on the careful selection of patient to receive the treatment because of the heterogeneity and complexity of the tumours. Progress with one patient cannot be always assured in another. Pharmacodynamics and pharmacokinetics of a drug will be affected by individual genetic variations (Wheeler et al. 2013). These limitations urged scientists to develop an alternative therapy that is less harmful as well as effective to treat the cancer cells. Use of nanotherapeutics, nanotechnology-based drug formulation, is one of the best choices for controlled and targeted drug delivery in the affected area. One of the unique characteristics of nanoparticles is enhanced permeability and retention (EPR) in which the nanoparticles could penetrate into tumour tissues, but due to the ineffective lymphatic system of the cancerous tumour, they cannot exit the cancer cells (Maeda et al. 2000, Talekar et al. 2011). Clinical studies proved that nanotherapeutics could reduce the adverse effects of chemotherapeutic agents, improve the bioavailability of drugs, and enhance drug tolerance thereby improving the overall survival rate of cancer patients (Uskokovic 2009). Substantial scientific success has been attained in this field but there are significant challenges to translate the technology into clinical practice and commercialization.

6.2.1 Complications in Diagnosis and Treatment of Cancer

Cancer diagnosis usually begins with the exposure of symptoms that related to the disease. Most of the initial symptoms are nested with other common health issues in daily life and hence can be easily ignored. The ultimate reason of low survival rates from cancer is the late stage discovery and diagnosis. Even when the symptoms are detected, there are several limitations (e.g., size of tumour and inadequate imaging period) in the traditional biomedical imaging methods such as ultrasound and magnetic resonance imaging (MRI) (Jain 1987, Talekar et al. 2011). Cancer diagnosis requires more information to outline the precise treatment for the patient. However, the questions answered about a tumour are limited due to the inadequate availability of the samples for testing. In order to overcome these limitations, it is important to develop more reliable and effective diagnostic methods. Medical research on improving cancer treatments has been prioritized in recent years, but concrete explanations on how best to diagnose the disease are still evolving. Precise diagnosis and staging of the cancer are vital for the treatment plan.

The treatment for cancer is usually a multifaceted approach. Depending on the type of cancer, the treatment involves chemotherapy, radiotherapy, surgery, or the combination of these therapies. Nevertheless, the issues such as long-term complications, impact on survival, quality of life, and physiological conditions have to be reviewed again on the existing cancer treatments. Numerous anticancer therapies are identified to induce pulmonary damage, with minor to severe clinical consequences (Jain 1987, Talekar et al. 2011). Drug-induced pneumonitis is a common chemotherapy-related lung injury. Cancer therapies could also cause cardiac diseases such as congestive heart failure or accelerated arterial hypertension (Vyskocil et al. 2017). Consequently, in some instances, a patient may survive from cancer but die due to cardiac dysfunction of anticancer treatments (Li et al. 2017). Moreover, haematological complications are common for cancer patients undergoing active treatment (Jain 1987, Talekar et al. 2011). Present therapeutic drugs inadequately penetrate the tumours owing to their poor pharmacokinetics and unreliable biodistribution. As a result, they may accumulate and induce toxicity towards various healthy organs. Hence, there is an urgent need to develop a therapeutic method that can expedite the diagnosis of cancer before the symptoms appear. This would be beneficial for early-stage diagnosis, design of precise treatment plans, high survival, and better quality of life.

6.2.2 Nanotherapeutics in Diagnosis and Treatment of Cancer

Nanotherapeutics is a rapidly developing field to resolve the numerous shortcomings of conventional drug delivery methods. Nanotherapeutic drug delivery system is considered to avoid the drawbacks of conventional oral medicines in cancer chemotherapy. Most of the present cancer therapies are nonspecific. Hence, it leads to the damage of rapidly proliferating normal cells and develops various side effects. Similarly, the water solubility of many anticancer drugs is very poor, which leads to insufficient release of drugs. Severe adverse effects such as hypersensitivity, fluid retention, and peripheral neuropathy are caused by the use of solvents to dissolve the poor water-soluble drugs (Shepherd 2003, Ten Tije et al. 2003). Nanotherapeutics can target specific cancerous sites, either actively or passively, which makes them ideal for diagnostic imaging. The high surface area of the nanoparticles could increase the dissolution velocity and saturation solubility. Nanotherapeutics exhibit flexible functions and properties that make them a promising alternative for cancer therapy. Recent developments in the drug-release performance are further improved by stimuli-responsive nanoparticles by either internal or external stimuli. Stimuli can be combined to create two or more signals such as pH/temperature or pH/magnetic field (Figure 6.1).

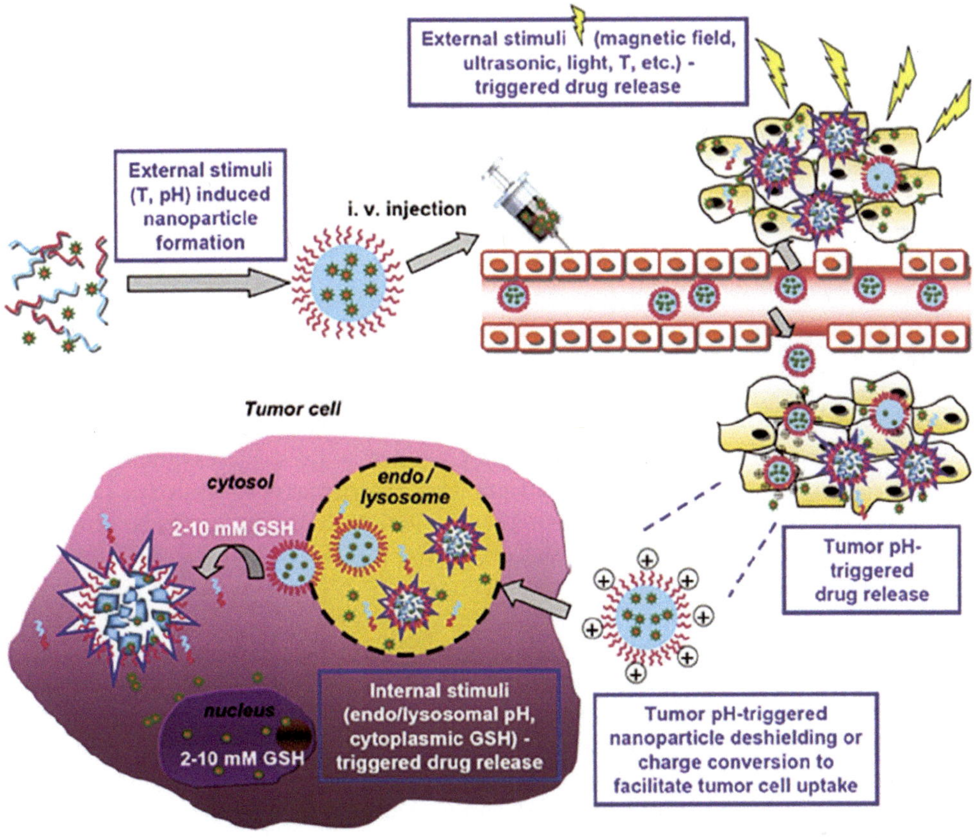

FIGURE 6.1 Schematic for the various steps involved in internal and external stimuli-responsive drug delivery for cancer therapy. (Reprinted from Cheng, R. et al., *Biomaterials*, 34, 3647–3657. Copyright 2013, with permission from Elsevier.)

Site-specific drug delivery and release of stimuli-responsive nanoparticles are beneficial for better *in vitro* and/or *in vivo* anti-cancer efficacy (Qiao et al. 2018).

6.3 Nanostructured Materials for Cancer Therapy

6.3.1 Carbon Nanotubes

Carbon nanotubes (CNTs) are one of the promising candidates in cancer diagnosis and therapy owing to their distinctive physicochemical properties. CNTs could be used in cancer treatments in numerous ways. CNTs act as bio-carriers without producing any cytotoxicity by shuttling different biomolecules such as proteins (Kam and Dai 2005), peptides (Pantarotto et al. 2004), drugs (Feazell et al. 2007, Liu et al. 2007a), and small interfering RNA (Kam et al. 2005a, Liu et al. 2007b) through endocytosis. Due to the high aspect ratio and specific surface area, CNTs can be conjugated or adsorbed onto several therapeutic molecules.

Another possible non-invasive way to destroy cancerous cells with limited side effects is using CNTs as a mediator. Platinum anticancer agents with short blood circulation times have problems in tumour uptake efficiency and intracellular DNA binding. In order to overcome these limitations a folate mediated drug delivery system was developed (Dhar et al. 2008). In that system, a platinum (IV) complex comprising a folate component (FA) was attached to the surface of an amine-functionalized SWCNT (SWCNT-PL-PEG-NH_2). This SWCNT acted as a delivery vehicle for platinum drugs and released the drug to the cancerous cells upon intracellular reduction of Pt(IV) to Pt(II) (Dhar et al. 2008). Kam et al. (2005b) developed SWCNTs to recognize and target cancerous cells with specific ligands. They used the concept of optical stimulation of CNTs inside the living cells in NIR spectral region (700 nm–1100 nm), where the biological systems are highly transparent. The tumour cells were internalized with folic acid (FA) and CNTs, which are specifically bound with cancer cells that overexpress folate receptors. These cancer cells were selectively destroyed when the CNTs were irradiated with NIR radiation by generating significant amount of energy (photo-thermal therapy [PTT]). *In vitro* analysis revealed that endosomal rupture and molecular cargo release occurred at the intended targets due to the local heating of SWCNTs by NIR pulses. NIR irradiation was not harmful to the normal cells that had not internalized the FA–bound CNTs. The cancer cell destruction was accomplished with low laser power (3.5 W/cm^2) and short irradiation time (2 min) (Kam et al. 2005b). This is ascribed to high NIR absorbance of the SWCNTs.

Photodynamic therapy (PDT) cytotoxicity is created by activating the photosensitizer using a specific wavelength of light. The activated photosensitizer transfers the energy to the oxygen molecule, which will be excited to singlet oxygen ($^1O_2$) from the ground state. The singlet oxygen is cytotoxic. The photochemical property of CNTs and its potential to act as a carrier for photosensitizers could be used to destroy cancer cells by the cytotoxic oxygen. Although PDT-based cancer therapy is currently in an early stage of development, PDT has shown potential effects for selective and controllable treatments. Different studies have been reported for the utilization of SWCNTs and MWCNTs for cancer treatment (Kam and Dai 2005, Kam et al. 2005b, Madani et al. 2011, Sanginario et al. 2017, Son et al. 2016). However, due to their poor solubility in water, it is difficult to use them in cancer therapy. In order to improve the bioavailability, they must be functionalized with specific ligands. Apart from their promising results and benefits, evaluation of its potential effects on normal tissues are important since CNTs are toxic at high concentrations. Various reports in the literature revealed the intrinsic cytotoxicity of CNTs (Madani et al. 2013, Tian et al. 2006). Specifically, it was reported that MWCNTs were toxic and developed DNA damage eventually leading to cell death in human dermal fibroblast cells (Patlolla et al. 2010). Other claims for human epithelial cells (Khaliullin et al. 2015, Mittal et al. 2011), macrophages (Cheng et al. 2009, Khaliullin et al. 2015), and lymphocytes (Kim and Yu 2014) also accounted for the CNT cytotoxicity.

6.3.2 Core Shell Nanoparticles

Core shell nanoparticles have distinctive properties such as high chemical and thermal stability, low cytotoxicity, excellent biocompatibility, and high dispensability (Lungu et al. 2016). Optically tuneable core shell nanoparticles can be designed to scatter and/or absorb light by tuning the shell thickness and core radius over a broad spectral range (including NIR). The scattering of light has prospective applications for cancer imaging whereas the absorbance has potential for treatment. Menon et al. developed a multifunctional dual drug-loaded nanoparticle (MDNP) containing a poly(N-isopropylacrylamide)-carboxymethyl chitosan shell and poly lactic-co-glycolic acid (PLGA) core to treat lung cancer. The nanoparticles showed low lung cytotoxicity during the treatment (Menon et al. 2017).

In *in vivo* systems, bimetallic core/shell nanoparticles are used as sensors and biomarkers for tumour detection (e.g., bimetallic Au/Ag core/shell nanoparticles) (Fu et al. 2015, Ghosh Chaudhuri and Paria 2012, Yang et al. 2017). Polymer-based core–shell nanoparticles helps to detoxify the widely employed anticancer drug, cis-Diamminedichloridoplatinum(II) (CDDP, cisplatin) for the treatment of glutathione (GSH) over-expressed breast cancer cells. The polymer–cisplatin drug conjugates are selectively accumulated at the cancer tissues (Fang et al. 2011, Surnar et al. 2015). This is attributed to the EPR effect. Elbaz et al. reported the Ag/polymeric core shell nanoparticle-based therapy for breast cancer. The cytotoxicity effect of different core shell (Ag/polymeric) nanoparticles were tested against breast cancer cells (MCF-7). It was found that the nanoparticles are more cytotoxic to MCF-7 cells as compared to normal fibroblast. A low dose of Doxorubicin (DOX)-loaded core shell nanoparticles displayed high anticancer activity (Elbaz et al. 2016) when compared to pure DOX or Ag/polymer core shell nanoparticles.

Core shell magnetic nanoparticles (MNPs) are usually made up of chemotherapeutic drugs which are proposed to be delivered by a pH-sensitive mechanism. A particular ligand that can relate to the specific tumour cells and iron oxide nanoparticles is used as a core (Lungu et al. 2016). Organic-inorganic hybrid core shell nanoparticles have the potential to incorporate considerable amounts of drug. Poly-2-dimethylaminoethylmethacrylate (PDMAEMA) is an ideal candidate due to its hydrophilicity, biocompatibility, and pH sensitivity. PDMAEMA-coated Fe_3O_4 core shell nanoparticles were reported for targeted drug delivery and controllable release. The super-paramagnetism of Fe_3O_4 was used for targeting with drug release at the targeted site controlled by changing the pH of the environment. Hence, this kind of hybrid core shell nanoparticles can be utilized for targeted drug delivery and controlled release (Zhou et al. 2009).

6.3.3 Quantum Dots

Recently, quantum dots (QD) have been widely used in biomedicine (Sun et al. 2015). QDs are semi-conducting tiny luminous nanocrystals usually in the range of 2–10 nm (Seeta Rama Raju et al. 2015). QDs generally have narrow emission spectra in the NIR region and can be tuned based on their size. Because of their powerful fluorescent properties, they can be used in photo-based therapies. QDs can fluoresce for several months in a living animal, and they are much more resistant to photo-bleaching (Sun et al. 2015). Hence, it is advantageous over radioactive tags and organic fluorophores. Presently QDs are being used as probes for high-resolution molecular imaging and tracking of cell activities inside the body. It is also possible to attach different proteins and receptors to QDs to track interaction with molecules. Black phosphorous QDs exhibited excellent NIR photo-thermal efficiency that leads to cancer cell death (Sun et al. 2015). Disappearance of tumour was reported when CdTe QD-coated SiO_2 (CdTe(710)/SiO_2 QDs) was injected in mouse melanoma tumours and irradiated with laser (Chu et al. 2012). Toxicity studies on QSG-7701 liver cell and A375 melanoma cell revealed that the side effects were minimized by the SiO_2 coating controlling the release rate of Cd^{2+} ions from CdTe(710)/SiO_2 QDs.

FA-targeted Mn:ZnS QDs were studied for cancer cell imaging and therapy. The cell viability and proliferation analysis did not exhibit any toxicity towards human breast cell line (MFC-10) and the breast cancer cell lines (MCF-7 and MDA-MB-231). In this system, Mn:ZnS was used as QDs, which was further encapsulated with chitosan (as a stabilizer and active binding site) and FA (as targeting agent towards the cells overexpressing folate receptors). Due to its nontoxic nature, it could be used as a favourable candidate for targeted delivery and cellular imaging (Bwatanglang et al. 2016). Dual extracellular and intracellular targeting and imaging of cancer markers (nucleolin and mRNA) using DNA-templated heterobivalent QD nanoprobes were reported to bypass the lysosomal sequestration and accomplish the imaging of intracellular biomolecules in cancer cells (Wei et al. 2014). The toxicity of QDs is a major problem in using them as a candidate for molecular imaging and it can be converted as a traceable chemotherapeutic compound (Sala-Rabanal et al. 2013, Zhao et al. 2016). A pair of human liver cells (HepG2, a well-differentiated hepatocellular carcinoma, and QSG-7701, a human hepatocyte line) were tested for QD cellular uptake. HepG2 cells uptake substantially higher than QSG-7701 cells, which is possibly due to upregulated PAT. *In vitro* cell growth inhibition ability of these QDs indicates preferential cancer cell killing. *In vivo* studies in tumour mouse model revealed that the tumour mass was reduced by 63.5% and the survival time was also increased by 2.5 fold as compared to control group (Zhao et al. 2016).

6.4 Combining Diagnosis and Treatment in a Nanotherapeutics

At present, cancer diagnosis and treatments are accomplished by discrete diagnostic and therapeutic modalities. Additionally, all treatment methods involve invasive surgical procedures or therapies that will be accompanied with major side effects, high treatment cost, and poor clinical outcome. If both diagnosis and treatment procedures can be provided by a single method/technology this would provide substantial savings in cost and time and minimize the patient discomfort. Hence, additional investigations to look out for innovative approaches of cancer diagnosis and treatments are crucial. One of the innovative strategies is theranostics (combination of diagnosis and treatment) which provides great assistance in cancer research. Nanotheranostics can be fabricated by combining the diagnostic probes and therapeutic drugs. It can be customized by changing probes and drugs. Different kinds of materials such as core-shells, QDs, (such as silver, gold, polymer), and magnetic NPs (such as iron oxide) can be used as nanotheranostics. These agents can be functionalized with drugs, ligands, and antibodies to enhance targeted drug delivery, diagnostic imaging, and therapy (Figure 6.2) (Mura and Couvreur 2012).

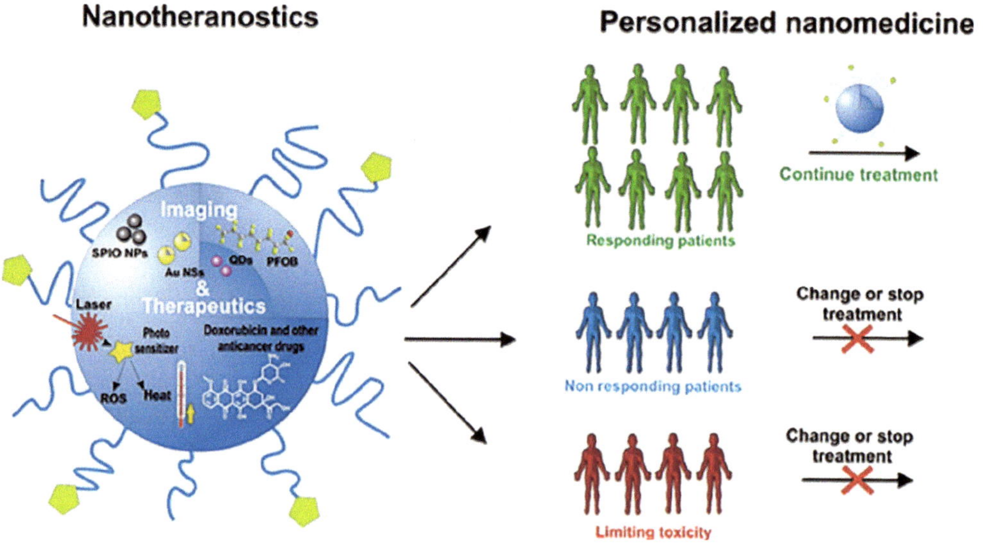

FIGURE 6.2 Schematic of the applications of nanotheranostics. (Reprinted from Mura, S. and Couvreur, P., *Adv. Drug Deliv. Rev.*, 64, 1394–1416. Copyright 2012, with permission from Elsevier.)

6.4.1 Magnetic Nanotheranostics

In the last few decades, magnetic nanotheranostics have attracted particular interest in biomedicine, especially in cancer treatment. MNPs are functionalized using different methods for their application in diagnosis and therapeutics. These include various coatings or embedment within polymers or biological molecules to improve their biocompatibility as well as reduce the toxicity within the biological systems. MNPs, with their small size and magnetic properties, are used in diagnostic methods such as MRI, single photon emission computed tomography (SPECT), and positron emission tomography (PET). These techniques are the main diagnostic tools for cancer. However, they possess some shortcomings such as poor spatial resolution and low sensitivity. Hence, the diagnostic techniques are continuously advanced to increase their accuracy. MNPs are extensively applied in therapy because of their ability to control them remotely (Xiong et al. 2013). One of the prominent advantages is the possibility to increase their concentration in a specific target, which will ultimately increase their efficacy.

A cluster of four individual iron oxide was capped with oleic acid and dioleate-modified PEG. It was formed as a rubik-like magnetic nano-assemblies (MNAs), which was further loaded with hydrophobic anticancer drugs (Xiong et al. 2013). Paclitaxel (PTX)-loaded MNAs showed high drug loading ability and magnetism, along with quick and prolonged drug release performance. Cells and animal experiments along with the optical imaging analysis of this new assembly showed significant amount of MNAs in the tumour even after removing the external magnetic field. The tumour size also reduced with the combination of the rubik-like MNAs with PTX as compared to pure PTX. This kind of design may be a prospective 'all-in-one' scheme for identification, diagnosis, treatment, and monitoring of the cancerous tumours (Xiong et al. 2013).

Magnetite functionalized with FA as theranostics for various types of cancer was reported by Azcona et al. (2018). Theranostic characteristics were assessed using the oncological drug DOX. The possibility of these designs to interact with simulated physiological fluid show that protein corona was formed around all the verified formulations providing more stable nano-devices (Azcona et al. 2018). A 20–30 nm diameter of magnetic carrier was prepared by Fe_3O_4 NPs. The nanoparticles were further coated with an active bio-glass layer and functionalized with hyper-branched polyglycerol through ring opening polymerization of glycidol. This carrier did not induce any cytotoxicity but yielded a better prospective contrast agent for MRI (Mousavi et al. 2015).

6.4.2 Core Shell Nanotheranostics

Polyethylene glycol-modified polypyrrole-coated bismuth core shell nano-hybrids (Bi@PPy-PEG) were studied for multifunctional X-ray computed tomography/photoacoustic (CT/PA) dual-modal imaging and PTT. This core shell structure showed excellent physiological stability with low cytotoxicity. They also exhibited strong NIR absorbance with the photo-thermal conversion efficiency of around 46 (Yang et al. 2018).

Prussian blue and gold (PB@Au) core–satellite nanoparticles (CSNPs) modified with PEG were reported as a multifunctional nanotheranostic for magnetic resonance-computed tomography (MR-CT) imaging and synergistic photo-thermal and radiosensitive therapy (PTT–RT). In this system, the core PBNPs assist T_1– and T_2–weighted MR contrast and strong photo-thermal effect, while satellite AuNPs assist CT enhancement and radio-sensitization. After intravenous injection, these CSNPs had efficient tumour localization. This was confirmed by CT and MR imaging. High photo-thermal conversion efficiency was observed from PBNPs core and efficient radio-sensitization from AuNP satellites. Tumour growth was completely suppressed after PTT-RT treatment using CSNPs without regrowth over the 14-day therapeutic period. Further *in vivo* analysis confirmed its biocompatibility without any evidence of cytotoxicity (Dou et al. 2017).

Loo et al. (2005) demonstrated integrated cancer imaging and therapy using immune-targeted Au nanoshells, which can scatter light in the NIR for cancer imaging while demonstrating sufficient absorption to selectively destroy the targeted cells by PTT. The performance of the nanoshell as nanotheranostic was confirmed by *in vitro* testing for the detection and destroying of breast cancer cells. No cytotoxicity was observed on the Au nanoshells (Loo et al. 2005). Multifunctional theranostic schemes by the combination of two different metal-organic-frameworks (MOFs) were proposed. A novel core-shell dual-metal-organic-frameworks (CSD-MOFs) nanotheranostic was prepared using ZIF-8 MOFs as shell and Prussian blue (PB) MOFs as core for simultaneous diagnosis, synergistic tumour therapy, and therapy monitoring using MRI and fluorescence optical imaging. The core shell nanoparticles were loaded with DOX for pH and NIR stimuli responsive. PB MOFs produce heat during NIR irradiation that will kill the cancer cells. *In-vivo* anti-tumour efficiency of CSD-MOFs@DOX+NIR was 7.16 and 5.07 times higher than chemo and thermal therapies, respectively (Wang et al. 2017a).

IR806 photothermal sensitizer (PTS) integrated into core shell NaYF4:Yb, Er@NaYF4:Yb@NaYF4:Yb, Nd up-conversion nanoparticles (UCNPs) were demonstrated for PTT and luminescence imaging by a single-wavelength NIR irradiation (Lin et al. 2018). The mechanism behind this new approach is as follows: When the UCNPs were irradiated with NIR laser at 793 nm, the ND^{3+} ions were sensitized and emitted visible light at 540 nm and 654 nm due to up-convert energy from Yb^{3+} to Er^{3+} ions and emitted NIR light at 980 nm by down-convert energy to the Yb^{3+} ions. IR806 dye could be excited by 793 nm NIR radiation for effective PTT and the same 793 nm NIR radiation is suitable for luminescence imaging, which can also penetrate deeply into tissues. Stable loading of the IR806 dye was achieved by further surface modification by mesoporous silica ($mSiO_2$) and polyallylamine (PAH). Tumour targeting was achieved by modification with polyethylene glycol-FA (PEG-FA) (Lin et al. 2018). *In vitro* analysis revealed that this nanocomposite could be accepted by the cancer cells (MDA-MB-231) but the cell viability was efficiently reduced when irradiated with the 793 nm NIR laser radiation (Lin et al. 2018).

6.5 Design and Synthesis of Nanotherapeutics

Nanotherapeutics are available in various forms such as polymeric nanoparticles (PNPs), MNPs, theranostics, emulsions, solid lipid nanoparticles, lipid carriers, gene carriers, dendrimers, QDs, carbon or graphene materials, capsules, sprays, and sponges for bio-medical applications (Kang et al. 2017). Numerous techniques such as precipitation, milling, ultra-sonication, polymerization, emulsification, salting out, high-pressure homogenization, chemical vapour deposition, and supercritical fluid have been used to synthesize nanotherapeutics (Kang et al. 2017). The key findings on the design and synthesis of some important nanotherapeutics are highlighted briefly in the following section.

6.5.1 Magnetic Nanoparticles

MNPs (e.g., iron oxide) have been widely employed for different bio-medical applications (such as drug delivery, medical imaging, and biosensor) under the guidance of an external magnetic field. The US Food and Drug Administration has accepted iron oxide nanoparticles (e.g., Fe_3O_4) for its clinical use as a magnetic resonance contrast agent (Kim et al. 2011). The magnetism and functionality of the nanoparticles are influenced by their shape, size, particle size distribution, and surface characteristics (Lee et al. 2015). Hydrothermal (Chen et al. 2008), precipitation (Rajamohan et al. 2017, Vignesh et al. 2014), thermal decomposition (Unni et al. 2017), and electro-chemical (Starowicz et al. 2011) methods are commonly used to synthesize MNPs. The biocompatibility and colloidal stability of MNPs are significantly determined by the nature of surface ligands. The surface of Fe_3O_4 is generally decorated by a polymer with required surface ligands for drug delivery. The surface modification is performed using ligand exchange (e.g., sulfonates, phosphonates, thiols, catechol, and carboxylic acids) and encapsulation (e.g., polymers, silica, DNA, and inorganic metal/metal oxide) techniques. Stimuli-responsive (internal and external) ligands have received much attention in recent years to improve the drug delivery of MNPs (Kang et al. 2017). The interaction of stimuli-responsive MNPs with internal/external factors results in a wide range of structural changes as shown in Figure 6.3a (Kang et al. 2017).

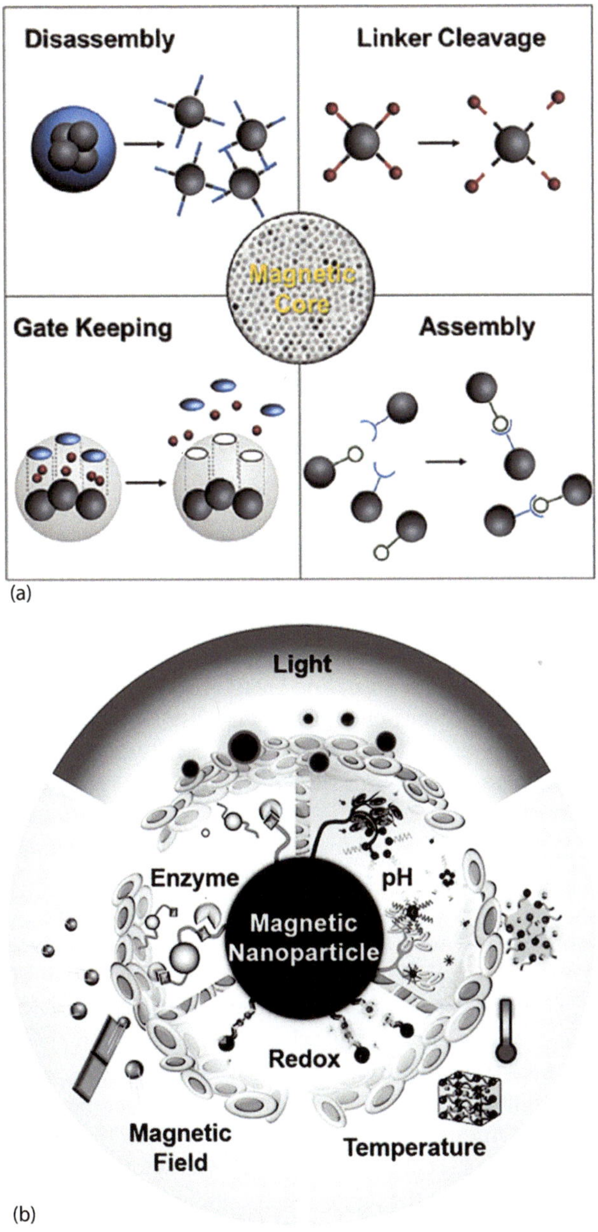

FIGURE 6.3 (a) Schematic representations of stimuli response structural changes of MNPs. (b) Schematic illustration of internal/external stimuli-responsive MNPs for the drug delivery. (Reprinted from Kang, T. et al., *Biomaterials*, 136, 98–114. Copyright 2017, with permission from Elsevier.)

The structural changes are related to the physico-chemical properties, suggesting improved drug delivery with negligible side effects *via* stimuli-driven drug release at the target receptors (Figure 6.3b). The internal stimuli MNP employs the features such as pH, enzyme concentrations, and hypoxia for drug delivery. Similarly, the external stimuli MNP utilizes the factors such as light intensity, temperature, and magnetic field for drug delivery.

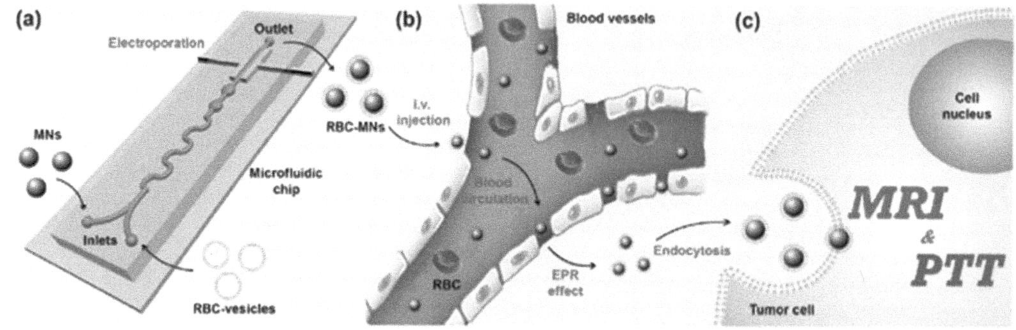

FIGURE 6.4 Schematic of (a) microfluidic electroporation synthesis of RBC- Fe_3O_4, (b) intravenous injection of RBC- Fe_3O_4 into the blood circulation, (c) bio-mimetic *in vivo* cancer therapy (MRI and PTT). (Reprinted with permission from Rao, L. et al., *Acs Nano*, 11, 3496–3505, 2017. Copyright 2017, American Chemical Society.)

Recently, multifunctional drug delivery of superparamagnetic nanoparticles was reported for cancer therapy (Malekzadeh et al. 2017). Fe_3O_4 nanoparticles were prepared through co-precipitation technique. Then, Fe_3O_4 was coated with bio-degradable poly citric acid (PCA) to avoid the aggregation and it was further surface functionalized with PEG and FA. Quercetin was used as an anti-cancer drug in this study. Human cervical cancer cells (HeLa) and breast cancer cell lines (MDA-MB-231) were used as cell culture models. *In vitro* cytotoxicity of quercetin-loaded MNPs was measured by the MTT assay.

The anti-cancer drug is selectively delivered to the tumour cells *via* FA-receptor-mediated endocytosis. Significant cytotoxicity is observed for Hela cells and MDA-MB-231 cells for quercetin-loaded Fe_3O_4@PCA-PEG-FA nanoparticles. However, no apparent cytotoxicity is noted for pure Fe_3O_4@PCA-PEG-FA nanoparticles. MRI studies revealed that Fe_3O_4@PCA-PEG-FA nanoparticles could be effectively used as a negative contrast agent when compared to commercial MRI contrast agents (Malekzadeh et al. 2017).

Microfluidic electroporation was used to synthesis red blood cell (RBC) membrane-coated Fe_3O_4 nanoparticles to enhance the MRI-guided cancer therapy (Rao et al. 2017). The electric pulses can promote the entry of Fe_3O_4 nanoparticles into the RBC vesicles when the reaction mixture is passed through the electroporation zone in a microfluidic chip (Figure 6.4). A complete cell membrane coating of MNPs is attained by the electroporation technique as compared to the conventional extrusion. The magnetic and photo-thermal properties of as-synthesized RBC membrane-coated Fe_3O_4 (RBC–Fe_3O_4) nano-particles were applied to study the *in vivo* cancer therapy. The PTT and MRI properties of RBC–Fe_3O_4 are enhanced *via* the synergistic effects of Fe_3O_4 and RBC. Cytotoxicity of RBC–Fe_3O_4 was studied in RAW 264.7 and MCF-7 cells using the nanoparticles synthesized by both conventional extrusion (RBC–Fe_3O_4–C) and electroporation (RBC–Fe_3O_4–E) techniques. Under prolonged laser irradiation, pure and RBC–Fe_3O_4 showed remarkable cytotoxicity against cancer cells.

6.5.2 Nano-emulsion

Design and development of lipophilic bioactive drugs using emulsion-based drug delivery systems are more attractive in recent years. Nano-emulsion (droplet diameter < 500 nm) is a biphasic dispersion of two immiscible liquids in the presence of suitable surfactants. If the surfactant is soluble in water then it is called oil-in-water (O/W) nano-emulsion while if the surfactant is soluble in oil then it is named water-in-oil (W/O) nano-emulsion (Figure 6.5a). Based on the number of constituents, the nano-emulsion is classified as bi-phasic or multiple. The drug molecules are generally dissolved in the internal phase (oil/lipid/water) (Singh et al. 2017).

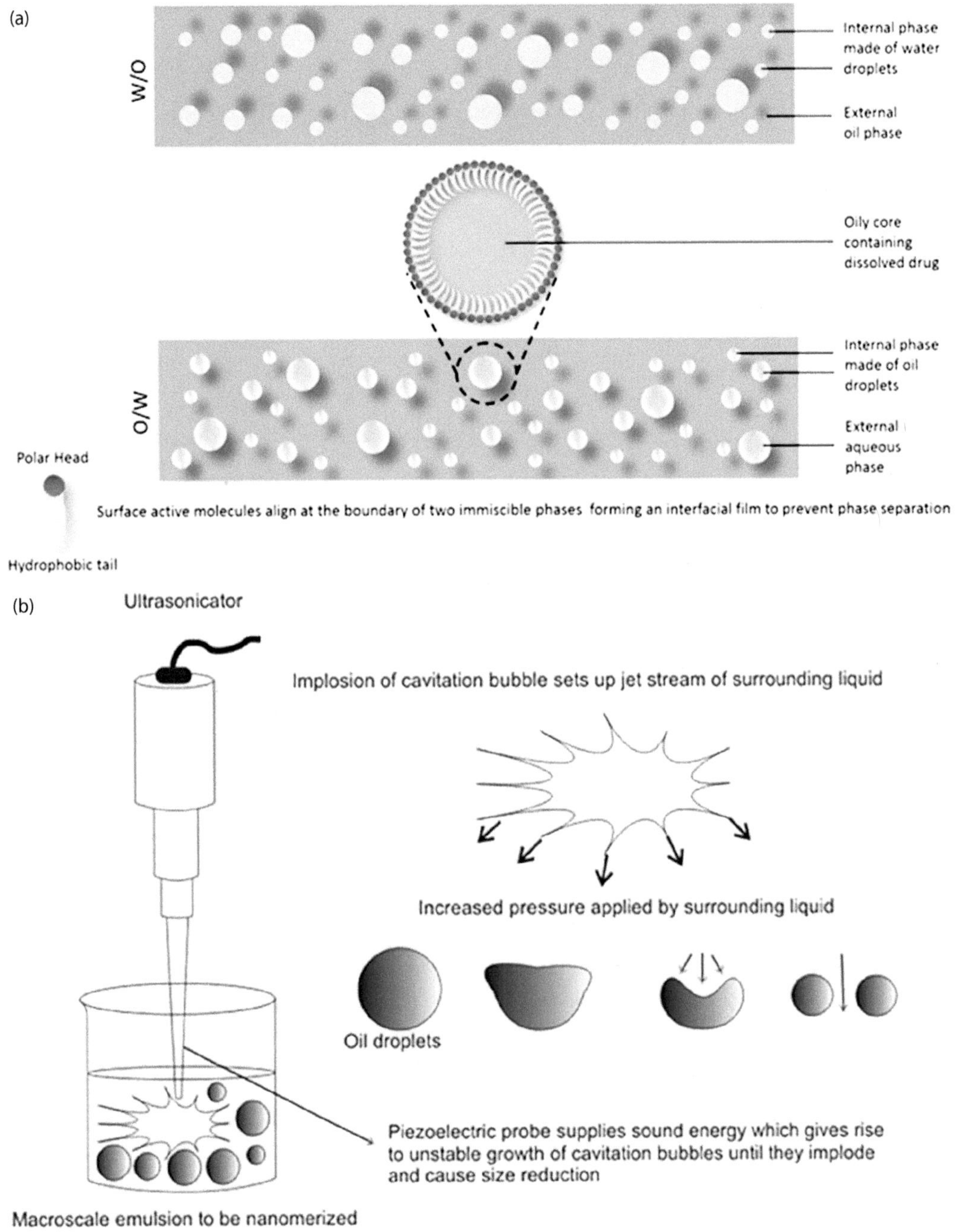

FIGURE 6.5 (a) Schematic for the structure of an oil in water or water in oil nano-emulsion. (b) Schematic for the synthesis of nano-emulsion by ultra-sonication. *(Continued)*

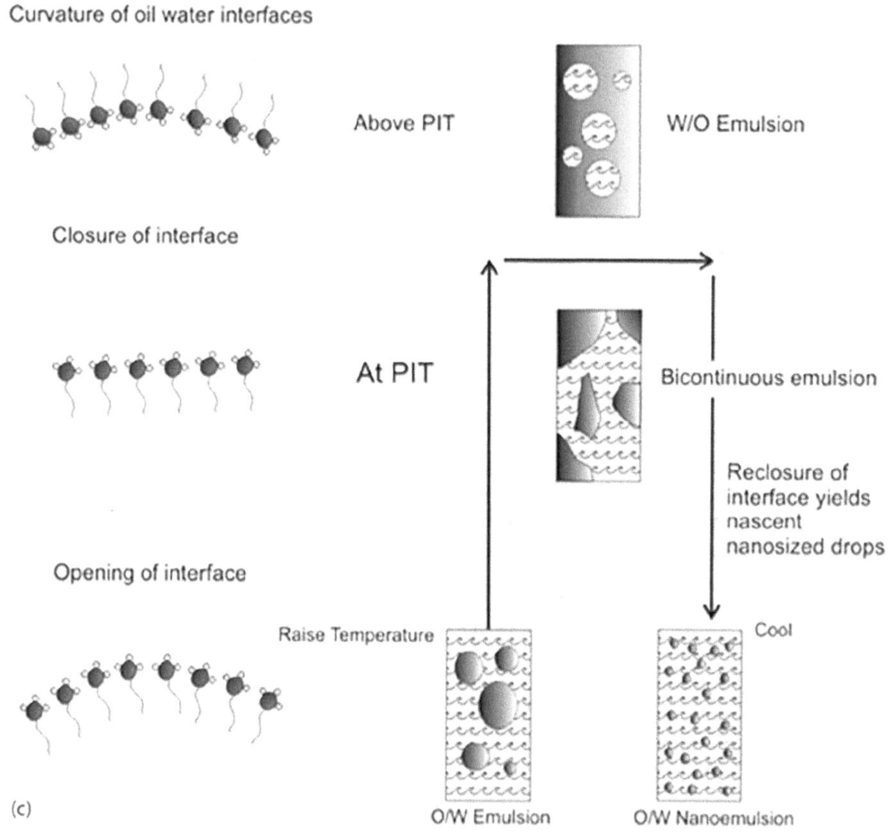

FIGURE 6.5 (Continued) (c) Schematic for the synthesis of nano-emulsion by phase-inversion temperature (PIT) method. (Reprinted from Singh, Y. et al., *J. Control Release*, 252, 28–49. Copyright 2017, with permission from Elsevier.)

Nano-emulsions can be synthesised through three methods: high-energy (ultra-sonication [Figure 6.5b], micro-fluidization, and piston gap homogenization), low-energy (phase inversion temperature and spontaneous emulsification [Figure 6.5c]), and the combination of low and high energy (Singh et al. 2017) techniques. The most important principle in the synthesis of nano-emulsions is to achieve the preferred droplet size with monomodal distribution.

High-energy methods are commonly used in industry and almost any oil can be converted into nano-emulsion. Low-energy methods are limited to a specific type of oil. Factors such as nature of surfactant or co-surfactant or aqueous component, hydrophilic lipophilic balance, type of oil, and temperature greatly influence the emulsification process in low-energy methods. Nano-emulsions are available in various forms such as gel, aerosol, foam, spray, and cream. It can be administered into the body *via* different routes including intravenous, oral, ocular, and intranasal routes.

Recently, nano-emulsion-filled alginate hydrogel was fabricated to study the digestion behaviour of hydrophobic nobiletin (3′,4′,5,6,7,8-hexamethoxyflavone) in the gastro intestinal tract (Lei et al. 2017). The nano-emulsion was synthesized by PIT method. Tween 80 and span 20 with a weight ratio of 3:1 were used as surfactants. Medium-chain triglyceride was used as an oil phase. The final formulation of nano-emulsion had 4 g of organic phase, 16 g of aqueous phase, and various amounts of nobiletin (0.0900 g, 0.1200 g, 0.1500 g). The particle size of the nano-emulsion was calculated as 205.3 ± 2.3 nm. Ca^{2+} cross-linked alginate hydrogel matrix was synthesized using the internal gelling method of calcium carbonate-D-glucono-δ-lactone ($CaCO_3$-GDL) system. The encapsulation of the hydrogel matrix can be beneficial to attain the maximum dissolution and steady release of nobiletin drug from the nano-emulsion. The gelation between Ca^{2+} and alginate is induced by GDL. Molar ratios of $CaCO_3$ and GDL were 0.5 and 0.135, respectively. The blank hydrogel is transparent in colour with a porous morphology. The nano-emulsion filled hydrogel is milky white in colour. SEM images revealed that the porous structure is completely filled by the nano-emulsion. *In vitro* experimental studies showed that the bioavailability of nobiletin in the nano-emulsion is higher (67.2 $\pm$ 0.4% at 4.5 mg/mL) than that available in the blank (44.7 $\pm$ 0.4% at 4.5 mg/mL).

In another recent study, cefuroxime (a broad-spectrum second-generation cephalosporin antibiotic)-loaded nano-emulsion was studied to enhance the drug penetration in the brain during parenteral administration (Harun et al. 2018). The multi-component nano-emulsion was synthesized through the high-pressure homogenization of the reaction mixture. The final nano-emulsion contained cefuroxime (0.3% w/w), lecithin (3% w/w), tween 80 (0.8% w/w), α-tocopherol (1% w/w), sodium oleate (0.1% w/w), glycerol (2.30% w/w), soybean oil (5% w/w), sunflower seed oil (5% w/w), and deionized water (82.50% w/w). The as-synthesized nano-emulsion was physico-chemically stable for 6 months (storage at 4°C). *In vivo* experimental studies in rats demonstrated that the concentration and biodistribution of cefuroxime drug are higher in the nano-emulsion-treated rats compared to control-treated rats. *In vitro* cytotoxicity studies were carried out in human brain endothelial cells (hCMEC/D3) using MTT assay. No toxic effects are observed for CLN up to 10 μg/mL. However, pure cefuroxime did not show any toxicity up to 100 μg/mL.

6.5.3 Nano-sponges

Nano-sponges are composed of sub-microscopic cross-linked polymers (e.g., cyclodextrin nano-sponges) with cavities in the range of a few nano-meters (Sherje et al. 2017). They have applications in *in vivo* or *in vitro* drug delivery. Most of the cyclodextrin nano-sponges have hydrophobic interior cavity with hydrophilic exterior surface. The polymerization of cyclodextrin is carried out using cross linkers such as diisocyanate, epichlorohydrin, acid anhydride, diphenylcarbonate, dicarboxylic acid, and alkyl dihalides. Schematics for the polymerization of cyclodextrin using various cross linkers are shown in Figures 6.6 and 6.7. The solubility and molecular interactions of cyclodextrin nano-sponges are influenced by the cross linkers. The drug is loaded into the cavities *via* solvent evaporation, microwave synthesis, ultrasound

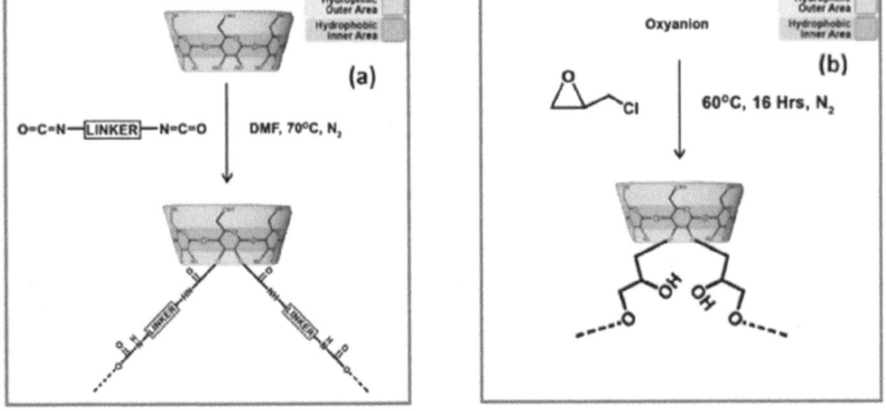

FIGURE 6.6 Schematic for the cyclodextrin polymerization using (a) diisocyanate and (b) epichlorohydrin cross-linkers. (Reprinted from Sherje, A. et al., *Carbohydr. Polym.*, 173, 37–49, 2017. Copyright 2017, with permission from Elsevier.)

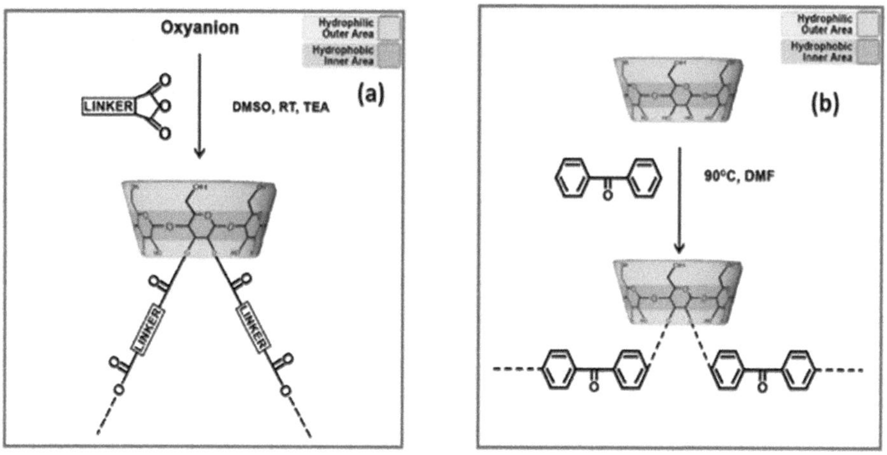

FIGURE 6.7 Schematic for the cyclodextrin polymerization using (a) acid anhydride and (b) diphenylcarbonate cross-linkers. (Reprinted from Sherje, A. et al., *Carbohydr. Polym.*, 173, 37–49, 2017. Copyright 2017, with permission from Elsevier.)

sonication, or dispersion with freeze-drying technique. The crystalline nano-sponges have more drug-loading capability when compared to the paracrystalline materials (Tejashri et al. 2013). Drug molecules are attached into the nano-cavities through Van der Waals, dipole-dipole, dispersion force, electrostatic, and hydrophobic interactions.

Recently, a simple, fast, and one-step fabrication method was reported for the glutathione-responsive cyclodextrin nano-sponges to host and release an anticancer drug (Trotta et al. 2016). The nano-sponge was synthesized using precursors such as β-cyclodextrin 2-hydroxyethyl disulphide, and pyromellitic dianhydride. A maximum of 95% yield was achieved.

Reduced graphene oxide (rGO)-supported protein lipid bi-layer nano-sponges were synthesized for photolytic cancer therapy (Su et al. 2016). The synergistic effects of chemo- and thermo-therapy was studied for anti-cancer drug delivery. Hydrophobic anti-cancer drugs docetaxel (DTX) and gasified perfluorohexane (PFH) were used in this study. The detailed schematic for the synthesis of graphene supported lipid nano-sponges is shown in Figure 6.8. At first, the rGO was treated with styrene, cetyltrimethylammonium bromide (CTAB), and tetraethylorthosilane (TEOS) to fabricate a core-shell structure. The role of rGO is to avoid the aggregation of micelles. The rGO and TEOS are adsorbed on the surface of CTAB *via* electrostatic interaction and Coulombic force. The as-synthesized compound was calcined under N_2 atmosphere. The styrene units were cross-linked by sulphide and sulphonyl bonds during sulfonation in the presence of sulphuric acid. The product was pyrolyzed at 800°C with Ar atmosphere to create a thin carbon layer on the porous surface. Then the hydrophobic anti-cancer drugs and lipids were attached on the carbon layer.

The as-synthesized drug was tested in mouse models through intravenous injection. Under near infrared radiation, the gasification of PFH is initiated with an increase of local temperature and the DTX is released, suggesting the rupture of tumour cells by chemo- and thermo-therapy. Cytotoxicity was studied in a brain cancer cell line (RG2 cells). The results revealed that the pure nano-sponges without DTX or PFH displayed low cytotoxicity against the cancer cells. However, the cytotoxicity is enhanced after the introduction of drugs and NIR radiation.

A supra-molecular peptide nano-sponge of (cholesterol-$(K/D)_n$ $DEVDGC)_3$-trimaleimide units with a trigonal maleimide linker was designed for the rapid uptake by leukocytes and neural stem cells (Yapa et al. 2018). Here, 'K' and 'D' are lysine and aspartic acid. 5(6)-carboxyfluorescein was used as a model drug. The therapeutic efficacy of the nano-sponge was tested for the caspase-6 mediated release of 5(6)-carboxyfluorescein drug. The N-terminal of the peptides were capped with cholesterol and the

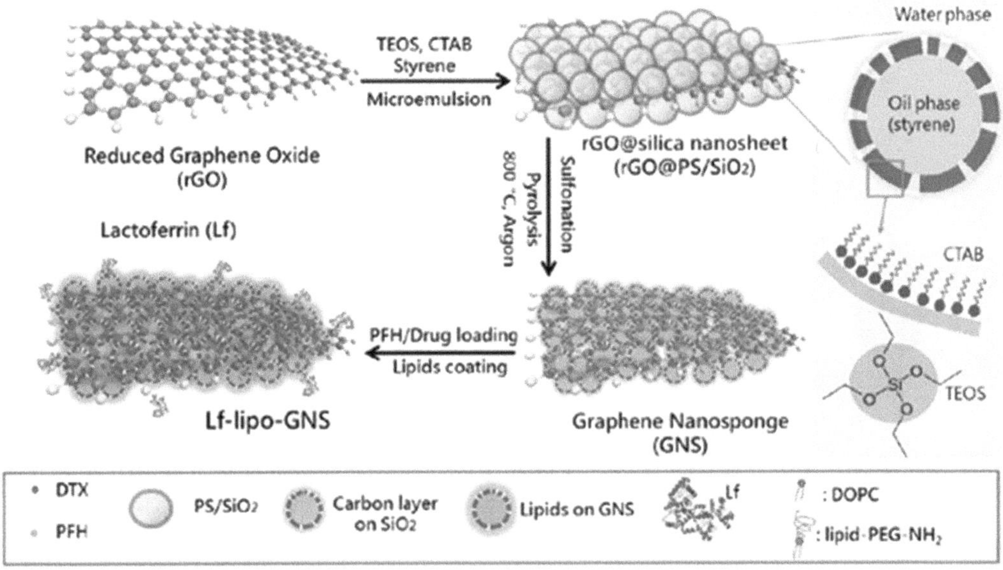

FIGURE 6.8 Schematic for the synthesis of reduced graphene oxide (rGO)-supported lipid nano-sponges. (Reprinted with permission from Su, Y. et al., *ACS Nano*, 10, 9420–9433, 2016. Copyright 2016 American Chemical Society.)

peptides were further connected to a trimaleimide scaffold through Michael-addition (Wang et al. 2017b). It was found that the nano-sponges were efficiently taken up by the leukocytes in the peripheral blood flow. The cytotoxicity was assessed in neural progenitor cells (C17.2) by the MTT assay. The results revealed that DK20 nano-sponges are non-toxic up to 100 μM. However, a slight increase in cell proliferation is noted at low concentrations of DK20.

6.5.4 Nano-theranostics

Nano-theranostics (e.g., QDs, silver, gold, iron oxide, polymer, silica, and carbon-based nanoparticles) are recently employed in drug delivery to detect and treat the disease in a single step with minimum side effects (Wang et al. 2012). Nanotheranostics can be easily localized at the specific sites of diseases to improve the permeability and retention effect. Tumour multimodal imaging and thermo-radiotherapy was studied using bismuth sulphide (Bi_2S_3) nanotheranostic agents (Wang et al. 2016). Bi_2S_3 nanoparticles were synthesized *via* a bovine serum albumin (BSA)-mediated bio-mineralization technique. Bi_2S_3 nanoparticles were synthesized through the following steps: (i) BSA was treated with bismuth nitrate ($Bi(NO_3)_3$) in acidic pH, (ii) BSA was bonded with Bi^{3+} ions through its –SH, $-NH_2$ and –COOH functional groups, and (iii) Bi_2S_3 nanoparticles were formed by adjusting the pH of BSA-Bi^{3+} complex mixture to 12. At high pH, BSA was denatured to release cysteine residues. Cysteine is one of the best sulphur sources to synthesize metal sulphide nanoparticles. Bi_2S_3 nanoparticles exhibited low *in vivo* toxicity and long circulation time during intravenous administration in mice.

Gadolinium (III)-complex-grafted lead sulphide (GCGLS) nanoparticles were examined for CT and magnetic resonance dual-modality-imaging-guided PTT (Zou et al. 2018). The schematic representation of synthesis and theranostic functions (PTT, MRI, and CT) of GCGLS nanoparticles is displayed in Figure 6.9. At first, 3-chloropropionic acid (CPA) modified lead sulphide (PbS) was synthesized using the precursors such as lead acetate, 2-mercaptoethanol, sodium sulphide, CPA, and water under N_2 atmosphere. The obtained product was centrifuged and washed with deionized water.

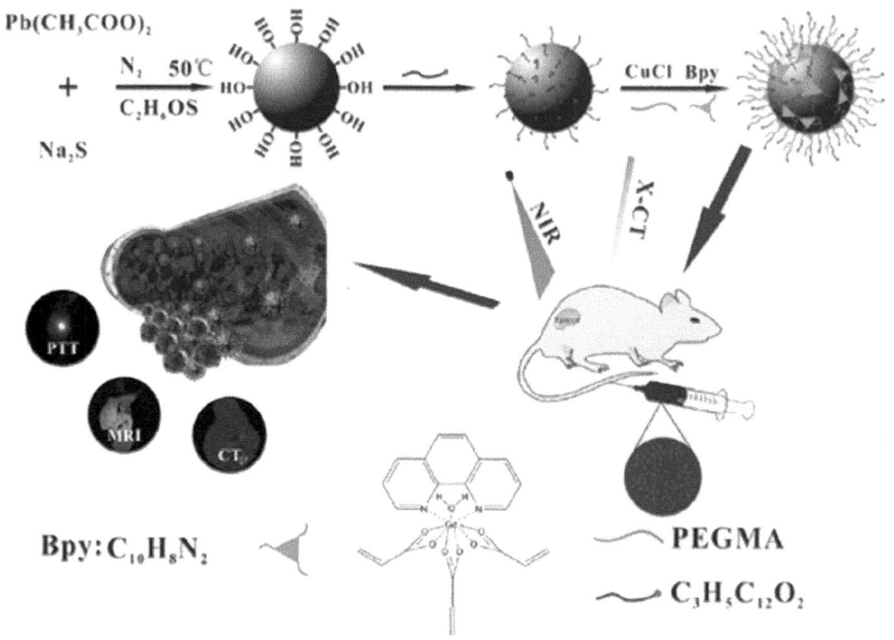

FIGURE 6.9 Schematic representation of the synthesis and theronastic functions (PTT, MRI, and CT) of GCGLS nanoparticles. (Reprinted with permission from Zou, Y. et al., *ACS Appl. Nano Mater.*, 1, 2294–2305, 2018. Copyright 2018, American Chemical Society.)

Then, CPA-PbS and $Gd(AA)_3Phen$ were dispersed in polyethylene glycol monomethacrylate (PEGMA) monomer under N_2 atmosphere. This mixture was further reacted with copper chloride (CuCl), bipyridine (Bpy), and tetrahydrofuran (THF). The product was centrifuged and washed with deionized water. The as-synthesized nanoparticles were injected into mice to detect and treat the tumour sites. PbS was responsible for the CT and the Gd complex was accountable for the MR imaging. GCGLS nanoparticles showed high stability and biocompatibility *in vitro/vivo*. The stability and dispersibility of GCGLS nanoparticles were influenced by the concentration of Gd complex. The cytotoxicity of GCGLS nanoparticles was evaluated using MTT assay in B16 cells. After 24 hours of incubation, the cell viability is up to 80%. However, the nanoparticles showed slight cytotoxicity at high concentrations during 48 hours of incubation. The photo-thermal effect was also tested after 24 hours of incubation. Under light irradiation, approximately 80% of the B16 cells were killed by 40 μg/mL of GCGLS nanoparticles. In contrast, the cytotoxicity is negligible under normal conditions (without light irradiation).

Theranostic application of sulphur-containing hyperbranched polyester (HBPE-S) nanomaterials was examined for the encapsulation of a bismuth complex, bimodal imaging, and cancer treatment (Heckert et al. 2017). The as-synthesized HBPE-S was encapsulated with different cargos such as Dil dye (for optical imaging), taxol (drug for the treatment), and Bi-DOTA (for enhanced X-ray attenuation) through solvent-diffusion method. The cytotoxicity of HBPE-S was studied using MTT assay (NSCLC lung A549 cells and cardiomyocyte H9c2 cells). The carboxylate- and folate-functionalized PNPs showed minimum toxicity for NSCLC lung A549 cells and cardiomyocyte H9c2 cells when compared to taxol-functionalized PNPs. This might be attributed to the existence of Bi-DOTA complex in the polymer cavities. Moreover, the affinity of Bi complex towards S atoms is beneficial for the encapsulation of Bi-DOTA complex at high concentration.

6.5.5 Nano-dendrimers

Nano-dendrimers are hyper-branched polymers with 3-dimensional (3D) structure, more active sites, and high solubility (García-Gallego et al. 2017). The low polydispersity of dendrimers can provide more reproducible pharmacokinetic behaviour as compared to the linear polymers. The *in vivo* application profile of dendrimers is better than that of linear polymers. A nano-dendrimer is composed of three parts such as central core, interior layer with branching units, and terminal functional group. A schematic representation of a bio-degradable nano-dendrimer is shown in Figure 6.10a (Huang and Wu 2018).

DNA, carbohydrates, polyamidoamines (PAMAM), polyesters, polyamines, poly(aryl ethers), and polypeptides are the most commonly used dendrimers for various biological applications including drug delivery, imaging, and gene therapy. Divergent and convergent approaches were initially used to

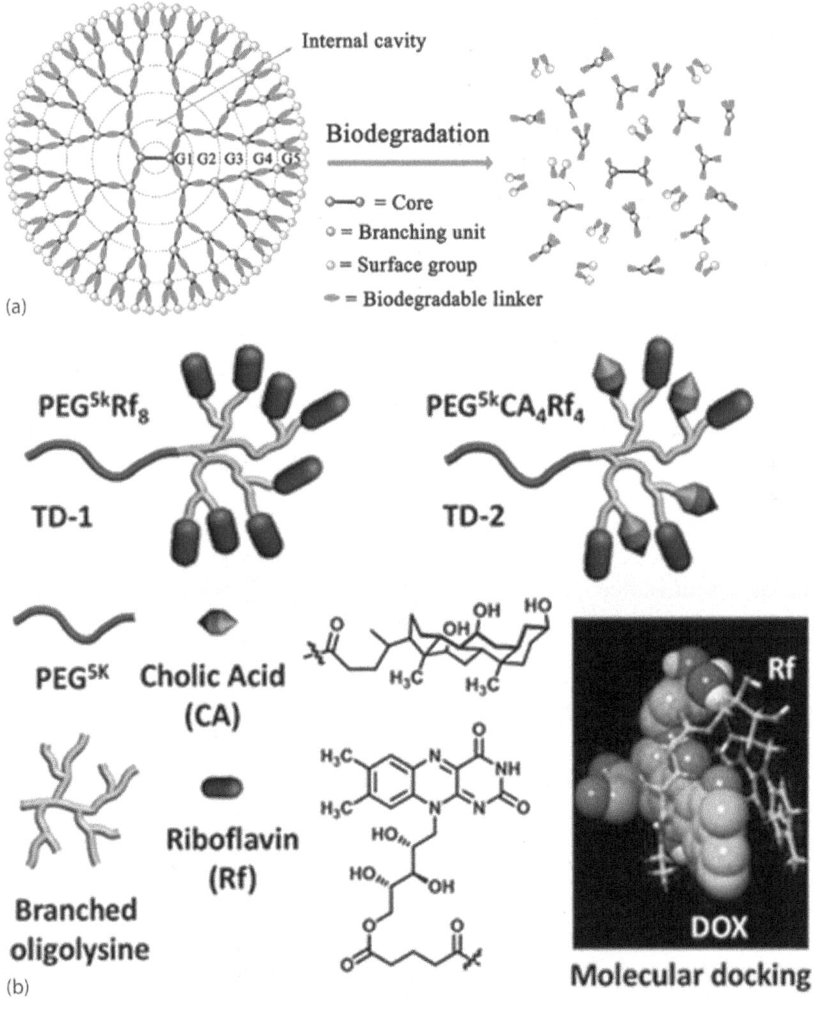

FIGURE 6.10 (a) Schematic representation of a bio-degradable nano-dendrimer. (Reprinted from *Mater. Sci. Eng. C.*, Huang, D. and Wu, D., Biodegradable dendrimers for drug delivery. Copyright 2018, with permission from Elsevier.) (b) The chemical architectures of TD1 (with riboflavin) and TD2 (with riboflavin and cholic acid). (Reprinted from Guo, D. et al., *Biomaterials*, 141, 161–175. Copyright 2017, with permission from Elsevier.)

synthesise dendrimers. These traditional techniques are time consuming and laborious. To address this, new 'accelerated strategies' such as double-stage conventional growth, double exponential growth, and hyper-monomer strategy have been introduced. Two or more robust and chemo-selective reactions can be introduced during the orthogonal growth to further expedite the synthesis of dendrimers. High-density functional groups are particularly preferred for the targeted drug delivery. In recent years, nano-dendrimers were used to delivery various drugs such as MTX, CPT, cisplatin, and paclitaxel.

Guo et al. (2017) examined the delivery of DOX using riboflavin-containing telodendrimer (TD) nanocarriers. TD is a polyethyleneglycol(PEG)-dendritic block copolymer with well-defined structures and tuneable properties. TD was synthesized from MeO-PEG^{5k}-NH_2 through solution-phase peptide chemistry technique. The chemical architectures of TD1 (with riboflavin) and TD2 (with riboflavin and cholic acid) are displayed in Figure 6.10b. The cytotoxicity and cell viability of TD were studied in various cell models (MDA-MB-231 and SKOV-3). There is no appreciable change in the IC-50 values of TD-DOX and free DOX after 72 hours of incubation time. The blank TD-1 and TD-2 do not display any cytotoxicity against most of the tested cells up to 625 mg/mL. Nano-carriers such as Doxil, DOX-TD-1, and DOX-TD-2 showed remarkably higher cell viabilities compared to free DOX, indicating low toxicity of nano-formulations during blood circulation. Haemolytic toxicity results also revealed that TD-DOX exhibited a low haemolysis rate ($\leq$4%) compared to free DOX, indicating that the TD-DOX can be administered safely via intravenous injection.

6.6 Conclusions and Future Outlook

The development of nanotherapeutics has a tremendous impact in the bio-medical industry for their potential applications in various life-threatening diseases such as diabetes, cardiovascular diseases, cancer, and infectious diseases. The main principle for the use of nanotherapeutics in oncology is ascribed to the EPR effect of the tumour microenvironment, but it can also be an obstacle for the drug delivery. This could be controlled by the precise designing of nanotherapeutics. The evolution of nanotherapeutics would absolutely improve the survival of human beings. Various nanotherapeutic candidates and their possible diagnosis and/or treatment strategies, cytotoxicity, were discussed briefly. Stimuli-responsive therapy by nanoparticles such as CNTs, QDs, and core shell structures possess high efficiency with less cytotoxicity. In some approaches, the cytotoxicity of nanomaterials was used to kill the cancer cells. Nanotheranostics are beneficial in terms of cost and time, and more importantly they lead to less discomfort for patients compared to current treatment methods, which involve invasive surgical procedures, or therapies with major side effects. The design/synthesis, drug delivery, and cytotoxicity of nanotherapeutics such as dendrimers, MNPs, sponges, theranostics, and emulsions have been discussed briefly. Nanotherapeutics with diagnosis and the precise characterization of drug targets are the foremost choice to achieve high efficacy with reduced side effects. Crystalline nanotherapeutics have increased bioavailability at the target organs as compared to amorphous nanotherapeutics. In terms of saturation solubility, amorphous nanotherapeutics are superior when compared to crystalline nanotherapeutics. Hence, more studies are required to understand the physico-chemical properties of nanomaterials for drug delivery.

The commercialization of nanotherapeutics is still a challenging task. Hence, more novel nanotherapeutics should be commercialized in the market for devastating diseases. Designing of cost-effective nanotherapeutics with multi-functionalities (diagnosing, targeting, treating), small size, high solubility, and required surface characteristics would offer a significant solution for various bio-medical applications. Most of the nanotherapeutics available in the market are commonly synthesized *via* top-down approaches. High-energy requirements, time consumption, and inconsistency in the particle size distribution are the main constraints for the conventional top-down methods. Therefore, significant work should be performed to discover suitable low-cost bottom-up approaches to design nanotherapeutics. Future research works should also be focused on multidisciplinary approaches with academia-industry collaboration.

References

Azcona, P., López-Corral, I. and Lassalle, V. (2018) Fabrication of folic acid magnetic nanotheranostics: An insight on the formation mechanism, physicochemical properties and stability in simulated physiological media, *Colloids Surf A Physicochem Eng Asp*, 537, pp. 185–196.

Blanco, E., Shen, H. and Ferrari, M. (2015) Principles of nanoparticle design for overcoming biological barriers to drug delivery, *Nature Biotechnol*, 33 (9), p. 941.

Bwatanglang, I. B., Mohammad, F., Yusof, N. A., Abdullah, J., Hussein, M. Z., Alitheen, N. B. and Abu, N. (2016) Folic acid targeted Mn:ZnS quantum dots for theranostic applications of cancer cell imaging and therapy, *Int J Nanomedicine*, 11, pp. 413–428.

Chen, F., Gao, Q., Hong, G. and Ni, J. (2008) Synthesis and characterization of magnetite dodecahedron nanostructure by hydrothermal method, *J Magn Magn Mater*, 320 (11), pp. 1775–1780.

Cheng, C., Müller, K. H., Koziol, K. K. K., Skepper, J. N., Midgley, P. A., Welland, M. E. and Porter, A. E. (2009) Toxicity and imaging of multi-walled carbon nanotubes in human macrophage cells, *Biomaterials*, 30 (25), pp. 4152–4160.

Cheng, R., Meng, F., Deng, C., Klok, H. A. and Zhong, Z. (2013) Dual and multi-stimuli responsive polymeric nanoparticles for programmed site-specific drug delivery, *Biomaterials*, 34 (14), pp. 3647–3657.

Chu, M., Pan, X., Zhang, D., Wu, Q., Peng, J. and Hai, W. (2012) The therapeutic efficacy of CdTe and CdSe quantum dots for photothermal cancer therapy, *Biomaterials*, 33 (29), pp. 7071–7083.

Couvreur, P. (2013) Nanoparticles in drug delivery: Past, present and future, *Adv Drug Deliv Rev*, 65 (1), pp. 21–23.

Dhar, S., Liu, Z., Thomale, J., Dai, H. and Lippard, S. J. (2008) Targeted single-wall carbon nanotube-mediated Pt(IV) prodrug delivery using folate as a homing device, *J Am Chem Soc*, 130 (34), pp. 11467–11476.

Dou, Y., Li, X., Yang, W., Guo, Y., Wu, M., Liu, Y., Li, X., Zhang, X. and Chang, J. (2017) PB@Au core-satellite multifunctional nanotheranostics for magnetic resonance and computed tomography imaging in vivo and synergetic photothermal and radiosensitive therapy, *ACS Appl Mater Interfaces*, 9 (2), pp. 1263–1272.

Elbaz, N. M., Ziko, L., Siam, R. and Mamdouh, W. (2016) Core-shell silver/polymeric nanoparticles-based combinatorial therapy against breast cancer in-vitro, *Sci Rep*, 6, p. 30729.

Elsabahy, M. and Wooley, K. L. (2012) Design of polymeric nanoparticles for biomedical delivery applications, *Chem Soc Rev*, 41 (7), pp. 2545–2561.

Fang, J., Nakamura, H. and Maeda, H. (2011) The EPR effect: Unique features of tumor blood vessels for drug delivery, factors involved, and limitations and augmentation of the effect, *Adv Drug Deliv Rev*, 63 (3), pp. 136–151.

Feazell, R. P., Nakayama-Ratchford, N., Dai, H. and Lippard, S. J. (2007) Soluble single-walled carbon nanotubes as longboat delivery systems for Platinum(IV) anticancer drug design, *J Am Chem Soc*, 129 (27), pp. 8438–8439.

Fu, Q., Liu, H. L., Wu, Z., Liu, A., Yao, C., Li, X., Xiao, W., Yu, S., Luo, Z. and Tang, Y. (2015) Rough surface Au@Ag core-shell nanoparticles to fabricating high sensitivity SERS immunochromatographic sensors, *J Nanobiotechnology*, 13, p. 81.

García-Gallego, S., Franci, G., Falanga, A., Gómez, R., Folliero, V., Galdiero, S., De La Mata, F. J. and Galdiero, M. (2017) Function oriented molecular design: Dendrimers as novel antimicrobials, *Molecules*, 22 (10), p. 1581.

Ghosh Chaudhuri, R. and Paria, S. (2012) Core/shell nanoparticles: Classes, properties, synthesis mechanisms, characterization, and applications, *Chem Rev*, 112 (4), pp. 2373–2433.

Guo, D., Shi, C., Wang, X., Wang, L., Zhang, S. and Luo, J. (2017) Riboflavin-containing telodendrimer nanocarriers for efficient doxorubicin delivery: High loading capacity, increased stability, and improved anticancer efficacy, *Biomaterials*, 141, pp. 161–175.

Harun, S. N., Nordin, S. A., Gani, S. S. A., Shamsuddin, A. F., Basri, M. and Basri, H. B. (2018) Development of nanoemulsion for efficient brain parenteral delivery of cefuroxime: Designs, characterizations, and pharmacokinetics, *Int J Nanomedicine*, 13, p. 2571.

Heckert, B., Banerjee, T., Sulthana, S., Naz, S., Alnasser, R., Thompson, D., Normand, G., Grimm, J., Perez, J. M. and Santra, S. (2017) Design and synthesis of new sulfur-containing hyperbranched polymer and theranostic nanomaterials for bimodal imaging and treatment of cancer, *ACS Macro Letters*, 6 (3), pp. 235–240.

Huang, D. and Wu, D. (2018) Biodegradable dendrimers for drug delivery, *Mater Sci Eng C*, 90 (2018), pp. 713–727.

Jain, R. K. (1987) Transport of molecules in the tumor interstitium: A review, *Cancer Res*, 47 (12), pp. 3039–3051.

Kam, N. W. and Dai, H. (2005) Carbon nanotubes as intracellular protein transporters: Generality and biological functionality, *J Am Chem Soc*, 127 (16), pp. 6021–6026.

Kam, N. W., Liu, Z. and Dai, H. (2005a) Functionalization of carbon nanotubes via cleavable disulfide bonds for efficient intracellular delivery of siRNA and potent gene silencing, *J Am Chem Soc*, 127 (36), pp. 12492–12493.

Kam, N. W., O'Connell, M., Wisdom, J. A. and Dai, H. (2005b) Carbon nanotubes as multifunctional biological transporters and near-infrared agents for selective cancer cell destruction, *Proc Natl Acad Sci USA*, 102 (33), pp. 11600–11605.

Kang, T., Li, F., Baik, S., Shao, W., Ling, D. and Hyeon, T. (2017) Surface design of magnetic nanoparticles for stimuli-responsive cancer imaging and therapy, *Biomaterials*, 136, pp. 98–114.

Khaliullin, T. O., Fatkhutdinova, L. M., Zalyalov, R. R., Kisin, E. R., Murray, A. R. and Shvedova, A. A. (2015) In vitro toxic effects of different types of carbon nanotubes, *IOP Conf Ser Mater Sci Eng*, 98 (1), p. 012021.

Kim, B. H., Lee, N., Kim, H., An, K., Park, Y. I., Choi, Y., Shin, K., Lee, Y., Kwon, S. G. and Na, H. B. (2011) Large-scale synthesis of uniform and extremely small-sized iron oxide nanoparticles for high-resolution T 1 magnetic resonance imaging contrast agents, *J Am Chem Soc*, 133 (32), pp. 12624–12631.

Kim, J. S. and Yu, I. J. (2014) Single-wall carbon nanotubes (SWCNT) induce cytotoxicity and genotoxicity produced by reactive oxygen species (ROS) generation in phytohemagglutinin (PHA)-stimulated male human peripheral blood lymphocytes, *J Toxicol Environ Health A*, 77 (19), pp. 1141–1153.

Lee, N., Yoo, D., Ling, D., Cho, M. H., Hyeon, T. and Cheon, J. (2015) Iron oxide based nanoparticles for multimodal imaging and magnetoresponsive therapy, *Chem Rev*, 115 (19), pp. 10637–10689.

Lei, L., Zhang, Y., He, L., Wu, S., Li, B. and Li, Y. (2017) Fabrication of nanoemulsion-filled alginate hydrogel to control the digestion behavior of hydrophobic nobiletin, *LWT-Food Sci Technol*, 82, pp. 260–267.

Li, X., Liu, M., Sun, R., Zeng, Y., Chen, S. and Zhang, P. (2017) Cardiac complications in cancer treatment—A review, *Hellenic J Cardiol*, 58 (3), pp. 190–193.

Lin, S. L., Chen, Z. R. and Chang, C. A. (2018) Nd^{3+} sensitized core-shell-shell nanocomposites loaded with IR806 dye for photothermal therapy and up-conversion luminescence imaging by a single wavelength NIR light irradiation, *Nanotheranostics*, 2 (3), pp. 243–257.

Liu, Z., Sun, X., Nakayama-Ratchford, N. and Dai, H. (2007a) Supramolecular chemistry on water-soluble carbon nanotubes for drug loading and delivery, *ACS Nano*, 1 (1), pp. 50–56.

Liu, Z., Winters, M., Holodniy, M. and Dai, H. (2007b) siRNA delivery into human T cells and primary cells with carbon-nanotube transporters, *Angew Chem Int Ed Engl*, 46 (12), pp. 2023–2027.

Loo, C., Lowery, A., Halas, N., West, J. and Drezek, R. (2005) Immunotargeted nanoshells for integrated cancer imaging and therapy, *Nano Lett*, 5 (4), pp. 709–711.

Lungu, I., Radulescu, M., Mogosanu, G. D. and Grumezescu, A. M. (2016) pH sensitive core-shell magnetic nanoparticles for targeted drug delivery in cancer therapy, *Rom J Morphol Embryol*, 57 (1), pp. 23–32.

Madani, S. Y., Mandel, A. and Seifalian, A. M. (2013) A concise review of carbon nanotube toxicology, *Nano Rev*, 4.

Madani, S. Y., Naderi, N., Dissanayake, O., Tan, A. and Seifalian, A. M. (2011) A new era of cancer treatment: Carbon nanotubes as drug delivery tools, *Int J Nanomedicine*, 6, pp. 2963–2979.

Maeda, H., Wu, J., Sawa, T., Matsumura, Y. and Hori, K. (2000) Tumor vascular permeability and the EPR effect in macromolecular therapeutics: A review, *J Control Release*, 65 (1–2), pp. 271–284.

Malekzadeh, A., Ramazani, A., Rezaei, S. and Niknejad, H. (2017) Design and construction of multifunctional hyperbranched polymers coated magnetite nanoparticles for both targeting magnetic resonance imaging and cancer therapy, *J Colloid Interface Sci*, 490, pp. 64–73.

Menon, J. U., Kuriakose, A., Iyer, R., Hernandez, E., Gandee, L., Zhang, S., Takahashi, M., Zhang, Z., Saha, D. and Nguyen, K. T. (2017) Dual-drug containing core-shell nanoparticles for lung cancer therapy, *Sci Rep*, 7 (1), p. 13249.

Mittal, S., Sharma, V., Vallabani, N. V., Kulshrestha, S., Dhawan, A. and Pandey, A. K. (2011) Toxicity evaluation of carbon nanotubes in normal human bronchial epithelial cells, *J Biomed Nanotechnol*, 7 (1), pp. 108–109.

Mousavi, H., Movahedi, B., Zarrabi, A. and Jahandar, M. (2015) A multifunctional hierarchically assembled magnetic nanostructure towards cancer nano-theranostics, *RSC Adv*, 5 (94), pp. 77255–77263.

Mura, S. and Couvreur, P. (2012) Nanotheranostics for personalized medicine, *Adv Drug Deliv Rev*, 64 (13), pp. 1394–1416.

Pantarotto, D., Briand, J. P., Prato, M. and Bianco, A. (2004) Translocation of bioactive peptides across cell membranes by carbon nanotubes, *Chem Commun*, (1), pp. 16–17.

Patlolla, A., Knighten, B. and Tchounwou, P. (2010) Multi-walled carbon nanotubes induce cytotoxicity, genotoxicity and apoptosis in normal human dermal fibroblast cells, *Ethn Dis*, 20 (1 Suppl 1), pp. S1-65-72.

Prasad, M., Lambe, U. P., Brar, B., Shah, I., Manimegalai, J., Ranjan, K., Rao, R., Kumar, S., Mahant, S. and Khurana, S. K. (2018) Nanotherapeutics: An insight into healthcare and multi-dimensional applications in medical sector of the modern world, *Biomed Pharmacother*, 97, pp. 1521–1537.

Qiao, Y., Wan, J., Zhou, L., Ma, W., Yang, Y., Luo, W., Yu, Z. and Wang, H. (2018) Stimuli-responsive nanotherapeutics for precision drug delivery and cancer therapy, *Wiley Interdiscip Rev Nanomed Nanobiotechnol*, p. e1527.

Rajamohan, S., Kumaravel, V., Muthuramalingam, R., Ayyadurai, S., Abdel-Wahab, A., Kwak, B. S., Kang, M. and Sreekantan, S. (2017) Fe_3O_4–Ag_2WO_4: Facile synthesis, characterization and visible light assisted photocatalytic activity, *New J Chem*, 41 (20), pp. 11722–11730.

Rao, L., Cai, B., Bu, L., Liao, Q., Guo, S., Zhao, X., Dong, W. and Liu, W. (2017) Microfluidic electroporation-facilitated synthesis of erythrocyte membrane-coated magnetic nanoparticles for enhanced imaging-guided cancer therapy, *ACS Nano*, 11 (4), pp. 3496–3505.

Sala-Rabanal, M., Li, D. C., Dake, G. R., Kurata, H. T., Inyushin, M., Skatchkov, S. N. and Nichols, C. G. (2013) Polyamine transport by the polyspecific organic cation transporters OCT1, OCT2, and OCT3, *Mol Pharm*, 10 (4), pp. 1450–1458.

Sanginario, A., Miccoli, B. and Demarchi, D. (2017) Carbon nanotubes as an effective opportunity for cancer diagnosis and treatment, *Biosensors*, 7 (1).

Seeta Rama Raju, G., Benton, L., Pavitra, E. and Yu, J. S. (2015) Multifunctional nanoparticles: Recent progress in cancer therapeutics, *Chem Commun*, 51 (68), pp. 13248–13259.

Shepherd, G. M. (2003) Hypersensitivity reactions to chemotherapeutic drugs, *Clin Rev Allergy Immunol*, 24 (3), pp. 253–262.

Sherje, A., Dravyakar, B., Kadam, D. and Jadhav, M. (2017) Cyclodextrin-based nanosponges: A critical review, *Carbohydr Polym*, 173, pp. 37–49.

Singh, R. and Lillard Jr, J. W. (2009) Nanoparticle-based targeted drug delivery, *Exp Mol Pathol*, 86 (3), pp. 215–223.

Singh, Y., Meher, J., Raval, K., Khan, F., Chaurasia, M., Jain, N. and Chourasia, M. (2017) Nanoemulsion: Concepts, development and applications in drug delivery, *J Control Release*, 252, pp. 28–49.

Son, K. H., Hong, J. H. and Lee, J. W. (2016) Carbon nanotubes as cancer therapeutic carriers and mediators, *Int J Nanomedicine*, 11, pp. 5163–5185.

Starowicz, M., Starowicz, P., Zukrowski, J., Przewoznik, J., Lemanski, A., Kapusta, C. and Banas, J. (2011) Electrochemical synthesis of magnetic iron oxide nanoparticles with controlled size, *J Nanopart Res*, 13 (12), pp. 7167–7176.

Su, Y., Chen, K., Sheu, Y., Sung, S., Hsu, R., Chiang, C. and Hu, S. (2016) The penetrated delivery of drug and energy to tumors by lipo-graphene nanosponges for photolytic therapy, *ACS Nano*, 10 (10), pp. 9420–9433.

Sun, Z., Xie, H., Tang, S., Yu, X. F., Guo, Z., Shao, J., Zhang, H., Huang, H., Wang, H. and Chu, P. K. (2015) Ultrasmall black phosphorus quantum dots: Synthesis and use as photothermal agents, *Angew Chem Int Ed Engl*, 54 (39), pp. 11526–11530.

Surnar, B., Sharma, K. and Jayakannan, M. (2015) Core-shell polymer nanoparticles for prevention of GSH drug detoxification and cisplatin delivery to breast cancer cells, *Nanoscale*, 7 (42), pp. 17964–17979.

Talekar, M., Kendall, J., Denny, W. and Garg, S. (2011) Targeting of nanoparticles in cancer: Drug delivery and diagnostics, *Anticancer Drugs*, 22 (10), pp. 949–962.

Tejashri, G., Amrita, B. and Darshana, J. (2013) Cyclodextrin based nanosponges for pharmaceutical use: A review, *Acta Pharm*, 63 (3), pp. 335–358.

Ten Tije, A. J., Verweij, J., Loos, W. J. and Sparreboom, A. (2003) Pharmacological effects of formulation vehicles: Implications for cancer chemotherapy, *Clin Pharmacokinet*, 42 (7), pp. 665–685.

Tian, F., Cui, D., Schwarz, H., Estrada, G. G. and Kobayashi, H. (2006) Cytotoxicity of single-wall carbon nanotubes on human fibroblasts, *Toxicol In Vitro*, 20 (7), pp. 1202–1212.

Trotta, F., Caldera, F., Dianzani, C., Argenziano, M., Barrera, G. and Cavalli, R. (2016) Glutathione bioresponsive cyclodextrin nanosponges, *ChemPlusChem*, 81 (5), pp. 439–443.

Unni, M., Uhl, A., Savliwala, S., Savitzky, B., Dhavalikar, R., Garraud, N., Arnold, D., Kourkoutis, L., Andrew, J. and Rinaldi, C. (2017) Thermal decomposition synthesis of iron oxide nanoparticles with diminished magnetic dead layer by controlled addition of oxygen, *ACS Nano*, 11 (2), pp. 2284–2303.

Uskokovic, V. (2009) Challenges for the modern science in its descend towards nano scale, *Curr Nanosci*, 5 (3), pp. 372–389.

Vignesh, K., Suganthi, A., Min, B. and Kang, M. (2014) Photocatalytic activity of magnetically recoverable $MnFe_2O_4$/g-C_3N_4/TiO_2 nanocomposite under simulated solar light irradiation, *J Mol Catal A Chem*, 395, pp. 373–383.

Vyskocil, J., Petrakova, K., Jelinek, P. and Furdek, M. (2017) Cardiovascular complications of cancers and anti-cancer therapy, *Vnitr Lek*, 63 (3), pp. 200–209.

Wang, D., Zhou, J., Shi, R., Wu, H., Chen, R., Duan, B., Xia, G. et al. (2017a) Biodegradable core-shell dual-metal-organic-frameworks nanotheranostic agent for multiple imaging guided combination cancer therapy, *Theranostics*, 7 (18), pp. 4605–4617.

Wang, H., Yapa, A., Kariyawasam, N., Shrestha, T., Kalubowilage, M., Wendel, S., Yu, J. et al. (2017b) Rationally designed peptide nanosponges for cell-based cancer therapy, *Nanomedicine*, 13 (8), pp. 2555–2564.

Wang, L.-S., Chuang, M.-C. and Ho, J.-A. A. (2012) Nanotheranostics—A review of recent publications, *Int J Nanomed*, 7, p. 4679.

Wang, Y., Wu, Y., Liu, Y., Shen, J., Lv, L., Li, L., Yang, L., Zeng, J., Wang, Y. and Zhang, L. W. (2016) BSA-mediated synthesis of bismuth sulfide nanotheranostic agents for tumor multimodal imaging and thermoradiotherapy, *Adv Funct Mater*, 26 (29), pp. 5335–5344.

Wei, W., He, X. and Ma, N. (2014) DNA-templated assembly of a heterobivalent quantum dot nanoprobe for extra- and intracellular dual-targeting and imaging of live cancer cells, *Angew Chem Int Ed Engl*, 53 (22), pp. 5573–5577.

Wheeler, H. E., Maitland, M. L., Dolan, M. E., Cox, N. J. and Ratain, M. J. (2013) Cancer pharmacogenomics: Strategies and challenges, *Nat Rev Genet*, 14 (1), pp. 23–34.

Wong, P. T. and Choi, S. K. (2015) Mechanisms of drug release in nanotherapeutic delivery systems, *Chem Rev*, 115 (9), pp. 3388–3432.

Xiong, F., Chen, Y., Chen, J., Yang, B., Zhang, Y., Gao, H., Hua, Z. and Gu, N. (2013) Rubik-like magnetic nanoassemblies as an efficient drug multifunctional carrier for cancer theranostics, *J Control Release*, 172 (3), pp. 993–1001.

Yang, L., Gao, M. X., Zhan, L., Gong, M., Zhen, S. J. and Huang, C. Z. (2017) An enzyme-induced Au@Ag core-shell nanostructure used for an ultrasensitive surface-enhanced Raman scattering immunoassay of cancer biomarkers, *Nanoscale*, 9 (7), pp. 2640–2645.

Yang, S., Li, Z., Wang, Y., Fan, X., Miao, Z., Hu, Y., Li, Z., Sun, Y., Besenbacher, F. and Yu, M. (2018) Multifunctional Bi@PPy-PEG core-shell nanohybrids for dual-modal imaging and photothermal therapy, *ACS Appl Mater Interfaces*, 10 (2), pp. 1605–1615.

Yapa, A., Wang, H., Wendel, S., Shrestha, T., Kariyawasam, N., Kalubowilage, M., Perera, A. et al. (2018) Peptide nanosponges designed for rapid uptake by leukocytes and neural stem cells, *RSC Advances*, 8 (29), pp. 16052–16060.

Zhao, M.-X., Zhu, B.-J., Yao, W.-J. and Chen, D.-F. (2016) Therapeutic effect of quantum dots for cancer treatment, *RSC Advances*, 6 (114), pp. 113791–113795.

Zhou, L., Yuan, J., Yuan, W., Sui, X., Wu, S., Li, Z. and Shen, D. (2009) Synthesis, characterization, and controllable drug release of pH-sensitive hybrid magnetic nanoparticles, *J Magn Magn Mater*, 321 (18), pp. 2799–2804.

Zou, Y., Jin, H., Sun, F., Dai, X., Xu, Z., Yang, S. and Liao, G. (2018) Design and synthesis of a lead sulfide based nanotheranostic agent for computer tomography/magnetic resonance dual-mode-bioimaging-guided photothermal therapy, *ACS Appl Nano Mater*, 1 (5), pp. 2294–2305.

7

Identifying Nanotoxicity at the Cellular Level Using Electron Microscopy

Kerry Thompson, Alanna Stanley, Emma McDermott, Alexander Black, and Peter Dockery

7.1 Introduction

Nanotoxicity is a relatively new term which describes the toxicological effects which nanomaterials and nanoparticles can exert on a biological system. Engineered nanomaterials can be defined as particles that are within the size range of 0.1–100 nanometres (nm) or having one dimension less than 100 nm in diameter (Report, 2007, Schrand et al., 2012). Nanoparticles are in this same size range and have all three dimensions of their shape in the nanoscale (Kumar et al., 2014). Nanoparticles can be engineered or may be found free in nature. They can be further classified according to their size, shape, and properties, and sometimes as organic or inorganic. For the purposes of this chapter we will continue to focus on nanoparticulate objects and will refer to them as nanoparticles. The past decade has seen an explosion of interest in the potential medical applications of nanoparticles. The need to understand the sometimes deleterious effects associated with their use has been heightened due to the increasing risk of exposure on living systems and potential negative impact on human health and the environment, so the nanotoxicological effects of these particles are being investigated more vigorously by academia and industry alike. These effects and overall fate of the nanoparticles are dependent on their distribution and interaction with the biological system (Krpetic et al., 2014).

Nanomaterials and nanoparticles have been used extensively as additives for paint, in water treatment systems, cosmetics, as antimicrobials, as vehicles for drug delivery, and more recently as theranostic, diagnostic, and imaging probes for cells and tissues (Schrand et al., 2012, Lynch et al., 2015). To fully understand the potential effects of nanoparticles on an *in vitro* or *in vivo* system, the type and size of nanoparticle under investigation must be considered. Much of the work cited in this chapter refers to the interaction of nanoparticles with epithelial cells – the cells in the body which come into contact with the exterior (external environment) and line the body cavities. Electron microscopy is the sole technique or research tool capable of fully elucidating the ultrastructural alterations that may occur in a biological system after exposure to nanoparticles, therefore helping to shed light on the possible ultrastructural hallmarks of nanotoxicity. As the size of many of these nanomaterials and nanoparticles is similar to or below the wavelength of light, the micron scale of light microscopical observations is often inadequate to resolve and clearly demonstrate their features. The spatial resolution delivered by electron microscopy allows thorough identification and characterisation at the nanoscale via the production of high-magnification high-resolution data.

This chapter will describe the various forms of electron microscopy, its associated techniques, and its usefulness in identifying nanotoxicity at the cellular level.

7.2 Introduction to Electron Microscopy

Electron microscopy is a form of microscopical imaging whereby accelerated electrons and a series of electromagnetic lenses are used to visualise the specimen under study. In the early twentieth century it was discovered that electrons behaved much in the same way as light, and when accelerated and contained in a vacuum, could be controlled by electric and magnetic fields.

One must first consider the concept of *resolution* to truly understand the power of the electron microscope. Resolution can be defined as the ability of an optical microscope to distinguish detail and is related to the numerical aperture, or light gathering ability of the lens, and the wavelength of light used (Weakley, 1972). Unaided, the human eye can distinguish clearly two points that are approximately 0.2 millimetres (mm) apart from one another. If the points are closer to one another, they appear as a single blurry dot. Light microscopy affords a resolution of approximately 0.2 micrometres (μm), whilst electron microscopy allows objects to be resolved down to the nanometre (nm) scale (approximately 1 nm) or, in a biological context roughly the size of some cellular organelles (Perkins et al., 2009). It is predominantly for this reason that electron microscopy is one of the most useful techniques to study nanotoxicity at the cellular level. The high-magnification, high-resolution images allow the observer to analyse the intricacies of tissue, cellular, and subcellular environments in detail.

In 1961, Cosslett noted in a foreword to Kays' *Techniques for Electron Microscopy* that 'Electron microscopy is an art as much as it is a science. Its applications extend into almost all branches of pure and applied science – biological as well as inorganic' (Kay, 1961). This still very much remains the case, and perhaps even to a greater extent nowadays with the resurgence in popularity and development of the technique. To best characterise structures, alterations, and interactions at the bio-nano interface a combinatorial approach may be adopted with each technique lending a particular strength to the workflow. The bio-nano interface can be described as 'the interface where artificial engineered nanomaterials or nanoparticle systems interact with biological systems at the nano-scale level' (Krpetic et al., 2014). Each technique in the workflow provides the necessary data for the establishment of reliable and robust characterisations of different element of the same sample. Over the course of this chapter, the various forms of electron microscopy will be discussed while emphasising the paramount importance of high standards of specimen preparation.

7.3 Electron Microscopy – Types, and the Modern Resurgence of the Technique

Conventionally it is regarded that there are two main types or modes of electron microscopy, transmission electron microscopy (TEM) and scanning electron microscopy (SEM). There have been many adaptations of these techniques, the base microscopes and their associated specimen preparations which include low temperature and cryo techniques. To ensure that the specimen is able to withstand the electron beam, various preparatory techniques (Figure 7.1) must be carried out and will be discussed in greater detail in Section 7.3. When gathering three-dimensional (3D) information about a specimen, the microscopist generally employs the SEM, where it is used to visualise the surface topography of a specimen. TEM images are generally two dimensional (2D) in nature but in recent years and with advances in technology, computational power, and software packages a series of 2D TEM images can be collected via electron tomography (ET) and reconstructed to create a 3D model of the structure of interest.

TEM is a technique that allows for the observation and imaging of the internal components of a sample and cellular ultrastructure. It is used to characterise structure of nanoparticles and the inherent interaction of these nanoparticles with the sample under study. Conventional TEM has long been regarded as the gold standard for characterisations of this nature (Costanzo et al., 2016). The high-resolution, high-magnification images afford the scientist detailed structural information which can be related and interpreted with regard to the functional context of the experimental scenario. Many laboratories, institutes, and universities have well-established electron microscopy units, providing relative ease of access to such technologies for interested groups.

SEM enables identification of samples down to length scales of roughly 10 nm, and images created in the SEM reveal information on the external structural arrangement, spatial distribution, and surface topography, along with the geometrical features of a structure. Micrographs display information about the exterior shape, size, orientation, and density of the sample under study. Deviations in the conventional SEM workflow are progressing with the expansion in development of the focused ion beam (FIB) and serial block face (SBF) microscopy techniques (Smith and Starborg, 2018). These processes lack the

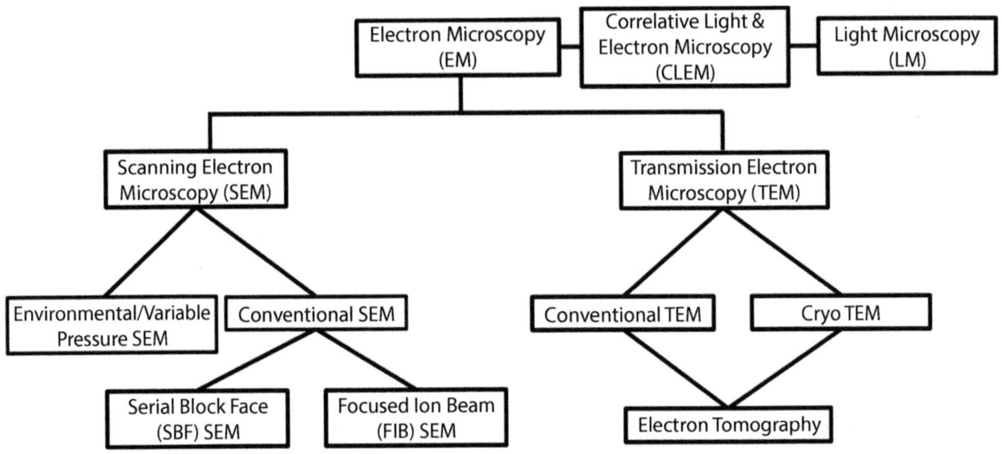

FIGURE 7.1 Overview of EM modalities. Electron microscopy can be classified into two main or parent modalities – scanning and transmission electron microscopy (SEM and TEM, respectively). Each modality has evolved and been modified in recent years with the resultant workflow described above. In addition, the correlative or CLEM workflows, combining any one of numerous light microscopy techniques with a form of electron microscopy, allow for a potential combination of dynamic, high-magnification, high-resolution data. The modifications of each of the EM modalities now incorporate the low-temperature and minimally damaging/invasive cryo and variable pressure or environmental techniques, along with the volume electron microscopy methodologies.

possible resolution that can be achieved with TEM but are much more time efficient and are capable of gathering much larger datasets. Another variation to the conventional use of SEM is the use of so called 'wet imaging' under high-vacuum conditions. This type of technique removes the need for the extensive sample dehydration and allows viewing of cell or tissue isolates in their fully hydrated state. Samples are contained on a membrane within a stable capsule adapted for SEM imaging.

The newer modes of electron microscopy have allowed for greater and renewed interest in the experimental usefulness of electron microscopy and the technique has undergone somewhat of a renaissance in recent years. These techniques and developments, along with the correlative workflows where a multimodal microscopical approach is adopted, are discussed further in Section 7.4 with the advantages of modern imaging techniques to assess nanotoxicity at the cellular level.

7.4 Interpretation of EM Images

The accurate interpretation of electron microscopy images is a learned skill. For reliable assessment of the toxicological impact of a nanoparticle on a sample the microscopist must first be confident with the expectations and procedures of the workflow. They will make a number of observations on the quality of the organism/tissue/cells at the initial inspection of the sample under the SEM or TEM. These observations will help to confirm if the sample has been prepared correctly with due care and attention. Other factors prior to this step in the workflow will be taken into account, such as the gross appearance of the prepared sample for SEM or the texture of the TEM block on sectioning and its tactile nature; for example, is its texture sticky/brittle when the sample is trimmed to a suitable size? Many of these questions tie directly back into the need for adequate sample preparation. All initial observations, as with any good experiment, should be made on an untreated control to establish a baseline from which to ascertain the 'normal' or expected. In our group we have benchmarked a number of simple questions thorough microscopists will ask themselves:

7.4.1 For SEM

1. Does the sample appear to be '*Charging*' (an accumulation of electrons under the beam where the image appears extremely bright or distorted)?
2. Was the sample chemically dried or critically point dried?
3. Are there cracks apparent on the sample surface which may have been induced by sample preparation?
4. Is there precipitate present on the sample surface – was this caused by inappropriate buffer use?
5. Are the features of interest apparent and present?

7.4.2 For TEM

1. Does the section of the sample in the resin on the grid hold up well under the beam of the microscope? Is the section starting to peel away or roll up? If the answer is yes, then perhaps the resin has been incorrectly prepared.
2. Does the untreated control sample appear, especially with regards to cellular membranes, intact? If the answer is no at this point, another sample from the group must be sectioned for comparative purposes. If membranes appear ruptured or the cell/tissue seems deformed and damage is also present, the fixation and perhaps osmolarity of the buffer used for the preparation must be questioned.
3. Intracellularly, does the nucleus demonstrate a clear double membrane and do the mitochondria appear to be adequately preserved? Are the membrane infoldings of the cristae apparent and clearly delineated. These features are regularly used as the major indicators on the level and quality of sample preservation and preparation.

4. Does the cytoplasm of the cell appear consistent, is the granular nature of the ribosomal content patchy, or is there a marked absence in contrast between cellular components?
5. Are there any tears, large holes, or unexplained structural damage in the sample – do cells or tissue components in close opposition touch as expected? Can unusual particulate be seen on the sample? Is it residue from the staining procedure or are there any crystalline structures present which may be attributable to the buffer used? Are there any knife or score marks present from the sectioning process? If so, move to an unobstructed area for analysis.

Satisfactory answers to the above questions will allow the microscopist to proceed to draw well-informed conclusions between the control and treated samples, knowing they have tried to exclude artefacts induced into the sample via sample preparation, processing, sectioning, or staining. The reader is directed towards Figure 7.5 for a summary of sample preparation procedures.

Electron microscopy, when properly performed, presents the scientist with a complex snapshot of a once dynamic structural system frozen in time. To understand the alterations which may be the result of treatment or exposure, the microscopist must first have a good understanding of the normal structure. The intracellular environment at the level of the electron micrograph is an intricate yet chaotic place. When confronted with an image in which similarly electron dense particles are present (e.g., titanium dioxide, ribosomes, or glycogen), the microscopist must be confident enough to be able to clearly distinguish the components and not allow for intracellular components to be confused with nanoparticles and vice versa (Muhlfeld et al., 2007b). In incidences like this, or should there be doubt about such a scenario, energy-dispersive X-ray (EDX) or X-Ray diffraction or crystallography may be useful to help identify the elemental composition of the queried object. Automated image processing and analysis techniques which employ machine learning are being developed to produce a more rapid and higher throughput investigation of the highly detailed images formed using electron microscopy (Laramy et al., 2015, Baichuan and Amanda, 2018, Madsen et al., 2018). In parallel, the automated identification of cells, intracellular compartments, and subcellular organelles is also being developed in this field using citizen science-type projects. These projects gather vast quantities of measurements via tracing or drawing around an object to contribute to characterisation of specified structures to help computers 'learn' to identify what the eye can distinguish easily (Heck et al., 2018). Identification of nanoparticles within a cellular environment using artificial intelligence, particularly when resulting images are composed of distinct electron densities and opacities, seems likely to be an impending reality which will vastly enhance analyses of this nature.

7.5 Nanotoxicology at the Cellular Level

Within the biological sample, tissue, or cell a number of detrimental or toxicity-driven processes can take place. These include but are not limited to

1. The generation of reactive oxygen species (ROS)
2. The initiation of inflammation like processes
3. Membrane and organelle membrane damage
4. DNA damage
5. A combination of the above with resultant cell death

These processes can be initiated by the introduction of the nanomaterial to the specimen and trigger a cascade of intracellular events which may cause an alteration in both structure and function of the sample under study.

When reviewing nanotoxicology at the cellular level, one must also first take into account the nature of the nanoparticle. A discussion on the full categorisation of nanoparticles is not within the remit of this chapter but we shall offer a brief synopsis. Nanoparticles can be deemed to be either free and unbound, or contained bound within material or associated with it (Hansen et al., 2008). Furthermore, nanoparticles can be classified or sorted according to their material, size, shape, surface coating, or charge and whether they are known to agglomerate, aggregate, or disperse when introduced to a biological system. Some are metallic in nature (gold, iron, copper) and others non-metal (carbon, silica, hydroxyapatite), whilst others are polymeric (Liu and Tang, 2017). Little is known about the long-term effects of nanomaterial and particle exposure due to the relatively recent nature of the field. Current research has identified the ability of these particles to utilise existing cellular mechanisms, particularly where the particle is of a size less than 50 nm (Kumari and Yadav, 2011, Elsaesser and Howard, 2012). Alterations to the surface chemistry of these particles can also make them more or less favourable for uptake into the cell. This surface chemistry can often help to predict what type of interaction the nanoparticle will have. In *in vitro* systems where cultured cells are incubated with a given particle, there are a number of cellular mechanisms which will determine the uptake, internalisation, or fate of this particle. Particle size is also known to have an impact on the distribution of nanoparticles in both cells and tissues, where particles of a smaller size are generally thought to exert a greater effect (Johnston et al., 2010, Kettler et al., 2014, Williams et al., 2016). This is thought to be related to the sheer quantity of particles that enter the cell and therefore the overall surface area they are capable of occupying within the cell or tissue. Little is still known about the overall interactions of various types and sizes of particles with specific organ systems, especially in the human.

Epithelial tissue, or *epithelium* (which comes from the Latin meaning '*upon sheets*') is found covering surfaces of the body that either come into contact with the exterior or line the internal tubes and cavities. The initial interaction of nanoparticles with cells in an *in vivo* system tend to be primarily with epithelial cells. Moreover, many nanotoxicological studies at the ultrastructural level using EM will use epithelial cells in culture to classify and investigate the interaction at the bio-nano interface (Muhlfeld et al., 2007b, Ye et al., 2015). All epithelia are avascular but adhere to a vascularised bed of connective tissue, with the two layers being separated by an intermediate layer known as the basement membrane, which is generally manufactured by and secreted from the epithelial cells themselves. Epithelial cells display three key characteristics: cells are tightly bound together to form these sheet-like structures through junctional complexes; there are functionally different membrane domains within the cell (apical, basal, and lateral), and cells adhere to the underlying basement membrane (Gray et al., 1995, Ross, 1995, Young and Wheater, 2006). Therefore, they present the ideal model system for studying nanoparticle uptake.

The primary function of epithelial tissue is to act as a selective barrier which either aids or prevents substances from traversing the surface that it presents. This barrier function is mediated by the presence of an intricate arrangement of junctional complexes, cytoskeletal networks, and domain-specific protein receptors (Rodriguez-Boulan et al., 2005, Shin et al., 2006). The cell membrane can also be thought of as a barrier, a delicately structured yet robust phospholipid bilayer which infers an intelligent selectivity on the components that try to pass into and out of the cell. The ability of the cell to maintain homeostasis is regulated primarily by the membrane and it also plays a major role in many signalling pathways, whose receptors are often found protruding from the membrane's surface (Behzadi et al., 2017). This tightly regulated and maintained zone is the first barrier that many nanoparticles must pass to gain entry to the cell. The phospholipid bilayer is studded with transmembrane proteins, responsible for signal transduction mechanisms (which trigger endo- and exocytosis), and cholesterol inclusions which provide strength to the structure. The signalling component on the cell surface, receptors, or transmembrane proteins have been described as possessing the ability to group together to initiate internalisation of particles of up to 50 nm in size. This specific particle size is thought to be linked to the clustering of receptors and is very important to consider when investigating nanoparticle internalisation processes, i.e., smaller particles would display more favourable uptake but can exert a much greater toxic effect (Missaoui et al., 2018). On crossing the cell membrane and gaining entry into the cell, the nanoparticle is then confronted with a chaotic cytoplasmic environment crowded with organelles including mitochondria, lysosomes, endoplasmic reticulum, nucleus, and a dense cytoskeletal network (actin microfilaments, microtubules, and intermediate filaments), amongst others. Historically, electron microscopy has been

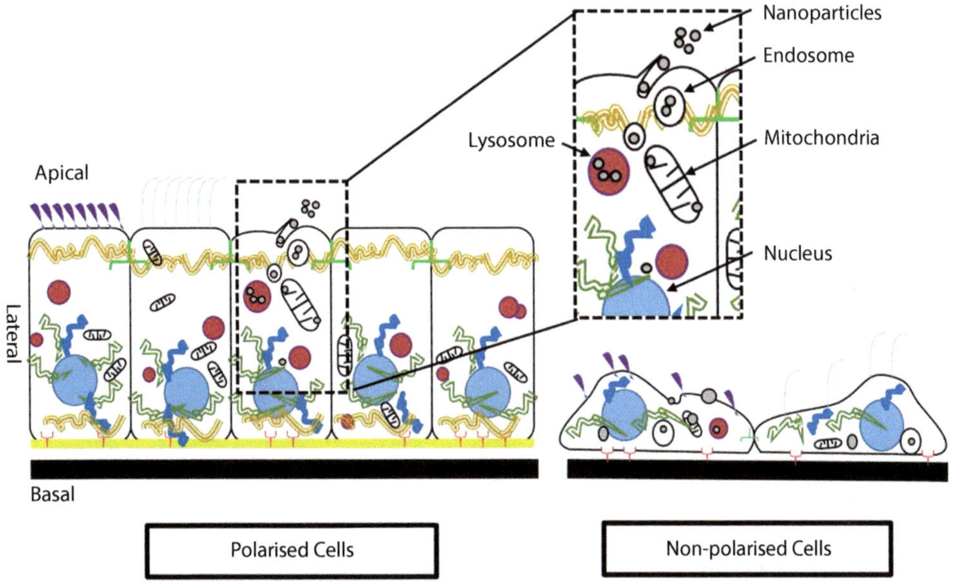

FIGURE 7.2 Polarised cell vs cell schematic. This figure highlights the phenotypical differences between polarised and non-polarised cells in schematic form. In the polarised cells, tight junctions and other cellular junctions (green and red brackets) assist with the formation of clearly distinguishable apical, basal, and lateral surfaces. These cells display a higher level of complexity in the distribution of the intracellular environment and its contents. For cells in routine culture to become polarised, they must be grown on an extracellular matrix (represented with yellow line in schematic). Uptake of nanoparticles into the cell is highlighted at the apical most surface of the polarised cell with particles tracking from the extracellular space, internally into the endocytic pathway. Endosomes can be seen to fuse with lysosomes, in mitochondria, near the nucleus, and free in the cytoplasm. These are the common locations in which particles have been observed after uptake. In the non-polarised cells, which are not cultured on an extracellular matrix, the structural delineation between apical and later surfaces is less apparent, even though cellular junctions are still formed. Cells display a larger surface area apically and uptake can occur at any location therein. The distance from the apical to basal surface is reduced and therefore this represents a more simplified model of uptake. (Purple and blue apical lines – microvilli and cilia, respectively; cytoskeletal components – microtubules (green), actin microfilaments (yellow), and intermediate filaments (blue lines); tissue culture surface (black line).

used to characterise and describe the ultrastructure of cells and tissues and today we now use these technologies to validate nanomaterial or particle uptake and internalisation along with the potential local nanotoxicity that may occur as a result of exposure.

Epithelial cells that are cultured to develop a polarised phenotype (i.e., displaying a distinct apical and basolateral surface) are most useful when studying trafficking mechanisms of nanoparticles through the cell (Figure 7.2). Cells cultured in a non-polarised state also remain a very useful tool when assessing the general cell-nanoparticle interactions.

7.5.1 Endocytosis and Internalisation Pathways

The delivery of nanomaterials and particles into the internal cytosolic environment frequently requires the destination to be specific and targeted. Often, nanoparticles end up enclosed within a membrane-bound organelle or endosome and remain trapped, unable to perform their desired function. The process of internalisation is generally referred to by the umbrella term of endocytosis. The internal environment in these compartments has a lower pH than the extracellular environment of origin and drops even further to an even lower pH in the lysosome, which is generally regarded as the 'graveyard' of nanoparticles

(Krpetic et al., 2014). The nanotoxicity that can result from internalisation may occur as a result of the degradation of the nanoparticle or material by intracellular enzymes. This may alter their surface, not just their surface chemistry, and the acidic pH4.5 of the lysosome can etch away the exterior coating. This can lead to localised generation of ROS and affect homeostasis (Deng and Gao, 2016). To better understand the potential effects of intracellular degradation and resultant nanotoxicity within the cell, the endocytic process will be described.

Endocytosis is a process in which small particles or fluid, along with select areas of the plasma membrane, are drawn into the intracellular environment by the individual cell (Figure 7.3).

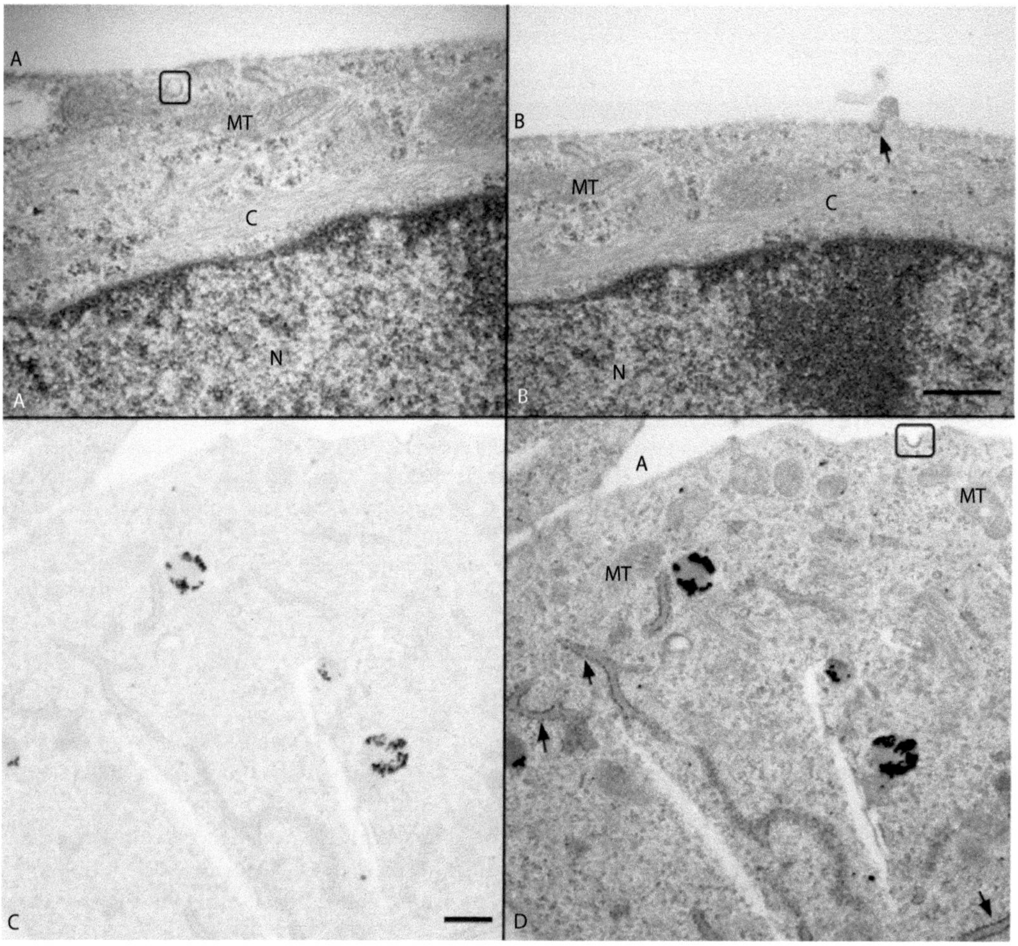

FIGURE 7.3 Endocytosis and nanoparticles within the cell. Control-untreated cells in this conventionally prepared TEM sample in A and B depict two stages of the endocytic mechanism – In A (black box), a completed endocytic event can be seen. A small membrane-bound vesicle is located just under the apical (A) surface of the cell. In B, the black arrow directs the reader towards an endocytic event in progress. Mitochondria (MT) and cytoskeletal filaments (C) can also be clearly seen superior to the nucleus (N). In C and D, the same image can be seen. In D, the contrast has been increased to demonstrate the ultrastructural detail within the cell. In C, the image clearly demonstrates the nanoparticles which have been taken up into the cell from the apical surface (A) via an endocytotic event (black box). In this instance, the mitochondria (MT) and rough endoplasmic reticulum (black arrow) appear unaffected by the uptake or inclusion of nanoparticles. Scale 500 nm for all.

This process may or may not be receptor-mediated. In this way the composition of the plasma membrane can be maintained which is important as it ultimately affects how the cell interacts with the external environment (Gray, 2008, Doherty and McMahon, 2009, Adjei et al., 2014). Endocytosis can vary in the speed at which it occurs, the amount of particulate that is endocytosed, and the size of endocytic vesicle formed (Rejman et al., 2004, Wu et al., 2009). The endocytic mechanism has been conserved in many biological systems and it remains the most prevalent route of entry to the body for viruses. Research into the uptake mechanisms of nanoparticles have shown that they too are taken into the cell by endocytic means (Adjei et al., 2014).

It is useful to note that the sheer mass of particles to reach the cell endocytotically may not be a useful indicator of the dose. Toxicological reactions generally have a linear relationship between the dose and the response, but this is not usually the case with nanotoxicology. When characterising nanoparticle exposure with resultant nanotoxicology, the surface area of the particle available to induce these subcellular alterations must also be taken into account. Furthermore, the route of exposure must be considered. In an *in vivo* context exposure may be via the following modes: inhalation, penetration of the skin, ingestion, or injection. A protein 'corona' generally forms on the exterior surface of the particles as soon as they come into contact with a biological system or in the case of *in vitro* experiments with proteinaceous cell culture medium (Lynch et al., 2013). This coating will further influence the interaction of the particle with the cell and is often thought to be a key player in determining whether the nanoparticle may be taken up by specific or nonspecific means into the cell (Monopoli et al., 2012).

Endocytosis can be roughly divided into two categories, depending on the nature of the internalised substance. *Phagocytosis* (literally, *cell-eating*) is the term used to refer to the mechanism by which the cell takes in particulate matter, whereas *pinocytosis* (literally, *cell-drinking*) is the term used to describe the intake of fluid. Endocytosis can be performed in one of four ways, depending on the type of cell in question and its functional requirement. The mechanisms are as follows: *clathrin-dependent endocytosis*, *caveolin-mediated endocytosis*, *macropinocytosis*, and *dynamin- and clathrin-independent endocytosis* (Seto et al., 2002, Liu and Shapiro, 2003). Clathrin-mediated endocytosis, also known as receptor-mediated endocytosis, takes place at specialised regions along the surface of the plasma membrane which are referred to as *clathrin-coated pits*. This form of endocytosis occurs at a much more rapid rate than that of phagocytosis. The protein clathrin has the ability to generate enclosed regions where the membrane invaginates to form an inward-facing hemisphere which contains the cell surface receptors and material being ingested (Gray, 2008). These enclosures then pinch off from the plasma membrane surface and are carried into the cell for further processing. Vesicles are generally sorted for either degradation by lysosomes or recycling to the plasma membrane. The endocytic system is linked functionally to lysosomal components of the cell (Apodaca, 2001). Lysosomes carry out both *heterophagy*, breakdown of exogenous contents imported into the cell, and *autophagy*, which is the breakdown of intracellular components (Dunn et al., 1980). This vesicular pathway is in turn functionally linked with the intracellular membrane system of both the Golgi apparatus and endoplasmic reticulum (Gray, 2008). As previously stated, the properties of the nanoparticle in question can determine the method by which it is internalised. Previous research has investigated whether cells display a preference for shape of particle and many studies corroborate that spherical particles are more readily taken up (Chithrani et al., 2006, Adjei et al., 2014).

The internalisation of nanoparticles into endosomes is often problematic for particle designers, who have had to incorporate surface modifications and size varieties to overcome this challenge. Inclusion into the cell can also be achieved via nanoparticle coatings composed of liposomal compounds or peptides which assist with penetration of the cell surface.

Endocytosis is generally a more complicated process in polarised *in vitro* epithelial cell systems in comparison to the non-polarised phenotype. This is due to the ability of the polarised cell to carry out internalisation of macromolecules from both the apical and basolateral surfaces (Apodaca, 2001). The actin microfilament and microtubule cytoskeletal components are thought to play pivotal roles in the endocytic process (Di Fiore and Scita, 2003, Kornilova, 2014). Not only do these filamentous structures provide the tracks along which endocytosed vesicles are shuttled, but the GTPase subfamilies which control them, Rho and Rab, are thought to provide integrated control of both membrane trafficking and cytoskeletal reorganisation (Feng et al., 1995, Di Fiore and Scita, 2003).

Calcium has also been implemented in triggering all forms of endocytosis in nerve cells in rodents (Wu et al., 2009). In this study, as calcium levels increased the rate of clathrin-mediated endocytosis increased from very slow to fast. It was suggested that this was as a result of an increase in speed of the membrane binding and invagination processes. Wu et al. have listed five speeds that endocytosis can occur at: very slow, slow, rapid, bulk, and overshoot endocytosis. Luzio et al. have also documented a need for calcium ions in aiding fusion events of endosomes to lysosomes and condensation of the lysosomal contents in non-neuronal cells (Luzio et al., 2007).

Low temperature is often used to selectively inhibit or slow down endocytosis and active transport along with the fusion of endosomes and liposomes (Haylett and Thilo, 1991, Johnston et al., 2010). At 4°C, which will from now on be referred to as low temperature, the active ATP-dependent transport mechanisms slow down within the cell, when compared to processes at 37°C. To try and elucidate the mechanism of uptake, Chithrani et al. carried out a comprehensive study in 2006 where a range of gold nanoparticle sizes (14, 30, 50, 74, and 100 nm) and shapes (spherical and rod) were placed in culture with HeLa epithelial cells in a culture medium which contained serum. They observed that the spherical particle uptake was greatest, and these particles appeared to be contained within membrane-bound vesicles. The 50 nm particles demonstrated the greatest level of uptake and it was suggested that perhaps particles of smaller diameter were too small to be recognised for endocytosis (Chithrani et al., 2006). To further study the uptake mechanism and to clarify whether these nanoparticles were in fact endocytosed, in 2007 Chithrani and Chan tested a temperature-driven experimental hypothesis. They demonstrated lowered rates of gold nanoparticle uptake at low temperatures in a number of cell lines (Chithrani and Chan, 2007). When cells were placed in low temperature and ATP-depleted environments, the rate of uptake of particles was significantly reduced. Moreover, to investigate if clathrin-mediated endocytosis was the mechanism by which the nanoparticles (both spherical and rod-shaped) were taken into the cell, the cellular environment was deprived of sucrose or depleted of potassium, which is likely to interfere with clathrin-mediated endocytosis. Again, the quantity of uptake was seen to reduce, suggesting the employment of the endocytic mechanism of entry into the cell. Dunn et al. suggest that the fusion process of endosomes with lysosomes is not only temperature but size dependant (Dunn et al., 1980). Their work demonstrates that small pinocytotic vesicles continued to be taken into the cell in temperatures under 20°C, with directionality towards the lysosome-rich perinuclear region, but no binding with lysosomes was evident. This process is related more specifically to lysosomes, as other smaller pinocytotic vesicles continued to fuse with one another at temperatures under 20°C.

7.5.2 Damage to the Cell

As with vast and varied nanoparticle composition, shape and structure, the nature of the specific damage that may arise in a sample after exposure is best characterised under a few umbrella headings as mentioned previously (Figure 7.4).

The nanotoxicological events will differ from cell line to cell line and one biological sample or system to another. Certain systems will of course be more predisposed to certain types of harm or alterations (such as respiratory epithelium after exposure to inhaled nanoparticles) and results may not completely reflect the true *in vivo* state due to employment of simplified models. Collecting and summarizing studies with different exposure methods, durations, nanoparticle type and composition, biological samples, and systems is problematic. For this reason, the ability to infer what differences are present after exposure or treatment relies on the ability to clearly interpret a 'normal' cellular ultrastructure (as previously discussed).

There are many other useful microscopic techniques in the light microscopy realm which can provide dynamic information necessarily missing from static electron microscopical images. Ultrastructural

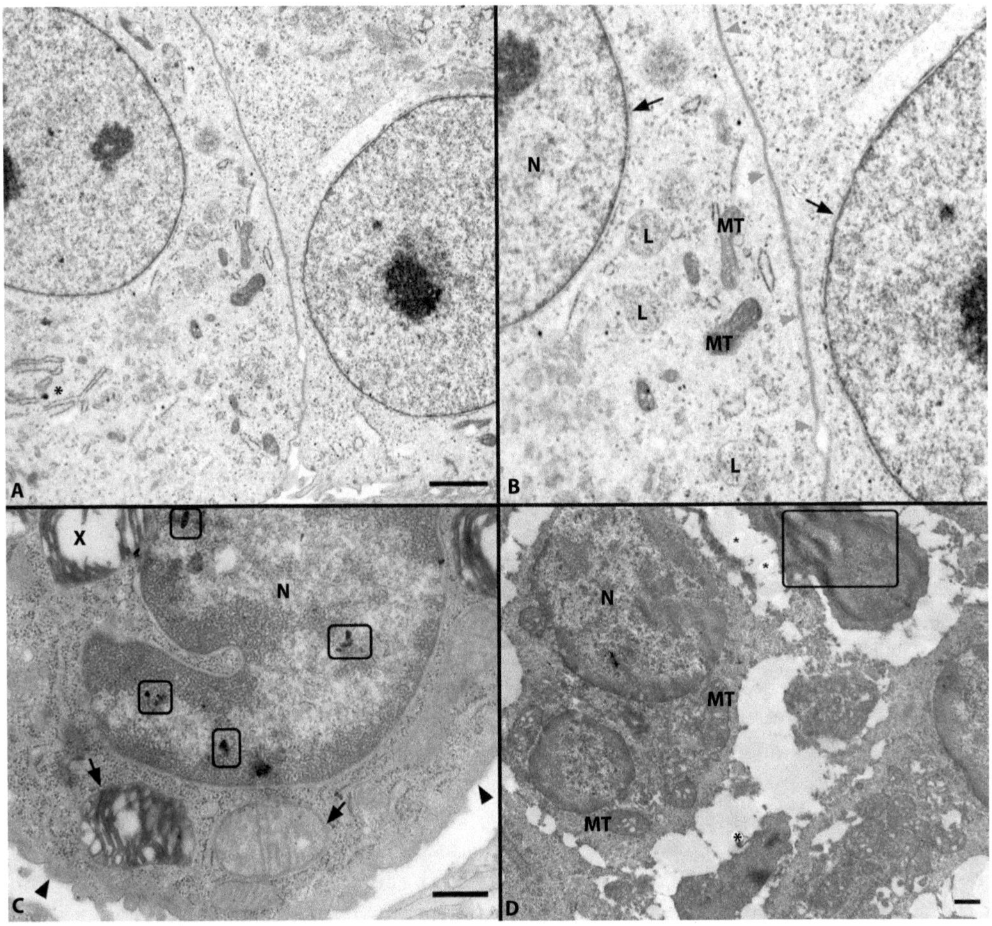

FIGURE 7.4 TEM micrograph of normal vs damaged cell samples. In these conventionally processed TEM images, well-preserved untreated control epithelial cells in A and B (enlarged portion of A) display double membrane-bound nuclei (N and black arrows). Also, clearly visible are the scattered rough endoplasmic reticulum clusters (RER – black asterisk), granular electron-dense membrane-bound lysosomal-like structures (L), and mitochondria (MT). The cells visible in A and B have a regular, well-structured cytoplasm with evenly distributed ribosomes, mitochondria with clearly discernible cristae, and other electron-dense membrane-bound organelles (nucleus and lysosomes). Scale 2 µm. The cell visible in C is distorted and is a good example of suboptimal preservation or sample preparation. The mitochondria (black arrows) are grossly enlarged and distended, whilst the nuclear material is punctate, and staining precipitate (an artefact of sample preparation) can be seen in the nucleus (N). The lysosomal like structures (L) are also enlarged with a deficit in the centre of the structure and appear to be swollen and distorted. Furthermore, the cell membranes appear to be 'soft' – not clearly distinguishable or discernible (black arrow heads). This would indicate poor sample preservation and or artefacts induced in the sample preparation workflow. Scale 500 nm. In D, a lower-magnification overview of an extremely poorly preserved sample can be seen. Enclosed within the black box, the sample appears wrinkled and puckered. This may indicate poor resin infiltration. Moreover, a large number of holes (astericks) can be seen in the section. Again, this would indicate poor sample preparation, where the section is unstable under the electron beam and has begun to deteriorate. The nuclear material (N) does not display a discernible double membrane-bound structure, and chromatin appears punctate and unevenly distributed. All cells in the images are damaged with distorted and altered intracellular content. Large spaces are present between cells where they appear to have come away from one another during sample processing. Scale 500 nm. Similarly, samples exposed to nanoparticles may also exhibit distortion in their intracellular makeup like that seen above. For this reason, sample preparation must be carried out with the utmost care and diligence to rule out these types of artefacts and damage.

and morphological analysis can be both qualitative and quantitative and the scientist can extract true phenotypical information from the sample due to the high magnification and high resolution of the micrographs and images formed. The limitations of quantification of nanotoxicology using electron microscopy will be discussed in Section 7.8. The means to deconstruct the mechanism of nanotoxicological damage to a biological sample after exposure relies heavily on knowledge of the intricate intracellular microenvironment and inherent closely regulated interactions between its organelles. The ability for one effect to have a 'knock on' to the next must be contemplated. The introduction of nanoparticles into the intracellular environment can be thought of as an assault to the cell. Damage can be induced through alterations of signalling mechanisms via uptake, the reorganisation of the intracellular environment, and the presence of actual particles themselves as foreign matter.

7.5.3 The Manifestation of ROS within the Cell at the EM Level

ROS are generated by the incomplete reduction of oxygen within the cell. When maintained at adequate levels these molecules can interact with and contribute to cell migration signalling pathways and specifically with the actin cytoskeleton (Stanley et al., 2014). When homeostasis fails and management of levels of ROS within the cell exceed the beneficial level, oxidative stress leads to damage within the sample which may ultimately trigger apoptosis or cell death (Ray et al., 2012). Nanoparticles have been implicated in both ROS-dependent and -independent apoptotic pathways (Yang et al., 2014, Zhu et al., 2016). Previous work from our group, cited above as Stanley et al., has focused on the delicate interaction and balance of oxidation and oxidative stress within the cell. As the actin cytoskeleton, and its regulators the GTPase family, are so closely implicated with the endocytic mechanism, the effects of an up- or downregulation of ROS levels within the cell may also affect the endocytosis of nanoparticulate matter.

In a study carried out by Abdal Dayem et al., the main sources of intracellular ROS are listed as the following organelles: mitochondria, endoplasmic reticulum, peroxisomes, and the plasma membrane in NOX complexes (Abdal Dayem et al., 2017). As key players in the endocytic mechanism for particle uptake, all are susceptible to the upstream and downstream effects of nanoparticle-induced oxidative stress. There is no one marker that signals to a microscopist a clear increase in the levels of ROS within the cell that may trigger oxidative stress; rather, subtle clues can be picked up during thorough investigation of the organelles mentioned above. The direct localisation of a cluster of nanoparticles near, or within, a mitochondrion may suggest to the microscopist that this organelle and cell may be starting to undergo change associated with this exposure. The alterations induced by trafficking of nanoparticles may also have indirect consequences on the cytoskeletal framework within the cell and subsequent interactions with the organelles shuttled along this intracellular highway.

7.5.4 Membrane and Organelle Membrane Damage

Lysosomal damage has recently emerged as a mechanism of nanotoxicity within cells which have accumulated large numbers of nanoparticles (Wang et al., 2018). To fully understand the impact of nanoparticles on cells, investigations should be performed to elucidate effects of nanoparticles on the lysosomal subcellular component itself. When considering the integrated nature of cellular functionality, this is not surprising. Much of the previous work carried on nanoparticles has indicated their ultimate final intracellular destination as being within the lysosome. The integration of the lysosome into the pathway is intrinsically linked to the endocytic uptake mechanism, temperature-dependent effects, and kinetics seen in cells capable of large levels of endocytosis. There is a lack of information regarding whether the nanoparticles ever escape the lysosome. Internally within the lysosome, the acidic pH environment and vast array of hydrolytic enzymes (esterases, proteases, phosphatases, nucleases, and lipases) are capable of at least commencing the degradation of even the most resilient nanoparticles (Stern et al., 2012). The endosomal escape model discusses the need for particles to get out of the membrane-bound organelle they are being shuttled along inside to reach the target destination in the case of nanoparticle-based therapy, or just to free themselves into the cytosol (Martens et al., 2014). The fusion of a lysosome with a late-stage endosome is one of the final stages in this process. In this instance, TEM images like those seen in Figure 7.3 can give a clear indication as to the type of organelle that the nanoparticle is in the vicinity of, attached to, or bound within.

Alterations in mitochondrial structure and function can also result from nanoparticle exposure. As previously stated, mitochondria are one of the main sources of intracellular ROS. The inherent links to the apoptosis pathway via nanotoxicological mechanisms manifests clearly in well-fixed and preserved mitochondria within cells. As a rule, they are one of the first organelles that are surveyed by our group in a 'control' or normal TEM preparation. The deposition of nanoparticles on or uptake within the mitochondria can also induce damage by just the process itself. This in turn would upregulate the oxidative stress activities within the cell (AshaRani et al., 2009).

7.5.5 DNA Damage and a Combination of the Above Factors with Resultant Cell Death

Alterations to the nucleus after coming into contact with nanoparticulate matter have major implications regarding the continuity of the cell cycle (Deng and Gao, 2016). The capacity for interaction with DNA within the nucleus in this manner has been recorded for gold nanoparticles and this mechanism has been exploited in the treatment of disease (Piperigkou et al., 2016). To reach the nucleus and perinuclear region and exert its effect, the nanoparticle must escape the endosome and lysosomal compartments as discussed above. Entry into the nucleus is mediated by nuclear pores which would exclude certain particles based on size. Molecules smaller than 45 kDa can freely pass into the nuclear structure. For studies relying on drug delivery to this organelle, deposition nearby may be sufficient to allow the active compounds on a modified nanoparticle to diffuse into the nucleus. A conformational change in the structure of the DNA is said to occur on binding, thereby preventing transcription from taking place. This can be thought of as a double-edged sword – both a positive and a negative effect of exposure. Should the interest of the study be DNA disruption as in the case of some therapies, this is a positive outcome. If not, then this unwanted and unsolicited alteration is a major detriment to a study. Again, the microscopist must have a sound knowledge of the variety of nuclear morphologies that exist, both in the normal and apoptotic state. During apoptosis the cell at the ultrastructural level can be thought of as though it were imploding. The nuclear material condenses, the membranes enclosing the distinct intracellular compartments eventually break down, and the cell disassembles. As the nanoparticle may be on its journey through the cell, it will in some instances be shuttled along the microtubule cytoskeletal network. The microtubule network is responsible for the formation of the mitotic spindle during mitosis and any perturbation in this process will affect the cell cycle (Adjei et al., 2014).

7.6 Considerations for Sample Preparation for Electron Microscopy

Currently no other technique other than electron microscopy is suitable to truly investigate morphological nanotoxicity at the cellular level. TEM, in particular, enables localisation of single nanoparticles inside cellular compartments and direct visualisation of the ultrastructural effects of these structures on the cell; therefore, this section will focus on sample preparation for this imaging method (Figure 7.5).

In order to be able to achieve representative ultrastructural images of a sample, extensive and complicated sample preparation must be carried out, including fixation, dehydration, resin embedding, sectioning, and staining. As each of the processing steps can have an impact on the sample, it is essential to have some degree of understanding of what is involved in each step.

The first consideration for sample preparation for electron microscopy is the sample itself. The quality of the biological sample is paramount to achieving the resolution capabilities of electron microscopy.

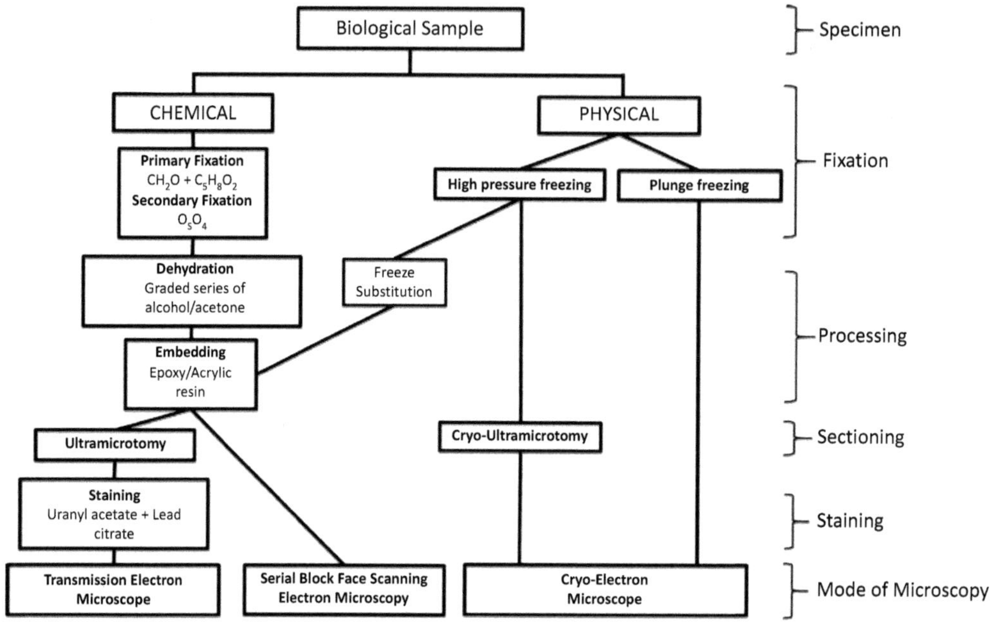

FIGURE 7.5 Sample preparation workflow. A brief summarised overview of some biological sample preparation workflows is described above, with each stage in the process indicated to the right-hand side.

The choice of preparation protocol is dependent on the type of biological sample, the particular area of interest, and here, the type of nanoparticle or nanomaterial. From a review of the extant literature, most studies utilising electron microscopy to investigate the interaction of nanoparticles or nanomaterials with biological samples utilise *in vitro* systems. The major influencing factor in this appears to be that localising nanoparticles in a whole animal or organism with this technique would be extremely difficult and overly time consuming. Indeed, one electron microscopy session will enable analysis of 1–10 μm^3 volume (Kempen et al., 2013). In addition, to improve the chances of being able to study the biodistribution of nanoparticles utilising electron microscopy, large volumes of nanoparticles have to be systemically administered to the organism under study (Jong et al., 2010, Kempen et al., 2013). However, studying nanoparticle biodistribution and nonotoxicity at the cellular level can be rendered more efficient and feasible by combining stereological methods with an effective sampling strategy. Advanced imaging techniques such as Serial Block Face (SBF) SEM and ET, as discussed below, help to overcome investigative limitations.

In vitro systems using adherent cells are often grown and treated in culture flasks and therefore will need to be enzymatically or mechanically removed and centrifuged in order to form a cell pellet for further processing. There is the option to grow adherent cells on glass coverslips and remove these later during the embedding process or utilise Thermanox coverslips, which can be embedded and sectioned. Non-adherent cells, however, require centrifugation prior to processing. As highlighted by Schrand, there is no well-defined speed for centrifuging cells for TEM and the centrifugal force may alter the localisation of nanoparticles leading to false interpretations (Schrand et al., 2010). This is an important point as determining if such particles are surface bound or internalised, and if internalised to which exact compartment, membrane bound, such as endosomes, lysosomes, the nucleus, or located within the cytoplasm is key to understanding the potential nanotoxicity at the cellular level. As there are many different preparation techniques depending on the cell or tissue culture type that can be utilised to circumvent the effect of processing, we refer the interested reader to excellent resources (Schrand et al., 2010, Bozzola, 2014a, 2014b).

The underlying problem with trying to view biological samples using electron microscopy is simply that they *are* biological. Cells consist of between 70%–80% water, it is the matrix for everything that happens within the cell and it provides structure and stability (Łukasz Mielańczyk, 2015). Given that the high vacuum of the electron microscope column would evaporate this water within the cell, it poses one of the fundamental challenges for any electron microscopy analysis - to view the complexity of the sample in as close to its native state as possible (Chlanda and Krijnse Locker, 2017). For conventional electron microscopy, this water therefore needs to be removed before the sample can be viewed, but in order to maintain the ultrastructural integrity of the sample, fixation is required. Fixation can be performed in a number of different ways depending on the sample type: for cell cultures *in situ* fixation is generally utilised; for tissue samples or cell pellets immersion fixation is used, and vascular perfusion can be used for animals.

Through fixation the overall volume, spatial relationships, and morphology of the sample are maintained by preserving structures of organelles and molecules in their native state, metabolism is halted thereby preventing any autolytic changes, and the sample is made stable for further processing (Wisse et al., 2010). Despite major advances in cryopreservation, which will be discussed later, traditional routine electron microscopy using chemical fixation still provides answers to most requirements needed to characterise cell interaction with nanoparticles while maintaining nanoparticle physiochemical properties (Schrand et al., 2010). Chemical fixatives act as crosslinking agents forming inter- and intramolecular chemical bonds with various macromolecules in the sample.

In order to maintain the quality of the biological sample and enable analysis of its morphological structure in as close to its native state as possible it is of utmost importance that fixation of the sample occurs rapidly. Chemical fixation is carried out using separate or combined reagents, the most common being osmium tetroxide (OsO_4) and aldehydes, namely formaldehyde and glutaraldehyde.

OsO_4 was the first fixative used for electron microscopy (Palade, 1952). This heavy metal salt is an excellent fixative primarily of unsaturated lipids through reaction with their C=C double bonds and is therefore very useful for membrane preservation. For more detailed information regarding its properties and specific interactions of OsO_4 we refer the reader to comprehensive books and reviews (Deetz and Behrman, 1981, Hayat, 1981b, Belazi et al., 2009). The use of OsO_4 should be considered carefully depending on the nanoparticle of interest; Chen, concluded that OsO_4 should be omitted when using silver as it causes a substantial morphological change to silver nanowires owing to its oxidation capability (Chen et al., 2016).

During the fixation process OsO_4 becomes reduced and therefore acts as an electron stain for biological tissue and therefore also any nanoparticles or materials containing unsaturated carbon bonds. Samples visibly blacken, which does have the added advantage of making it easier to see the samples during further processing and sectioning steps. This electron stain advantage can also be considered a disadvantage if low-contrast nanoparticles are being investigated as the increased contrast in the sample could make them difficult to visualise; this is also the case when using nanoparticles that are of similar size, shape, and electron density to cellular structures, as distinguishing between them is rendered difficult. It is therefore advisable to combine elemental analysis when performing ultrastructural analysis (Brandenberger et al., 2010a). Its ability to permeabilise membranes and halt cytoplasmic changes within seconds, as well as it's solulibility in both polar and nonpolar solutions, allows fixation of both hydrophobic and hydrophilic cell components making it an attractive primary fixative. In practice, however, this is only useful for cell monolayers. Its rate of penetration into tissue is very slow, approximately 0.5 mm/hour, meaning that the standard recommended sample size of 1 mm cube will take 1 hour before the centre is fixed, leading to necrotic change of the sample core. Therefore, when using tissue samples, OsO_4, as is, is rendered unsuitable as a primary fixative.

The greatest disadvantage of OsO_4 is its limited ability for crosslinking many proteins and, of the cellular macromolecules, fixation of proteins first and foremost is required to achieve excellent ultrastructural preservation. A huge advance in achieving this came in the early 1960s following investigations into the use of aldehydes as fixatives (Sabatini et al., 1963, Karnovsky, 1965). This led to the most commonly used electron microscopy fixation method today – a mixture of formaldehyde and glutaraldehyde as a primary fixative, to be followed by OsO_4 as a secondary fixative.

Formaldehyde (CH_2O) on its own will not yield good ultrastructural fixation as, while it does form crosslinks in cellular macromolecules, particularly proteins, these crosslinks are weak and are partly or completely reversible during subsequent sample preparation steps. As it is a monomer and the smallest and simplest aldehyde, it penetrates samples quickly. Glutaraldehyde ($C_5H_8O_2$), a five-carbon dialdehyde with two hydroxyl groups, is an extremely good fixative of cellular ultrastructure introducing irreversible intra- and intermolecular cross-links between proteins, in addition to having weak reactions with lipids, carbohydrates, and nucleic acids. Its penetration rate, however, is very slow, particularly into tissue samples. Therefore, by combining these two aldehydes the advantages of both can be utilised, the formaldehyde will penetrate the sample quickly, and while it will only form weak cross-links, it stabilises ultrastructural integrity until glutaraldehyde has had sufficient time to form more permanent fixation. The aldehydes primarily target proteins, so subsequent sample preparation steps result in loss of lipids which can be counteracted with the secondary fixation with OsO_4 providing excellent ultrastructural fixation for morphological analysis to the scale of 10 nm.

Choice of buffer for fixatives is vital in order to maintain the pH, tonicity, and osmolarity of the sample. Without appropriate buffering the pH of the sample will significantly decrease, resulting in the formation of artefacts (Hayat, 1981a). There is a huge range of different buffers available depending on the sample type and it is worth reviewing the relevant literature for any particular sample. The most commonly used buffers for electron microscopy today are phosphate and cacodylate buffers. Phosphate buffers are most commonly used as they are non-toxic, so can be used as washes for living cells, and they maintain pH extremely well when used at physiological levels of slightly alkaline (pH 7.2–7.4). However, this buffer can cause precipitation of the stains used for TEM: uranyl acetate and lead citrate (Hayat, 1981a, Dykstra and Reuss, 2003). Sodium cacodylate results in fewer precipitation artefacts than phosphate buffers. It works very well at physiological levels but can maintain a pH between 6.4–7.4 and therefore can be used with a wider range of sample types. Safety precautions must be observed during use as it contains arsenic and can form arsenic gas when exposed to acids (Hayat, 1981a, Dykstra and Reuss, 2003).

Once the sample has been appropriately fixed, the water is then removed by dehydrating the sample through a graded series of solvents, generally alcohol or acetone. These not only remove the water from the sample but also act as a solvent for embedding resins that are generally immiscible with water. Depending on the type of resin, an intermediate solvent, such as propylene oxide, may be required to achieve better infiltration of the sample. However, most resins available today mix well in alcohol or acetone and achieve excellent infiltration using these agents.

Embedding in resin involves the infiltration of the sample with, most commonly, acrylic or epoxy resins, followed by polymerisation into a block so that ultrathin sections can be obtained. Acrylic resins are hydrophilic making them useful when immunocytochemistry is required. These resins are prone to uneven polymerisation and are not as stable under the electron beam as epoxy resins. The infiltration and polymerisation of epoxy resins is more reliable, there is minimal shrinkage within the sample and more stable sections can be obtained (Dykstra and Reuss, 2003, Schrand et al., 2010). As noted by Schrand, there is no preferred resin for cell-nanoparticle interaction studies, but it is worth investigating the different resins as they can be formulated to match the hardness of the nanoparticle of interest, therefore reducing potential sectioning artefacts.

In order to view the internal structure of the samples at a high resolution in the TEM, ultrathin sections, ideally ≈50–100 nm in thickness, must be obtained from the polymerised resin block. Firstly, the block needs to be trimmed to expose the sample and form a block face from which the sections will be taken. In order to reduce artefacts from stresses imposed by the knife, the block face is trimmed to a trapezoid shape, aiming to remove plain resin from around the sample and make the block face as small as possible (≈0.3 mm^3), as the smaller this is the better the section (Hayat, 1986). Sections are taken using a specialised piece of equipment, an ultramicrotome, which will cut the sections while advancing the block in increments of the required section thickness. Ultrathin sections are taken using a diamond knife mounted in a metal trough which is filled with water to float the sections, after which they can be picked up onto specimen support grids. The major issue during sectioning samples for analysis of nanotoxicity at the cellular level is nanoparticle pull-out by the knife, where either the particle is lost or it has been dragged by the knife, as it can be difficult to determine if the nanoparticle is embedded in, or seated on the surface of, the section (Hondow et al., 2012). Avoidance of this artefact is through adjustments to thickness of section, angle of knife, speed of sectioning, and size of block face. The more recent advances in imaging techniques, namely SBF SEM and ET, as discussed below, can also prevent misinterpretation from such artefacts, in addition to their main advantage of enabling 3D analysis.

To provide sufficient contrast to the sample, heavy metal salt staining is required. Unlike in light microscopy where different cellular components can be selectively stained, for electron microscopy stains are compounds or ions of high atomic number which scatter the electron beam and are only semi-selective. However, the interaction of such stains with certain chemical groups is known and should be considered depending on the nanoparticle material as they may not only enhance the contrast of the nanoparticles, but by attaching to them may increase their size. The standard TEM counterstains for biological samples are uranyl acetate followed by lead citrate. Uranyl acetate interacts with anionic compounds and binds strongly to phosphate groups, particularly those of nucleic acids and membrane phospholipids, in addition to sialic acid carboxyl groups of proteins and lipids. Lead citrate acts as a mordant for osmium tetroxide and uranyl acetate and therefore further enhances the contrast provided by them. It also binds to negatively charged amino acids of proteins, hydroxyl groups of carbohydrates, and phosphate groups of nucleic acids (Bozzola, 2014b).

It is important to note that the type of nanoparticles may affect the ability to use these stains due to their pH. For example, in a study investigating artefacts in characterisation of nanomaterial-cell interactions it was found that TiO_2 was unaffected by the pH of the stains but ZnO- and MgO-based nanoparticles were dissolved, leaving voids in the sample sections (Leung et al., 2017). It must be noted that staining may be omitted if it is thought to impede the visualisation of the nanoparticle; however, this comes at the cost of reduced resolution.

In order to overcome the potential artefacts and problems leading to misinterpretation of data that can arise from conventional chemical sample preparation, some investigative focus has shifted to physical fixation methods, namely cryo-techniques. Instead of using chemicals to preserve the ultrastructure of the cell, samples are frozen at such a rapid rate that the water molecules form amorphous vitreous ice (Dykstra and Reuss, 2003, Studer et al., 2008, Weston et al., 2009, Mielanczyk et al., 2014). This results in more native-state imaging as the rate of heat diffusion from a sample is much faster than chemical diffusion into a sample resulting in cessation of biological activity in 10 ms (Gilkey and Staehelin, 1986, Dykstra and Reuss, 2003). This technique requires careful planning and execution for if freezing occurs too slowly or the sample is warmed above −137°C ice crystal formation occurs, which will destroy the cellular ultrastructure (Dubochet et al., 1988).

The most widely used techniques for cryopreservation for electron microscopy are plunge and high-pressure freezing. Plunge freezing, where a biological sample is immersed in a suitable cryogen, e.g., ethane, propane, or a mixture of both, is only suitable for thin samples between 10–15 μm. In the context of analysing nanotoxicity at the cellular level, cells need to be grown on specimen support grids. The major advantage is that once vitrified, these can be viewed immediately with a cryo-electron microscope. However, while this sample preparation technique is probably the least invasive, the limitation of sample thickness means only thin cells, or the thinner outer periphery of cells, can be adequately analysed and it is not suitable for tissue samples.

With high-pressure freezing, thicker samples in the region of 200 μm can be analysed. This is achieved by pressurising the sample to 2048 bar, meaning that less heat is produced during vitrification and less heat needs to be diffused from the sample; using liquid nitrogen jets, the sample is vitrified in approximately 10 milliseconds (Studer et al., 2008). As we know, only thin sections can be viewed using the TEM, so these samples require sectioning. This is carried out either using a cryo-ultramicrotome by the CEMOVIS procedure, thinned using focused-ion beam (FIB) milling, or freeze substitution followed by resin embedding (Weston et al., 2009, Mielanczyk et al., 2014). With the two former methods there is no need for chemicals or stains, which can impact the interpretation of nanoparticle tissue interactions as previously discussed. Sectioning artefacts such as knife marks, compression, and chatter can occur with CEMOVIS and curtaining and rippling can occur due to the milling procedure FIB.

Freeze substitution uses conventional sample preparation methods as discussed above, but this is carried out as the sample is brought up in temperature, meaning the sample is fixed and processed in a solid phase rather than a liquid phase (Studer et al., 2008).

Freeze substitution may ultimately be the most appropriate method for analysing nanotoxicity at the cellular level. While the problems and artefacts of conventional electron microscopy sample preparation remain, the sample itself is closer to its native state. In addition, with this method elemental analysis remains possible.

7.7 Advances in Imaging Techniques to Quantify Nanotoxicity

7.7.1 3D Imaging

Electron microscopy is indispensable for high-resolution visualisation of cellular ultrastructure and has contributed significantly to our understanding of numerous cellular processes. As TEM and SEM are 2D imaging techniques used to image complex 3D structures, incorrect interpretation of the true 3D structure can occur. To truly understand the 3D structure of an object it is sometimes necessary to use techniques which allow the object to be visualised in 3D.

Over the past 20 years, novel advanced 3D electron microscopy imaging techniques have been developed. These techniques include ET and SBF SEM. By imaging a specimen in sections or slices, we are able to visualise the 3D nature of a specimen. Computed tomography (CT) and magnetic resonance imaging (MRI) have been used in medical imaging since the 1970s; both methods employ imaging by virtual cross-sections to visualise internal structures. This method of 'imaging by section' has also long been used to investigate the 3D morphology of tissues and cells but relatively recent advances in electron microscopy have now allowed for quicker acquisition of high-resolution 3D data sets.

A relatively recent key development in TEM has been ET. ET is a 3D imaging technique that allows for the capture of nanometre resolution of subcellular structures. Samples are processed as they are for standard 2D TEM imaging (post-fixation in OsO4, dehydration, resin infiltration, and resin embedding), but instead of obtaining ultrathin (~70 nm) sections of the sample, slightly thicker sections between 120–250 nm are used. To acquire the image series, the sample is tilted within the TEM around a central axis perpendicular to the electron beam. As the sample is tilted, an image is acquired at every degree or half degree tilt increment through a ±70-degree range. Following image acquisition, the image series is compiled into a tomogram by back-projecting the images to create a 3D tomogram. From the tomogram, 3D models of specific structures of interest can be generated using 3D modelling software (Subramaniam and Zhang, 2003, Bárcena and Koster, 2009, Gan and Jensen, 2012).

To date, ET has been applied to visualise numerous cellular processes including endocytosis (Murk et al., 2003, Kukulski et al., 2012) and exocytosis (Lenzi et al., 1999, Mourik et al., 2014) and the morphology of membranous organelles such as the Golgi apparatus (Ladinsky et al., 1999). ET has been used to investigate and visualise nanoparticle internalisation by liposomes and allowed for the stages of this process to be characterised in great detail (Le Bihan et al., 2009). This approach is therefore ideal for investigating nanoparticle uptake and intracellular trafficking, as well as visualising evidence of nanotoxicity induced by the nanoparticle. ET is currently a relatively low-throughput imaging technique which is suited for high resolution, 3D visualisation of individual cellular components rather than whole-cell visualisation. This may be considered a limitation of this method but alternative approaches for visualising larger volumes, which are rapidly approaching comparable resolution to ET, are currently being developed (Figure 7.6).

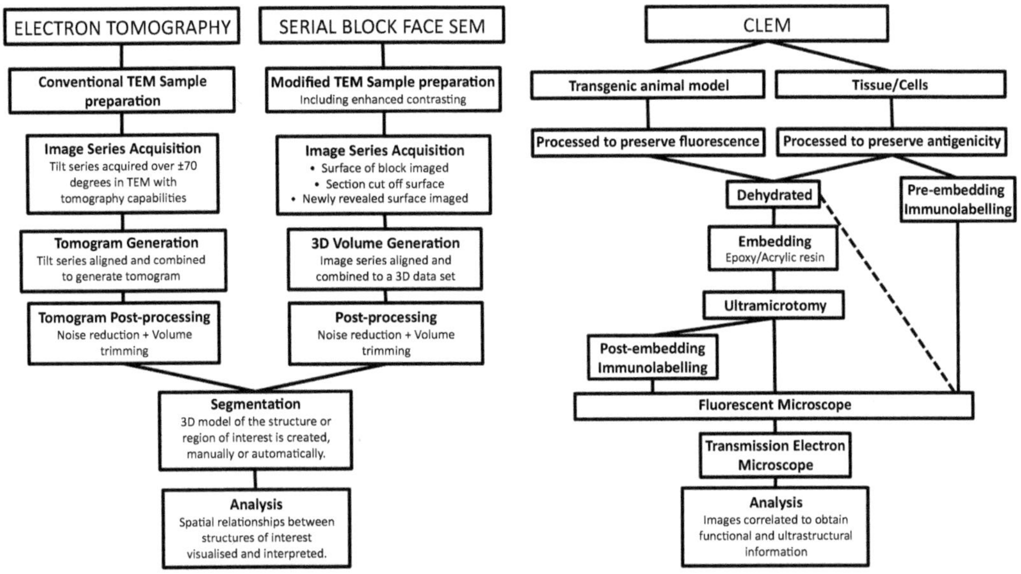

FIGURE 7.6 Modified sample prep workflows. Summarised steps in the workflows of the advanced volume electron microscopy methods, ET and SBF SEM. Furthermore, a brief CLEM workflow is described. Following pre-embedding immunolabelling and fluorescence imaging, samples can return to the electron microscopy workflow for processing (dashed line).

A second method of high-resolution, 3D electron microscopy is SBF SEM. Using this technique, large volumes of tissues/cells are imaged, in contrast to ET where subcellular structures are imaged. To prepare a sample for SBF SEM, the samples are processed using a modified TEM sample processing protocol optimised for enhancing contrast (Deerinck et al., 2010). Following processing, the blocks are trimmed down to the region of interest, placed onto a specialised stub, and placed in an SEM with an inbuilt ultramicrotome or focussed ion beam. The block face is then imaged using backscattered electron detection. A section of a set thickness (10–200 nm) is removed from the block face, discarded, and the newly exposed block face is imaged. This process is repeated until a data set of the desired volume is acquired. As with ET, the data set is combined to produce a 3D volume from which 3D models of structures of interest are generated (Denk and Horstmann, 2004, Hughes et al., 2014). SBF SEM has been applied in a variety of studies ranging from the determination of collagen fibril size and 3D organisation in tendon (Starborg et al., 2013), to elucidating the complex ultrastructure of synaptic structures within the hippocampus (Wilke et al., 2013). Pertaining to nanoparticles, SBF SEM has been applied to investigate trafficking of gold nanoparticles at the blood-brain barrier. In this study, as well as visualising uptake and distribution of the nanoparticles within the cells located at the blood-brain barrier, the true 3D morphology of endocytic vesicles containing the nanoparticles was realised. The vesicles were found to be branching and ellipsoid – contrary to previous 2D observations that had concluded the vesicles to be round (Cabezón et al., 2017).

7.7.2 Correlative Light and Electron Microscopy

Correlative light and electron microscopy (CLEM) is a multimodal imaging technique that allows for simultaneous, or near-simultaneous, visualisation and identification of cellular components using fluorescent and electron microscopy. There are numerous different types of workflows that can be used in order to achieve CLEM, which is dependent on several factors such as imaging equipment available and the nature of the sample (De Boer et al., 2015).

Fully integrated imaging systems have been developed that combine fluorescence microscopy with SEM (SECOM by DELMIC). Samples imaged using this novel method are processed in a way that allows for preservation of the fluorescent signal in resin-embedded specimens, for example using adapted high-pressure freezing protocols. Sections are then obtained and placed onto an indium-tin-oxide (ITO)-coated coverslip which is placed in the microscope. A cell/region of interest is then imaged both by fluorescence microscopy and SEM and both fluorescent and SEM images overlain (Baudoin et al., 2018).

Where access to such an integrated system is limited, CLEM is still achievable using a considerably more manual fashion. Samples are processed to retain in-resin fluorescence, but the ultrathin sections are then placed onto TEM finder grids. The grids can then be mounted onto a glass slide and imaged by fluorescence microscopy. The coordinates of the cell/region of interest on the finder grid is noted and the grid is then placed into the TEM. The position of the fluorescent signal can then be identified using the aforementioned coordinates and high-resolution images acquired. The fluorescent image can then be overlaid onto the TEM image and the precise location of the fluorescent signal can be identified (De Boer et al., 2015). Both of these approaches allow for combined functional and structural data to be obtained.

The second factor to consider when planning CLEM is the nature of the sample to be imaged. While many studies have focussed on imaging samples from transgenic animal models, dual probes such as fluoronanogold can also be used (Takizawa et al., 2015). These probes are used in the same manner as a secondary antibody and can be used pre- or post-resin embedding. Using the pre-embedding approach, the sample is imaged using conventional fluorescence microscopy and the localisation of the fluorescent signal is determined. For TEM, the fluoronanogold probe is enhanced using the silver enhancement technique, processed as normal, and imaged by TEM. Using the post-embedding approach, the sample is processed for TEM using an optimised antigenicity-preserving technique (such as HPF with progressive lowering of temperature processing into lowicryl resin or cryo-ultramicrotomy). Ultrathin sections are acquired and picked up onto TEM finder grids. Antibody labelling is performed directly onto the section, which is then subsequently imaged by fluorescence microscopy. The coordinates of ROI are recorded, and the grid is transferred to the TEM where the ROI is relocated and imaged. In both pre- and post-embedding approaches, when imaged by TEM, the probe is seen as electron-dense particles which indicate the precise location of the antibody-labelled protein of interest which places the label in the context of the cellular ultrastructure (Takizawa et al., 2015). While this technique may not be as directly correlative as the aforementioned techniques, it allows for high-resolution visualisation of specific protein localisation. In recent studies, CLEM has been utilised to investigate gold nanoparticle localisation and distribution *in vivo* in C. elegans (Gonzalez-Moragas et al., 2017) and gold nanoparticle uptake in human lung carcinoma cells *in vitro* (Böse et al., 2014).

7.8 Limitations of Electron Microscopy in Quantifying Nanotoxicity

Adequate quantification of nanoparticles in cells and tissues is becoming established as an essential element in studies of nanoparticle toxicology. Estimating the cellular dose helps in the assessment of both the effect of and possible risk from nanomaterials (Elsaesser and Howard, 2012). The wide range of microscopical and analytical methods that have been deployed in nanoparticle research studies include light microscopy (Huang et al., 2002, Alivisatos et al., 2005), electron microscopy (Muhlfeld et al., 2007a, Nativo et al., 2008), stereology (Brandenberger et al., 2010b), radioactive labelling (Jani et al., 1990, Kreyling et al., 2009), magnetic nanoparticle labelling (Sun et al., 2008, 2010), mass spectrometry (Chithrani et al., 2006, Gojova et al., 2007) and field flow fractionation (Deering et al., 2008).

The microscopic approach results in not only nanoparticle detection but also precise localisation in tissues, cells, and intracellular compartments. This precision has resulted in greater insight into the pathways involved in their interactions with living systems. Quantum dots, fluorescent dyes, and nanoparticle conjugates have been used to visualise nanoparticle cell dynamics by light microscopy and confocal microscopy. Flow cytometry has been employed to track cellular uptake (Elsaesser et al., 2010). The key advantage of adopting light microscopical approaches is that they can permit live cell imaging and analysis in a relatively quick and easy manner in terms of sample preparation with the significant benefit of not being sample destructive. The main disadvantage is that there is no resolution of nanoparticles therefore no absolute quantification and no imaging of bare particles. The recent emergence of super-resolution light microscopy techniques may provide sufficient resolution at the light microscopic level, promising a potential tool for dynamic *in vivo* studies (Owen et al., 2013).

SEM and TEM are the key enabling technologies that provide adequate resolution for nanoparticle visualisation. Although the chemical nature and subsequent electron contrast of specific nanoparticles may restrict visualisation to higher atomic number particles, elemental analysis techniques such as energy dispersive spectroscopy or electron energy loss spectroscopy provide data on the distribution of particles of low atomic number. It is important to consider the small volume of tissue that can be imaged via EM, in particular typical TEM section thickness of 50–70 nm. The use of three-dimensional reconstruction techniques such as ET and the SBF sectioning methods have greatly expanded imaging potential and offer new possibilities in the field of intracellular nanoparticle localisation research (Denk and Horstmann, 2004). A caveat on these approaches is that, despite providing detailed ultrastructure, they are time-consuming and result in huge datasets on limited sample sizes, which rightly raises concerns regarding sampling bias.

Irrespective of the investigative technique used, adequate sampling is essential for any serious quantitation. Use of the stereological toolkit in combination with electron microscopy provides a most useful approach in estimating nanoparticle uptake by cells. Stereological methods also enable the study of spatial distributions of nanoparticles within cells and tissues and, coupled with the high resolving power of electron microscopy, allows this distribution to be framed within the cellular/organelle context (Figure 7.7).

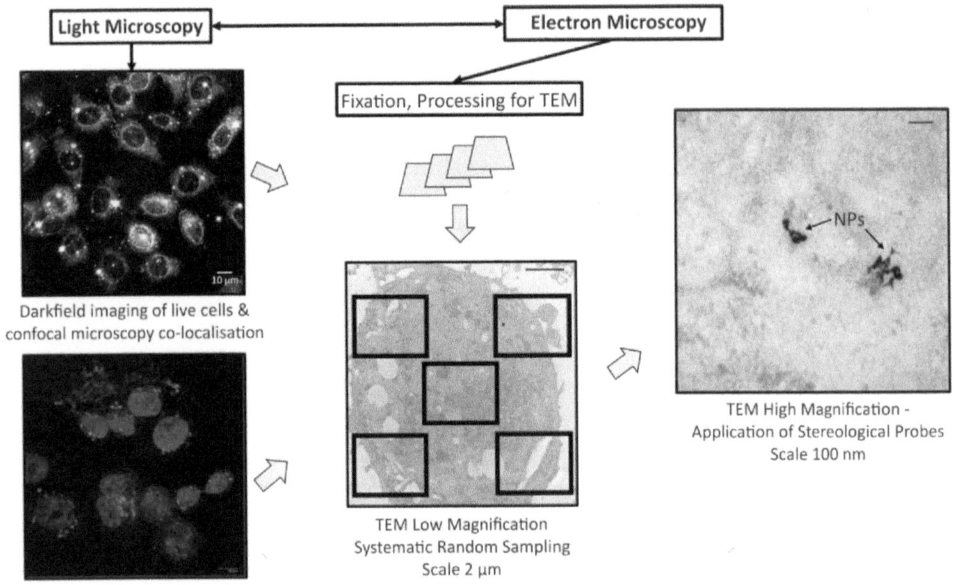

FIGURE 7.7 LM, dark field, Raman confocal, EM, and sampling schematic. A combination of both light and electron microscopy techniques can be used for efficient sample quantification and analysis via unbiased stereological probes. Nanoparticle uptake can be monitored and assessed using dark field and fluorescence microscopy. The nanotoxicological effects of exposure can then be definitively identified and structural alterations quantified using high-resolution micrographs and stereological probes.

If the nanoparticle being studied is electron dense then pre- or post-embedding immunolabelling methods as outlined previously can be used to advantage. These design-based stereological methods appear to be the most promising way of combining spatial resolution afforded by electron microscopy and uniform random sampling in order to quantify nanoparticle numbers. Due to their size, nanoparticles enter cells and tissue easily and in vast numbers; therefore, an efficient and effective sampling regime has to be employed (Mayhew et al., 2009, Muhlfeld et al., 2007a , Lucocq and Gawden-Bone, 2009, Mayhew et al., 2009, Mayhew and Lucocq, 2011, Lucocq and Hacker, 2013, Lucocq et al., 2015).

7.9 Conclusions

Interest in nanoparticles is increasing in both academia and industry and at their interfaces. The use of adequate tools to allow accurate reporting of the location, effect, and interactions of nanoparticles at all levels within a biological system(s) is of paramount importance.

Bio-nano research calls for a truly multidisciplinary approach between and across disciplines, from the chemists who manufacture nanoparticles to the physicists who examine their properties to the biologists who examine their uptake, distribution, and effects through to the medical and health-care industry who search for ways to safely incorporate them into therapeutic regimes. This disparate yet linked group of interested parties requires reliable and reproducible ways to visualise nanoparticles at all levels within systems.

Nanoparticle use within medicine and allied fields has at its core the need to be able to efficiently locate cellular and subcellular targets, whether this be containing or coated with pharmacological agents, as radiological contrast enhancers or potential sources of focal hyperthermia within tumours. In all cases the ultimate sequestration site of used nanoparticles remains to be elucidated; examination of initial effect and potential sequelae at the site of sequestration is an important area of future study.

The increasing prevalence of nanoparticles – both manufactured and natural – is of concern. Unwanted exposure and potential release into the food-chain on land and sea may prove to be a potential hazard. The World Health Organization in 2017 published guidelines on protecting workers from potential risks of manufactured nanomaterials (WHO, 2017). They offer preliminary guidelines and recommendations on the use of and exposure to nanoparticulate matter but emphasise the lack of reliable high-quality evidence regarding the potential toxicity, rates of exposure, and effect of manufactured nanomaterials.

Electron microscopy remains the only tool which can adequately provide high-quality data which permits precise location of nanoparticles at all levels within the organism. Its use, combined with other microscopical technologies, will ensure enhanced knowledge of the bio-nano interface, allowing safe and informed applications of nanoparticles to be developed.

Acknowledgements

The authors acknowledge the facilities and technical assistance of the staff from the Centre for Microscopy & Imaging at the National University of Ireland Galway (www.imaging.nuigalway.ie), a facility that is funded by NUIG and the Irish Government's Programme for Research in Third Level Institutions, Cycles 4 and 5, National Development Plan 2007–2013. They would also like to thank Dr Jennifer Connolly, Ross University School of Medicine, for contributing TEM images.

References

Abdal Dayem, A., Hossain, M. K., Lee, S. B., Kim, K., Saha, S. K., Yang, G.-M., Choi, H. Y. & Cho, S.-G. 2017. The role of reactive oxygen species (ROS) in the biological activities of metallic nanoparticles. *Int J Mol Sci*, 18, 120.

Adjei, I. M., Sharma, B. & Labhasetwar, V. 2014. Nanoparticles: Cellular uptake and cytotoxicity. *Adv Exp Med Biol*, 811, 73–91.

Alivisatos, A. P., Gu, W. & Larabell, C. 2005. Quantum dots as cellular probes. *Annu Rev Biomed Eng*, 7, 55–76.

Apodaca, G. 2001. Endocytic traffic in polarized epithelial cells: Role of the actin and microtubule cytoskeleton. *Traffic*, 2, 149.

AshaRani, P. V., Low Kah Mun, G., z, M. P. & Valiyaveettil, S. 2009. Cytotoxicity and genotoxicity of silver nanoparticles in human cells. *ACS Nano*, 3, 279–290.

Baichuan, S. & Amanda, S. B. 2018. Texture based image classification for nanoparticle surface characterisation and machine learning. *J Phy Materials*, 1, 016001.

Bárcena, M. & Koster, A. J. 2009. Electron tomography in life science. *Semin Cell Dev Biol*, 20, 920–930.

Baudoin, J. P., Hari, S., Sahmi-Bounsiar, D., Traore, S. I., Bou-Khalil, J., Andréani, J. & La Scola, B. 2018. Correlative light electron microscopy of giant viruses with the SECOM system. *New Microbes New Infect*, 26, 110–113.

Behzadi, S., Serpooshan, V., Tao, W., Hamaly, M. A., Alkawareek, M. Y., Dreaden, E. C., Brown, D., Alkilany, A. M., Farokhzad, O. C. & Mahmoudi, M. 2017. Cellular uptake of nanoparticles: Journey inside the cell. *Chem Soc Rev*, 46, 4218–4244.

Belazi, D., Sole-Domenech, S., Johansson, B., Schalling, M. & Sjovall, P. 2009. Chemical analysis of osmium tetroxide staining in adipose tissue using imaging ToF-SIMS. *Histochem Cell Biol*, 132, 105–115.

Böse, K., Koch, M., Cavelius, C., Kiemer, A. K. & Kraegeloh, A. 2014. A correlative analysis of gold nanoparticles internalized by A549 cells. *Part & Part Syst Char*, 31, 439–448.

Bozzola, J. J. 2014a. Conventional specimen preparation techniques for scanning electron microscopy of biological specimens. *Methods Mol Biol*, 1117, 133–150.

Bozzola, J. J. 2014b. Conventional specimen preparation techniques for transmission electron microscopy of cultured cells. *Methods Mol Biol*, 1117, 1–19.

Brandenberger, C., Clift, M. J., Vanhecke, D., Muhlfeld, C., Stone, V., Gehr, P. & Rothen-Rutishauser, B. 2010a. Intracellular imaging of nanoparticles: Is it an elemental mistake to believe what you see? *Part Fibre Toxicol*, 7, 15.

Brandenberger, C., Muhlfeld, C., Ali, Z., Lenz, A. G., Schmid, O., Parak, W. J., Gehr, P. & Rothen-Rutishauser, B. 2010b. Quantitative evaluation of cellular uptake and trafficking of plain and polyethylene glycol-coated gold nanoparticles. *Small*, 6, 1669–1678.

Cabezón, I., Augé, E., Bosch, M., Beckett, A. J., Prior, I. A., Pelegrí, C. & Vilaplana, J. 2017. Serial block-face scanning electron microscopy applied to study the trafficking of 8D3-coated gold nanoparticles at the blood–brain barrier. *Histochem Cell Biol*, 148, 3–12.

Chen, S., Goode, A. E., Skepper, J. N., Thorley, A. J., Seiffert, J. M., Chung, K. F., Tetley, T. D., Shaffer, M. S., Ryan, M. P. & Porter, A. E. 2016. Avoiding artefacts during electron microscopy of silver nanomaterials exposed to biological environments. *J Microsc*, 261, 157–166.

Chithrani, B. D. & Chan, W. C. W. 2007. Elucidating the mechanism of cellular uptake and removal of protein-coated gold nanoparticles of different sizes and shapes. *Nano Lett*, 7, 1542–1550.

Chithrani, B. D., Ghazani, A. A. & Chan, W. C. 2006. Determining the size and shape dependence of gold nanoparticle uptake into mammalian cells. *Nano Lett*, 6, 662–668.

Chlanda, P. & Krijnse Locker, J. 2017. The sleeping beauty kissed awake: New methods in electron microscopy to study cellular membranes. *Biochem J*, 474, 1041–1053.

Costanzo, M., Carton, F., Marengo, A., Berlier, G., Stella, B., Arpicco, S. & Malatesta, M. 2016. Fluorescence and electron microscopy to visualize the intracellular fate of nanoparticles for drug delivery. *Eur J Histochem*, 60, 2640.

De Boer, P., Hoogenboom, J. P. & Giepmans, B. N. G. 2015. Correlated light and electron microscopy: Ultrastructure lights up! *Nature Methods*, 12, 503–513.

Deerinck, T. J., Bushong, E. A., Thor, A. & Ellisman, M. H. 2010. NCMIR methods for 3D EM: A new protocol for preparation of biological specimens for serial block face scanning electron microscopy. *Microscopy*, 6–8.

Deering, C. E., Tadjiki, S., Assemi, S., Miller, J. D., Yost, G. S. & Veranth, J. M. 2008. A novel method to detect unlabeled inorganic nanoparticles and submicron particles in tissue by sedimentation field-flow fractionation. *Part Fibre Toxicol*, 5, 18.

Deetz, J. S. & Behrman, E. J. 1981. Reaction of osmium reagents with amino acids and proteins. Reactivity of amino acid residues and peptide bond cleavage. *Int J Pept Protein Res*, 17, 495–500.

Deng, J. & Gao, C. 2016. Recent advances in interactions of designed nanoparticles and cells with respect to cellular uptake, intracellular fate, degradation and cytotoxicity. *Nanotechnology*, 27, 412002.

Denk, W. & Horstmann, H. 2004. Serial block-face scanning electron microscopy to reconstruct three-dimensional tissue nanostructure. *PLoS Biology*, 2, e329–e329.

Di Fiore, P. P. & Scita, G. 2003. Endocytosis and Cytoskeleton. In: Bradshaw, R. & Dennis, E. A. (eds.) *Handbook of Cell Signalling*. London, UK: Academic Press, Elsevier.

Doherty, G. J. & McMahon, H. T. 2009. Mechanisms of endocytosis. *Annu. Rev Biochem*, 78, 857–902.

Dubochet, J., Adrian, M., Chang, J. J., Homo, J. C., Lepault, J., McDowall, A. W. & Schultz, P. 1988. Cryo-electron microscopy of vitrified specimens. *Q Rev Biophys*, 21, 129–228.

Dunn, W. A., Hubbard, A. L. & Aronson, N. N. 1980. Low temperature selectively inhibits fusion between pinocytic vesicles and lysosomes during heterophagy of 125I-asialofetuin by the perfused rat liver. *J Biol Chem*, 255(12), 5971–5978.

Dykstra, M. & Reuss, L. 2003. *Biological Electron Microscopy Theory, Techniques and Troubleshooting*. New York: Springer.

Elsaesser, A. & Howard, C. V. 2012. Toxicology of nanoparticles. *Adv Drug Deliv Rev*, 64, 129–137.

Elsaesser, A., Taylor, A., De Yanes, G. S., McKerr, G., Kim, E. M., O'Hare, E. & Howard, C. V. 2010. Quantification of nanoparticle uptake by cells using microscopical and analytical techniques. *Nanomedicine*, 5, 1447–1457.

Feng, Y., Press, B. & Wandinger-Ness, A. 1995. Rab 7: An important regulator of late endocytic membrane traffic. *J Cell Biol*, 131, 1435–1452.

Gan, L. & Jensen, G. J. 2012. Electron tomography of cells. *Q Rev Biophys*, 45, 27–56.

Gilkey, J. C. & Staehelin, A. 1986. Advances in ultrarapid freezing for the preservation of cellular ultrastructure. *J Electron Microsc Tech*, 3, 177–210.

Gojova, A., Guo, B., Kota, R. S., Rutledge, J. C., Kennedy, I. M. & Barakat, A. I. 2007. Induction of inflammation in vascular endothelial cells by metal oxide nanoparticles: Effect of particle composition. *Environ Health Perspect*, 115, 403–409.

Gonzalez-Moragas, L., Berto, P., Vilches, C., Quidant, R., Kolovou, A., Santarella-Mellwig, R., Schwab, Y., Stürzenbaum, S., Roig, A. & Laromaine, A. 2017. In vivo testing of gold nanoparticles using the Caenorhabditis elegans model organism. *Acta Biomater*, 53, 598–609.

Gray, H. 2008. Cell Structure: Exocytosis and Endocytosis. In: Standring, S. (ed.) *Gray's Anatomy: The Anatomical Basis of Clinical Practice*. 40th ed. Edinburgh, Scotland: Churchill Livingstone, Elsevier.

Gray, H., Williams, P. L. & Bannister, L. H. 1995. Cells and tissues. *Gray's anatomy*. 38th ed. New York: Churchill Livingstone.

Hansen, S. F., Michelson, E. S., Kamper, A., Borling, P., Stuer-Lauridsen, F. & Baun, A. 2008. Categorization framework to aid exposure assessment of nanomaterials in consumer products. *Ecotoxicology*, 17, 438–447.

Hayat, M. A. 1981a. 2—Factors affecting the quality of fixation. In: Hayat, M. A. (ed.) *Fixation for Electron Microscopy*. New York: Academic Press.

Hayat, M. A. 1981b. 4—Osmium tetroxide. In: Hayat, M. A. (ed.) *Fixation for Electron Microscopy*. New York: Academic Press.

Hayat, M. A. 1986. 3—Sectioning. *In:* Hayat, M. A. (ed.) *Basic Techniques for Transmission Electron Microscopy*. New York: Academic Press.

Haylett, T. & Thilo, L. 1991. Endosome-lysosome fusion at low temperature. *J Biol Chem*, 266, 8322–8327.

Heck, R., Vuculescu, O., Sørensen, J. J., Zoller, J., Andreasen, M. G., Bason, M. G., Ejlertsen et al. 2018. Remote optimization of an ultracold atoms experiment by experts and citizen scientists. *Proc Natl Acad Sci*, 115(48), E11231–E11237.

Hondow, N., Harrington, J., Brydson, R. & Brown, A. 2012. STEM mode in the SEM for the analysis of cellular sections prepared by ultramicrotome sectioning. *Electron Microscopy and Analysis Group Conference 2011*, 371.

Huang, M., Ma, Z., Khor, E. & Lim, L. Y. 2002. Uptake of FITC-chitosan nanoparticles by A549 cells. *Pharm Res*, 19, 1488–1494.

Hughes, L., Hawes, C., Monteith, S. & Vaughan, S. 2014. Serial block face scanning electron microscopy-the future of cell ultrastructure imaging. *Protoplasma*, 251, 395–401.

Jani, P., Halbert, G. W., Langridge, J. & Florence, A. T. 1990. Nanoparticle uptake by the rat gastrointestinal mucosa: Quantitation and particle size dependency. *J Pharm Pharmacol*, 42, 821–826.

Johnston, H. J., Hutchison, G., Christensen, F. M., Peters, S., Hankin, S. & Stone, V. 2010. A review of the in vivo and in vitro toxicity of silver and gold particulates: Particle attributes and biological mechanisms responsible for the observed toxicity. *Crit Rev Toxicol*, 40, 328–346.

Jong, W. H., Burger, M. C., Verheijen, M. A. & Geertsma, R. E. 2010. Detection of the presence of gold nanoparticles in organs by transmission electron microscopy. *Materials (Basel)*, 3, 4681–4694.

Karnovsky, M. J. 1965. A formaldehyde-glutaraldehyde fixative of high osmolality for use in electron microscopy. *J Cell Biol*, 27, 1A–149A.

Kay, D. 1961. *Techniques for Electron Microscopy*. Oxford, UK: Blackwell Scientific Publications.

Kempen, P. J., Thakor, A. S., Zavaleta, C., Gambhir, S. S. & Sinclair, R. 2013. A scanning transmission electron microscopy approach to analyzing large volumes of tissue to detect nanoparticles. *Microsc Microanal*, 19, 1290–1297.

Kettler, K., Veltman, K., Van De Meent, D., Van Wezel, A. & Hendriks, A. J. 2014. Cellular uptake of nanoparticles as determined by particle properties, experimental conditions, and cell type. *Environ Toxicol Chem*, 33, 481–492.

Kornilova, E. S. 2014. Receptor-mediated endocytosis and cytoskeleton. *Biochemistry (Moscow)*, 79, 865–878.

Kreyling, W. G., Semmler-Behnke, M., Seitz, J., Scymczak, W., Wenk, A., Mayer, P., Takenaka, S. & Oberdorster, G. 2009. Size dependence of the translocation of inhaled iridium and carbon nanoparticle aggregates from the lung of rats to the blood and secondary target organs. *Inhal Toxicol*, 21 Suppl 1, 55–60.

Krpetic, Z., Anguissola, S., Garry, D., Kelly, P. M. & Dawson, K. A. 2014. Nanomaterials: Impact on cells and cell organelles. *Adv Exp Med Biol*, 811, 135–156.

Kukulski, W., Schorb, M., Kaksonen, M. & Briggs, J. A. G. 2012. Plasma membrane reshaping during endocytosis is revealed by time-resolved electron tomography. *Cell*, 150, 508–520.

Kumar, A., Kumar, P., Anandan, A., Fernandes, T. F., Ayoko, G. A. & Biskos, G. 2014. Engineered nanomaterials: Knowledge gaps in fate, exposure, toxicity, and future directions. *J Nanomater*, 2014, 16.

Kumari, A. & Yadav, S. K. 2011. Cellular interactions of therapeutically delivered nanoparticles. *Expert Opin Drug Deliv*, 8, 141–151.

Ladinsky, M. S., Mastronarde, D. N., Mcintosh, J. R., Howell, K. E. & Staehlin, L. A. 1999. Golgi structure in three dimensions: Functional insights from the normal rat kidney cell. *J Cell Biol*, 144, 1135–1149.

Laramy, C. R., Brown, K. A., O'Brien, M. N. & Mirkin, C. A. 2015. High-throughput, algorithmic determination of nanoparticle structure from electron microscopy images. *ACS Nano*, 9, 12488–12495.

Le Bihan, O., Bonnafous, P., Marak, L., Bickel, T., Trépout, S., Mornet, S., De Haas, F., Talbot, H., Taveau, J. C. & Lambert, O. 2009. Cryo-electron tomography of nanoparticle transmigration into liposome. *J Struct Biol*, 168, 419–425.

Lenzi, D., Runyeon, J. W., Crum, J., Ellisman, M. H. & Roberts, W. M. 1999. Synaptic vesicle populations in saccular hair cells reconstructed by electron tomography. *J Neurosci*, 19, 119–132.

Leung, Y. H., Guo, M. Y., Ma, A. P. Y., Ng, A. M. C., Djurisic, A. B., Degger, N. & Leung, F. C. C. 2017. Transmission electron microscopy artifacts in characterization of the nanomaterial-cell interactions. *Appl Microbiol Biotechnol*, 101, 5469–5479.

Liu, J. & Shapiro, J. I. 2003. Endocytosis and signal transduction: Basic science update. *Biol Res Nurs*, 5, 117–128.

Liu, X.-Q. & Tang, R.-Z. 2017. Biological responses to nanomaterials: Understanding nano-bio effects on cell behaviors. *Drug Deliv*, 24, 1–15.

Lucocq, J. M. & Gawden-Bone, C. 2009. A stereological approach for estimation of cellular immunogold labeling and its spatial distribution in oriented sections using the rotator. *J Histochem Cytochem*, 57, 709–719.

Lucocq, J. M. & Hacker, C. 2013. Cutting a fine figure: On the use of thin sections in electron microscopy to quantify autophagy. *Autophagy*, 9, 1443–1448.

Lucocq, J. M., Mayhew, T. M., Schwab, Y., Steyer, A. M. & Hacker, C. 2015. Systems biology in 3D space—Enter the morphome. *Trends Cell Biol*, 25, 59–64.

Mielańczyk, L., Matysiak, N., Klymenko, O. and Wojnicz, R. 2015. Transmission electron microscopy of biological samples. *The Transmission Electron Microscope-Theory and Applications*. InTech.

Luzio, J. P., Bright, N. A. & Pryor, P. R. 2007. The role of calcium and other ions in sorting and delivery in the late endocytic pathway. *Biochem Soc Trans*, 35, 1088–1091.

Lynch, I., Ahluwalia, A., Boraschi, D., Byrne Hugh, J., Fadeel, B., Gehr, P., Gutleb Arno, C., Kendall, M. & Papadopoulos Manthos, G. 2013. The bio-nano-interface in predicting nanoparticle fate and behaviour in living organisms: Towards grouping and categorising nanomaterials and ensuring nanosafety by design. *BioNanoMaterials*, 14(3–4), 195–216.

Lynch, I., Feitshans, I. L. & Kendall, M. 2015. 'Bio-nano interactions: New tools, insights and impacts': Summary of the Royal Society discussion meeting. *Philos Trans Royal Soc B Biol Sci*, 370, 20140162.

Madsen, J., Liu, P., Kling, J., Wagner, J. B., Hansen, T. W., Winther, O. & Schiøtz, J. 2018. A deep learning approach to identify local structures in atomic-resolution transmission electron microscopy images. *Adv Theory Simulat*, 1, 1800037.

Martens, T. F., Remaut, K., Demeester, J., De Smedt, S. C. & Braeckmans, K. 2014. Intracellular delivery of nanomaterials: How to catch endosomal escape in the act. *Nano Today*, 9, 344–364.

Mayhew, T. M. & Lucocq, J. M. 2011. Multiple-labelling immune EM using different sizes of colloidal gold: Alternative approaches to test for differential distribution and colocalization in subcellular structures. *Histochem Cell Biol*, 135, 317–326.

Mayhew, T. M., Muhlfeld, C., Vanhecke, D. & Ochs, M. 2009. A review of recent methods for efficiently quantifying immunogold and other nanoparticles using TEM sections through cells, tissues and organs. *Ann Anat*, 191, 153–170.

Mielanczyk, L., Matysiak, N., Michalski, M., Buldak, R. & Wojnicz, R. 2014. Closer to the native state. Critical evaluation of cryo-techniques for transmission electron microscopy: Preparation of biological samples. *Folia Histochem Cytobiol*, 52, 1–17.

Missaoui, W. N., Arnold, R. D. & Cummings, B. S. 2018. Toxicological status of nanoparticles: What we know and what we don't know. *Chem Biol Interact*, 295, 1–12.

Monopoli, M. P., Åberg, C., Salvati, A. & Dawson, K. A. 2012. Biomolecular coronas provide the biological identity of nanosized materials. *Nature Nanotechnol*, 7, 779.

Mourik, M. J., Faas, F. G. A., Valentijn, K. M., Valentijn, J. A., Eikenboom, J. C. & Koster, A. J. 2014. Correlative light microscopy and electron tomography to study von Willebrand factor exocytosis from vascular endothelial cells. *Methods Cell Biol*, 124, 71–92.

Muhlfeld, C., Mayhew, T. M., Gehr, P. & Rothen-Rutishauser, B. 2007a. A novel quantitative method for analyzing the distributions of nanoparticles between different tissue and intracellular compartments. *J Aerosol Med*, 20, 395–407.

Muhlfeld, C., Rothen-Rutishauser, B., Vanhecke, D., Blank, F., Gehr, P. & Ochs, M. 2007b. Visualization and quantitative analysis of nanoparticles in the respiratory tract by transmission electron microscopy. *Part Fibre Toxicol*, 4, 11.

Murk, J. L. A. N., Humbel, B. M., Ziese, U., Griffith, J. M., Posthuma, G., Slot, J. W., Koster, A. J., Verkleij, A. J., Geuze, H. J. & Kleijmeer, M. J. 2003. Endosomal compartmentalization in three dimensions: Implications for membrane fusion. *Proc Natl Acad Sci*, 100, 13332–13337.

Nativo, P., Prior, I. A. & Brust, M. 2008. Uptake and intracellular fate of surface-modified gold nanoparticles. *ACS Nano*, 2, 1639–1644.

Owen, D. M., Magenau, A., Williamson, D. J. & Gaus, K. 2013. Super-resolution imaging by localization microscopy. *Methods Mol Biol*, 950, 81–93.

Palade, G. E. 1952. A study of fixation for electron microscopy. *J Exp Med*, 95, 285–298.

Perkins, G. A., Sun, M. G. & Frey, T. G. 2009. Chapter 2 correlated light and electron microscopy/electron tomography of mitochondria in situ. *Method Enzymol*, 456, 29–52.

Piperigkou, Z., Karamanou, K., Engin, A. B., Gialeli, C., Docea, A. O., Vynios, D. H., Pavão, M. S. G. et al. 2016. Emerging aspects of nanotoxicology in health and disease: From agriculture and food sector to cancer therapeutics. *Food Chem Toxicol*, 91, 42–57.

Ray, P. D., Huang, B.-W. & Tsuji, Y. 2012. Reactive oxygen species (ROS) homeostasis and redox regulation in cellular signaling. *Cell Signal*, 24, 981–990.

Rejman, J., Oberle, V., Zuhorn, I. S. & Hoekstra, D. 2004. Size-dependent internalization of particles via the pathways of clathrin- and caveolae-mediated endocytosis. *Biochem J*, 377, 159–169.

Report, B. 2007. Publicly Available Specification 136 Terminology for nanomaterials. British Standards Institute PSI 136:2007 ISBN: 9780580613210.

Rodriguez-Boulan, E., Kreitzer, G. & Maisch, A. 2005. Organization of vesicular trafficking in epithelia. *Nat Rev Mol Cell Biol*, 6, 233–247.

Ross, M. H., Romrell, L.J., & Kaye, G.I. 1995. Epithelial Tissue. *In:* Coryell, P. A. (ed.) *Histology: A Text and Atlas*. Third ed, Philadelphia, PA: LIppincott, Williams & Wilkins.

Sabatini, D. D., Bensch, K. & Barrnett, R. J. 1963. Cytochemistry and electron microscopy. The preservation of cellular ultrastructure and enzymatic activity by aldehyde fixation. *J Cell Biol*, 17, 19–58.

Schrand, A. M., Dai, L., Schlager, J. J. & Hussain, S. M. 2012. Toxicity testing of nanomaterials. *Adv Exp Med Biol*, 745, 58–75.

Schrand, A. M., Schlager, J. J., Dai, L. & Hussain, S. M. 2010. Preparation of cells for assessing ultrastructural localization of nanoparticles with transmission electron microscopy. *Nat Protoc*, 5, 744–757.

Seto, E. S., Bellen, H. J. & Lloyd, T. E. 2002. When cell biology meets development: Endocytic regulation of signaling pathways. *Genes Dev*, 16, 1314–1336.

Shin, K., Fogg, V. C. & Margolis, B. 2006. Tight junctions and cell polarity. *Annu Rev Cell Dev Biol*, 22, 207–235.

Smith, D. & Starborg, T. 2018. Serial block face scanning electron microscopy in cell biology: Applications and technology. *Tissue Cell*.

Stanley, A., Thompson, K., Hynes, A., Brakebusch, C. & Quondamatteo, F. 2014. Nadph oxidase complex-derived reactive oxygen species, the actin cytoskeleton, and Rho GTPases in cell migration. *Antioxid Redox Signal*, 20, 2026–2042.

Starborg, T., Kalson, N. S., Lu, Y., Mironov, A., Cootes, T. F., Holmes, D. F. & Kadler, K. E. 2013. Using transmission electron microscopy and 3View to determine collagen fibril size and three-dimensional organization. *Nat Protoc*, 8, 1433–1448.

Stern, S. T., Adiseshaiah, P. P. & Crist, R. M. 2012. Autophagy and lysosomal dysfunction as emerging mechanisms of nanomaterial toxicity. *Part Fibre Toxicol*, 9, 20.

Studer, D., Humbel, B. M. & Chiquet, M. 2008. Electron microscopy of high pressure frozen samples: Bridging the gap between cellular ultrastructure and atomic resolution. *Histochem Cell Biol*, 130, 877–889.

Subramaniam, S. & Zhang, P. 2003. Electron tomography: A powerful tool for 3D cellular microscopy. *ASM News*, 69, 240–245.

Sun, C., Du, K., Fang, C., Bhattarai, N., Veiseh, O., Kievit, F., Stephen, Z. et al. 2010. PEG-mediated synthesis of highly dispersive multifunctional superparamagnetic nanoparticles: Their physicochemical properties and function in vivo. *ACS Nano*, 4, 2402–2410.

Sun, C., Fang, C., Stephen, Z., Veiseh, O., Hansen, S., Lee, D., Ellenbogen, R. G., Olson, J. & Zhang, M. 2008. Tumor-targeted drug delivery and MRI contrast enhancement by chlorotoxin-conjugated iron oxide nanoparticles. *Nanomedicine*, 3, 495–505.

Takizawa, T., Powell, R. D., Hainfeld, J. F. & Robinson, J. M. 2015. FluoroNanogold: An important probe for correlative microscopy. *J Chem Biol*, 8, 129–142.

Wang, F., Salvati, A. & Boya, P. 2018. Lysosome-dependent cell death and deregulated autophagy induced by amine-modified polystyrene nanoparticles. *Open Biol*, 8.

Weakley, B. 1972. Simplified basic theory of transmission electron microscopy. In: *Biological Transmission Electron Microscopy*. 2nd ed. Edinburgh, Scotland: Churchill Livingstone.

Weston, A. E., Armer, H. E. & Collinson, L. M. 2009. Towards native-state imaging in biological context in the electron microscope. *J Chem Biol*, 3, 101–112.

WHO. 2017. WHO guidelines on protecting workers from potential risks of manufactured nanomaterials.

Wilke, S. A., Antonios, J. K., Bushong, E. A., Badkoobehi, A., Malek, E., Hwang, M., Terada, M., Ellisman, M. H. & Ghosh, A. 2013. Deconstructing complexity: Serial block-face electron microscopic analysis of the hippocampal mossy fiber synapse. *J Neurosci*, 33, 507–522.

Williams, K. M., Gokulan, K., Cerniglia, C. E. & Khare, S. 2016. Size and dose dependent effects of silver nanoparticle exposure on intestinal permeability in an in vitro model of the human gut epithelium. *J Nanobiotechnol*, 14, 62.

Wisse, E., Braet, F., Duimel, H., Vreuls, C., Koek, G., Olde Damink, S. W., Van Den Broek, M. A. et al. 2010. Fixation methods for electron microscopy of human and other liver. *World J Gastroenterol*, 16, 2851–2866.

Wu, X. S., McNeil, B. D., Xu, J., Fan, J., Xue, L., Melicoff, E., Adachi, R., Bai, L. & Wu, L. G. 2009. Ca^{2+} and calmodulin initiate all forms of endocytosis during depolarization at a nerve terminal. *Nat Neurosci*, 12, 1003–1010.

Yang, M., Zhang, M., Tahara, Y., Chechetka, S., Miyako, E., Iijima, S. & Yudasaka, M. 2014. Lysosomal membrane permeabilization: Carbon nanohorn-induced reactive oxygen species generation and toxicity by this neglected mechanism. *Toxicol Appl Pharmacol*, 280, 117–126.

Ye, D., Dawson, K. A. & Lynch, I. 2015. A TEM protocol for quality assurance of in vitro cellular barrier models and its application to the assessment of nanoparticle transport mechanisms across barriers. *Analyst*, 140, 83–97.

Young, B. & Wheater, P. R. 2006. *Wheater's Functional Histology: A Text and Colour Atlas; Drawings by Philip J. Deakin*. Ediburgh, Scotland: Churchill Livingstone.

Zhu, B., Li, Y., Lin, Z., Zhao, M., Xu, T., Wang, C. & Deng, N. 2016. Silver nanoparticles induce HePG-2 Cells apoptosis through ROS-mediated signaling pathways. *Nanoscale Res Lett*, 11, 198.

8

Ecotoxicology of Nanoparticles

Iain Murray and
Andy Fogarty

8.1 Introduction: The Problem Facing Nano-Ecotoxicology

Since the phrase nanotechnology was first coined in 1974 by Norio Taniguchi, nano-based technology has seen massive growth in development and applications (Taniguchi, 1974). This is evident from a simple key word search of peer-reviewed literature available on databases such as ISI Web of Science using the term 'nanomaterials'. The number of publications per year is shown in Figure 8.1 and demonstrates that the number of publications increased significantly every year from the late 1990s onwards with the increases peaking in the mid-2000s.

In 2011, the European Commission defined nanomaterials as *'natural, incidental or manufactured material containing particles, in an unbound state or as an aggregate or as an agglomerate and where, for 50% or more of the particles in the number size distribution, one or more of the external dimensions is in the size range 1–100 nm'*. Although this seems fairly definitive, the question of adverse effects on the environment remains unanswered. This lack of clarity surrounding nano-ecotoxicology is largely due to the plethora of nanomaterial forms as well as the substantial number of characteristics available to researchers for the classification of nanomaterials. These characteristics include: material, size,

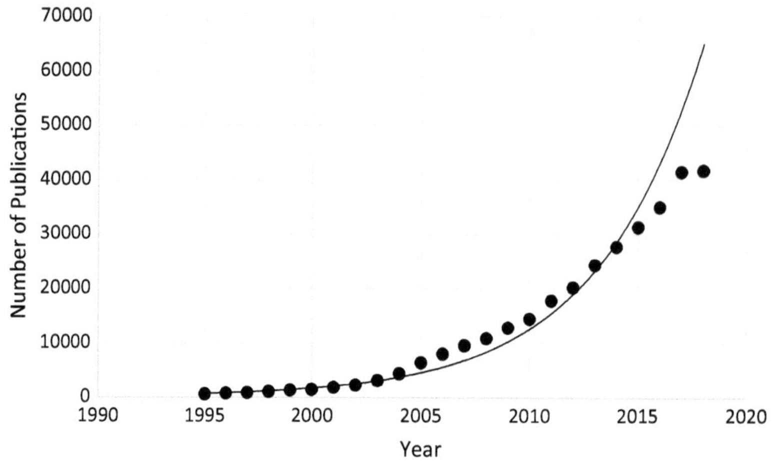

FIGURE 8.1 Number of publications using keyword search 'nanoparticles' listed on Science Direct by year.

shape, mass, coating, zeta-potential, optical properties, surface area, curvature, and concentration. The modern approach to ecotoxicology and chemical risk assessment under REACH legislation is to prioritize a known chemical of concern, sometimes based on modelled or computational toxicity data based on physico-chemical properties of similar chemical compounds. This approach creates a problem for nanotoxicology in that the chemicals of concern are not known. Manufacturers or users of these products in most jurisdictions are under no obligation to declare, register, or characterize the nanomaterials they use. This is often due to the proprietary nature of these materials. Databases of nanomaterials in use have been established (European Observatory for Nanomaterials, 2018) in an attempt to fill the knowledge gap surrounding the nanomaterials in common usage, but until it becomes mandatory for companies to register their materials, it is likely these databases will be sufficient.

In addition to deliberately manufactured nanomaterials, other nanomaterials can be produced incidentally as a result of other processes; for example, soot or carbon black often produced as a result of the partial combustion of fossil fuels can fall within the definition of nanomaterials (Nowack & Bucheli, 2007). This chapter focuses on anthropogenic nanomaterials, both metallic and carbon-based.

8.2 Ecotoxicology: Managing the Risk Assessment

Traditional ecotoxicological risk assessment of pollutants starts with a clear knowledge of the nature and chemical identity of the pollutant of concern. This is usually a known chemical or effluent from a known source or an observed effect such as a fish kill leading to an investigation and identification of the source. The composition of the effluent can be characterized chemically, respective concentrations measured, and appropriate ecotoxicity studies carried out, even taking synergistic effects into account where necessary. In the case of nanomaterials, this approach is not possible. Firstly, as previously mentioned, the nanomaterials in use are not known (McGillicuddy, Murray, Kavanagh, Morrison, Fogarty, Cormican, Rowan, & Morris, 2016). Secondly, the behaviour of nanomaterials in the freshwater ecosystem is highly variable and dependent not only on the physico-chemical properties of the nanomaterials themselves, but also the water chemistry in the receiving environment (Bury, Shaw, Glover, & Hogstrand, 2002; Ellis, Baalousha, Valsami-Jones, & Lead, 2018; Handy, Von Der Kammer, Lead, Hassellöv, Owen, Crane, 2008). Thirdly, the predicted environmental concentration (PEC) in the receiving environment is often so low as to be below the limit of detection using current standard methods and therefore immeasurable (Blaser, Scheringer, MacLeod, & Hungerbühler, 2008; Boxall, Chaudhry, Sinclair, Jones, Aitken, Jefferson, & Watts, 2007; Fadri Gottschalk, Sondere, Schols, & Nowack, 2009; Giese, Klaessig, Park, Kaegi, Steinfeldt, Wigger, Gleich, & Gottschalk, 2018).

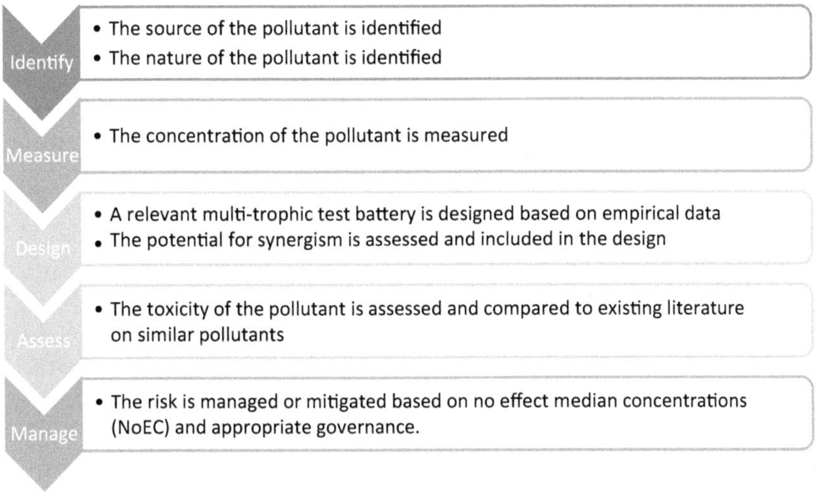

FIGURE 8.2 An overview of an ecotoxicological risk assessment methodology.

Figure 8.2 demonstrates a normal ecotoxicological risk assessment procedure. Following the identification of the risk, an appropriate risk assessment is designed and carried out to inform the necessary governance to mitigate further or increasing risk. Nanomaterials create problems for this process at every stage of the process. This is a five-stage process as depicted in Figure 8.2 and the stages are outlined in Subsections 8.2.1–8.2.5.

8.2.1 Stage (I) of Ecotoxicological Risk Assessment: Hazard Identification

In normal circumstances, chemical effluents are known as resultant effluents from a particular industry or process. In nanotoxicology, this may not always be straightforward. Even if the nano-metal effluent is known, downstream effects must be taken into account to determine the exact toxicant of concern. For example, silver nanoparticles when eluted from a source are reported to be influenced heavily by the water chemistry in the receiving body. (Bury et al., 2002; Nowack & Mueller, 2008).

These processes include: photochemical, redox, dissolution, aggregation/agglomeration, sedimentation, adsorption to organic matter or other surfaces such as soils, biodegradation, or biomodification (Fabrega, Luoma, Tyler, Galloway, & Lead, 2011; Klaine, Alvarez, Batley, Fernandes, Handy, Lyon, McLaughlin, & Lead, 2008). Different nanomaterials behave in different ways depending on several factors including water chemistry as previously mentioned, but also the form of the nanomaterial to begin with, i.e., silver/carbon/titanium/gold/copper/zinc, coatings, original size, and morphology (Ellis et al., 2018; He, Chen, Zeng, Peng, Huang, Shi, & Huang, 2017; Park, Woodhall, Ma, Veinot, Cresser & Boxall, 2014; Weinberg, Galyean, & Leopold, 2011). The effects of these processes on nanomaterial toxicity will be discussed later in this chapter.

This variability can be geographic in nature as well as temporal and as a result, researchers are now using specific waters from the receiving environment at a number of different seasons to determine the toxicity of a given nanomaterial (Ellis et al., 2018; Yin, Yang, Zhou, Wang, Yu, Liu, & Jiang, 2015). The variability in the characteristics of the nanomaterials is particularly true for silver nanomaterials. These materials may agglomerate into larger particles or ionize, resulting in smaller particles or soluble silver ions. These silver ions may then form other compounds with natural salts in the receiving water, each yielding different toxic effects (Kaegi, Voegelin, Sinnet, Zuleeg, Hagendorfer, Burkhardt, & Siegrist, 2011).

Typically, the temporal effects on the fate and behaviour of silver nanoparticles in the aquatic environment focus on seasonal changes often attributed to organic detritus, e.g., leaves or pine needles and concomitant forestry-associated chemical changes such as reduced pH following an influx of humic or fulvic acids (Akaighe, Maccuspie, Navarro, Aga, Banerjee, Sohn, & Sharma, 2011; Gunsolus, Mousavi, Hussein, Bühlmann, & Haynes, 2015); however, other seasonal changes may also have an effect. These seasonal variations can be photochemical in nature caused by increased sunlight during summer months and longer daylight hours, which has been shown to promote the reduction of ionic silver (Ag^+) to silver nanoparticles by organic matter (Zhang, Yang, Shen, Yin, & Liu, 2015).

Using the example of silver nanoparticles, Figure 8.3 shows some of the processes silver from consumer products may undergo on its route to the receiving environment.

In the example of silver nanoparticles in Figure 8.3, the dynamic nature of silver, particularly when influenced by water chemistry, is particularly evident.

The processes nanometallic materials undergo in the receiving environment are extensive, complex, and as will be discussed in this chapter, can have a significant effect on the toxicity of the materials to biota in the receiving environment.

Although empirical data can be generated from bench models (Benn & Westerhoff, 2008; Jarvie, Al-Obaidi, King, Bowes, Lawrence, Drake, Green, & Dobson, 2009) or *in silico* models (Gottschalk, Sun, & Nowack, 2013a), these models need a specified input, i.e., a known nanomaterial to work with at the start of the model. In order to do this, the specific nature and characteristics of the nanomaterials must be known. This includes details on the specific materials being produced, used commercially and in consumer products likely to end up in the environment at the end of their life cycle. The lack of this information has been noted by researchers for years (Giese et al., 2018; Hussain, Warheit, Ng, Comfort, Grabinski, & Braydich-Stolle, 2015; Maillard & Hartemann, 2012; Wijnhoven et al., 2009). A number of databases have been in existence since 2005 (Nanowerk, 2016); however, the details provided of the nanomaterials on the database are limited and not consistent. Many of these novel materials are the closely guarded secrets of their

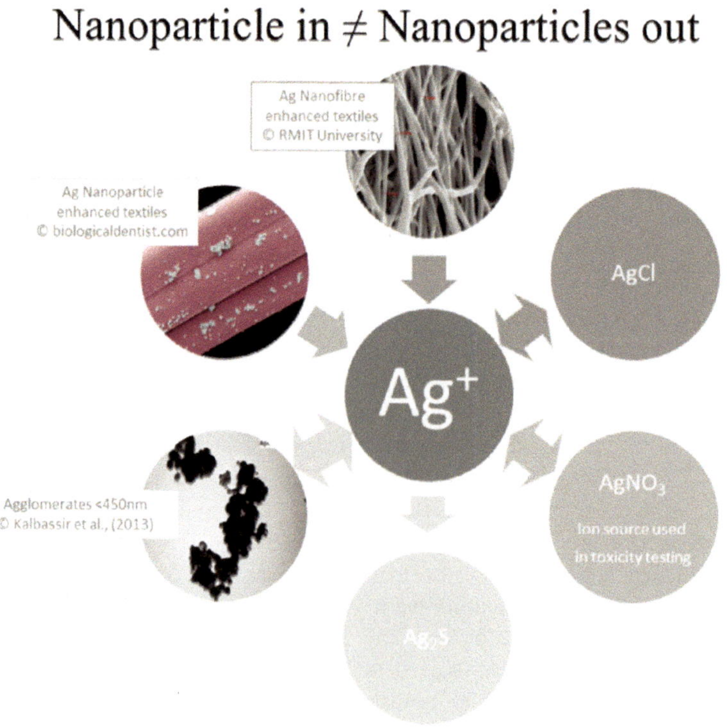

FIGURE 8.3 An overview of the dynamic water chemistry-influenced processes influencing the speciation and environmental fate of nano-silver desorbed from silver functionalised consumer products eluted to the aquatic environment.

manufacturers and developers to protect intellectual property. The introduction of two nanomaterial databases was announced in 2018, which is a welcome development (European Observatory for Nanomaterials, 2018), although unless mandatory reporting and registration of nanomaterial production and use is introduced across Europe it is likely to be of limited benefit to the field of ecotoxicology. Until then, the scientific community must continue to risk assess as many materials as possible and lobby the regulatory bodies into enforcing mandatory reporting so that when the necessary information becomes available it should be relatively straightforward to compare it with existing toxicity and ecotoxicity data.

8.2.2 Stage (II) of Ecotoxicological Risk Assessment: Measure

This step of the ecotoxicological risk assessment can involve both qualitative and quantitative measurements of the concentrations of toxicants of concern in the receiving environment. Once these typical environmental concentrations are known, the risk assessment is designed to ensure that the ecotoxicological study is environmentally relevant. This provides for a thorough understanding of the toxicological effects of the chemicals or effluents at the concentrations likely to be found in the environment. Again, the process becomes problematic and the problems are twofold. Firstly, in order to measure the concentration of something, the identity of what you are measuring must be known. This is not always possible with regards to nanomaterials as outlined in Section 8.2.2. If the identity of the chemical or metal is not known, it is not possible to measure their concentration in the environment. This problem can be further compounded with nanomaterials such as silver, which are known to be highly dynamic in nature and may exist as nanoparticles, ionic silver, or silver salts when influenced by the water chemistry, meaning it is not clear which species of silver needs to be assayed (McGillicuddy et al., 2016). However, efforts are ongoing to develop capture methods for silver nanoparticles (McGillicuddy, Morrison, Cormican, Dockery, & Morris, 2018). These methods, once developed and validated, may allow for the capture and concentration of the sample to levels above the limit of detection. This can, however, be complicated by polydisperse suspensions (Tomaszewska, Soliwoda, Kadziola, Tkacz-Szczesna, Celichowski, Cichomski, Szmaja, & Grobelny, 2013), whereby it may be difficult to ascertain the appropriate filter sizes as well as having difficulties with distortion of results when using methods such as dynamic light scattering (DLS) or UV-Vis spectroscopy.

Ecotoxicologists are therefore dependent on predictive models to determine the relevant ranges for toxicity assessments. Several such models have been developed and a review by Gottschalk, Sun, and Nowack (2013) summarized and graphically illustrated the wide range of findings and predictions from a number of studies of the environmental partitions available to nanomaterials, i.e., surface water, waste water treatment plant effluent, biosolids, sediments, soils, soils treated with biosolids, and the atmosphere. In surface water, it was predicted that concentrations could differ by a factor of 10^4 (3 ng/L up to 1.6 μg/L) but are likely to be in the ng/L range for TiO_2, Ag, ZnO, Fullerenes, and CeO_2. The same study suggested that concentrations in sediments will be higher and higher still in biosolids (0.001–100 μg/g). The risk of contamination of ground water and subsequently surface water is therefore considered to be substantial from soils treated with biosolids, i.e., sewage sludge. This and other studies again allude to substantial geographic and temporal variations in the PEC.

8.2.3 Stage (III) of Ecotoxicological Risk Assessment: Design

Given that the majority of traditionally employed ecotoxicity bioassays typically yield median effective concentrations in the μg/L–mg/L range, which is substantially higher than the PECs, it is difficult to design an environmentally relevant multi-trophic test battery within the confines of the traditional, validated, standard methods and biomarkers.

The ECHA provide guidance on the adopted and proposed test requirements for testing of aquatic pelagic toxicity (European Chemicals Agency [ECHA] 2014, updated June 2017) and a limited number of options for sediment-water toxicity testing using chironomids. The adopted guidelines prescribe a multi-trophic test battery to include algae, *Daphnia sp.*, acute immobilization and reproduction, and a series of *in vivo* piscine studies. The development of *in vitro* methods such as piscine cell lines is desirable, which may prove problematic for nanomaterials. Cyto-toxicological studies assessing nanomaterials have been investigated in a number of studies in recent years and the components of test matrices have been shown to interfere with the bioavailability and hence the toxicity of the test materials (Connolly, Fernandez-Cruz, Quesada-Garcia, Alte, Segner, & Navas, 2015; Minghetti & Schirmer, 2016). The formation of protein corona, the use of sera, and chloride concentrations are examples of complications which may contribute to variations in the toxicity observed with cytological toxicity assays.

The need to adapt, standardize, and validate a suite of repeatable, transferrable, and robust toxicity assay protocols for nanotoxicity assessment is critical in the near future. Until then, ecotoxicologists must continue to report their findings of adapted/non-standardized tests which they have optimized for the nanometallic toxicants of concern.

This remains a worthwhile exercise which will provide valuable data or a menu of options for regulators to choose from and establish standardized protocols going forward. However, in the meantime, the toxicity results and data as published might best be viewed as fragments contributing towards the design of future standardized protocols and not definitive assessments of nanomaterials in general given the plethora of adapted methods, matrices, and nanomaterial forms tested to date and again in the absence of relevant data of which nanomaterials are of particular concern in a specific geographical region.

8.2.4 Stage (IV) of Ecotoxicological Risk Assessment: Assess

The thorough risk assessment of any toxicant is a relatively straightforward process given an appropriate design and validated, standardized protocols. Results from standard methods can be compared and the toxicity of chemicals ranked as a result. The challenges for ecotoxicologists are immense in the area of nanotoxicity. It is necessary to try to choose nanomaterials of concern with limited data of which ones are in use, then trawl the extant literature to see what has been assessed already (if characterized thoroughly), choose an appropriate method, and potentially have to validate and justify adaptations to that method from first principles. As a result of the variety of methods, adaptations, and materials employed in these risk assessments, it is not possible for regulators to form an overall opinion on the toxicity of nanomaterials. The ecotoxicity of many nanomaterials has been assessed by a wide variety of methods, both standard and adapted, and the results published. The dynamic nature of these materials in solution leads to substantial variability and a lack of repeatability. The need to publish novel assessments is at odds with the scientific imperative of independent inter-laboratory repetition of assessments and comparison of results.

The finite resources in terms of time and funding of laboratory research further limits the assessment as a substantial amount of these resources go into validation of methods; thus, a reduction in replication is typical. This, when presented with the adaptation of standard methods and the lack of clarity surrounding the relevance of the materials being assessed, challenges the robustness of the assessment.

As described in Section 8.2.4, the modelled environmental concentrations and the dearth of definitive empirical data result in ecotoxicological assessments at concentrations several fold higher than the predictions when acute assessments are employed. As a result, more discrete sub-acute endpoints and longitudinal studies will be needed to determine if a true no effect concentration (NoEC) exists. The fidelity of assessments will need to be refined further, identifying more sensitive endpoints and bioindicators not currently possible with existing assessment protocols.

8.2.5 Stage (V) of Ecotoxicological Risk Assessment: Manage

The resultant data from risk assessments is normally managed and validated through replication before being compared and ranked with other similar but previously assessed toxicants. Regulators can then make decisions regarding permissible levels, mandatory reporting, discharge licensing criteria, etc. Due to the predictive approach ecotoxicologists are forced to employ with regards to the risk assessment of nanomaterials, the management of this data is challenging for industry and regulators. As a result, nothing decisive regarding nanomaterial regulation has been forthcoming from Europe or the relevant competent authorities and as such, it has not been possible thus far, to effectively manage the usage or elution of nanomaterials into our aquatic environment. Moves towards addressing the shortcomings in the management of ENMs in the environment took a big step forward in 2018 with the introduction of two searchable EU databases of ENMs and some associated toxicological and characterization data (European Observatory for Nanomaterials, 2018). The accessibility of this data will further the knowledge and validity of ENM toxicology by making data easily available to researchers for comparative analysis. However, there are substantial shortcomings such as limited ecotoxicological data and the lack of a registry of ENMs in use in industry or consumer products. Without this data, it remains impossible for ecotoxicologists to predict and risk assess nanometallic toxins of concern or prioritize these for toxicological assessment and hence regulation and effective hazard management remains problematic.

In a technical research report published by the Irish Environmental Protection Agency, these problems are highlighted (McGillicuddy et al., 2018) and urge regulators to adapt mandatory reporting on the use and import of ENMs and their characteristics to address the issues facing ecotoxicologists and the shortcomings in ENM management.

8.3 Ecotoxicity of 0-D, 1-D, 2-D, and 3-D Nanomaterials

Nanomaterials and nanoparticles are defined for the European Union as having one or more external dimensions (i.e., at least one dimension) in the range of 1–100 nm (The European Commission, 2011). The dimensionality of nanomaterials refers to the number of dimensions outside that range. Zero-dimensional (0-D) nanomaterials have no dimensions outside of that range. These would include small nanoparticles, but not nanowires, for example. One-dimensional (1-D) nanomaterials have one dimension outside the nanoscale as defined above. These materials include nanowires, nanofibers, nanotubes, and nanorods. Two-dimensional (2-D) nanomaterials have two dimensions outside the nanorange and tend to be sheet-like materials. Examples of 2-D nanomaterials include graphene, nanofilms, and nanolayers. These materials are frequently found to be used as coatings. All dimensions of three-dimensional (3-D) nanomaterials are outside the nanoscale, i.e., they essentially do not fit the strict European Commission definition. Nonetheless, because of the dynamic nature of nanomaterials in general, their tendency to agglomerate and their methods of synthesis being both bottom up and top down, it would be remiss to ignore them when talking about nanomaterials as they may be made up of agglomerates or layers of smaller particles or materials, e.g., multiple layers of nanosheets or plates, bundles of nanowires, fibres or tubes, agglomerates of nanoparticles, or of course bulk powders or dispersions.

The ecotoxicity of nanoparticles, i.e., 0-D and 1-D, can be two-fold: toxicity as a result of the material itself and toxicity as a result of the morphology of the presentation of the material. The former being mechanistic in nature and specific to the elements or compounds in the nanoparticle and its coatings and the latter being mechanical. Mechanical toxicity refers to toxicological effects caused by the size or shape of the materials, e.g., physical blocking of digestive tracts for example. The toxicological profile of the material can change throughout the product lifecycle, so the most relevant ecotoxicological studies may not include the native original form of the ENM. This is particularly important in the case of 2-D and 3-D materials. For example, in the case of anti-bacterial textiles utilizing silver nanowires woven into the fabric, the nanowire itself may not be a toxicant of concern as long as it remains *in-situ* on the fabric (Kwak & An, 2015). However, as part of the degradation process, the Ag nanowires may shed ionic silver, which then becomes the relevant toxicant. Similarly, bulk powders or agglomerates of copper or zinc may not technically be in the EU-defined nanoscale; however, as they begin to succumb to the pressures of water chemistry and other process, they too may begin to degrade potentially shedding metallic ions into the environment, or in the case of agglomerates, break down into smaller 0-D or 1-D nanomaterials each with its own distinct toxicological profile and concerns.

Essentially, what this means for ecotoxicologists is that they need to think outside the box with regards to choice of analytes. A full material life cycle analysis (not just product life cycle) is necessary to evaluate on the balance of probability, what the most likely effect the water chemistry is going to have on the materials, and hence identify the most likely toxicant species of concern, noting, that it may or may not be a nanomaterial at point of contact with the receiving environment or its biota.

8.4 Ecotoxicity of Graphene

8.4.1 Sources and Environmental Fate of Graphene Nanomaterials

Graphene, including the fullerenes, carbon nanotubes (CNT) and other presentations of graphene, is becoming ever more frequently used in the production of electronics due to its lightweight and conduction properties. Applications include graphical displays, capacitors, radio antennae, batteries, healthcare monitors, environmental remediation uses, filtration and water treatment, solar cells, electronic components, and microelectronics to name but a few. These materials are usually in sheet or rolled sheet forms and annual production for 2009 was estimated by Arvidsson, Molander, and Sandén (2013) at 15,000 kg, equivalent to a surface area of approximately 40,000 m^2. It is likely, given the rapid development in graphical interfaces and the advent of flexible graphics hardware and other applications of graphene and graphene oxide-related products, that this level of production will rise rapidly. The market value of graphene is expected to rise from under US$200 million in 2018 to US$1.3 billion by 2023 (Markovic, Kumar, Andjelkovic, Lath, Kirby, Losic, Batley, & McLaughlin, 2018).

Given the plethora of applications, it is not unreasonable to expect that some environmental infiltration will occur. Although significant improvements in the levels of electrical goods recycling have been ongoing, inevitably some end up in landfill or improperly disposed of. This has led to a substantial amount of materials, in particular from batteries and unrecyclable components, ending up in landfill leachate lagoons as they are washed through the landfill and then collected. Graphene, its derivatives, and indeed other carbon compounds are now seen as a method of dealing with other toxicants, particularly metals and heavy metals and these bioremediation usages further add to their likelihood of ending up in lagoons (McGillicuddy et al., 2018; Zhang, Hu, Deng, 2016).

Enroute to the receiving environment, graphene nanomaterials such as graphene oxide may undergo several transformations and exist in a number of iterations. This can be driven by numerous physico-chemical pressures such as UV light (sunlight), microbes, pH, ions, organic matter, leading to dispersion, or aggregation. These dynamic processes may alter the morphological, dimensional, or chemical properties of the materials and hence impact on their toxicity. Effects on the toxicological properties include partitioning of the material in the receiving environment, e.g., to benthic or pelagic regions, and hence accessibility and exposure levels to specific organisms and of with resultant bioavailability implications. Alterations to surface chemistry may also have an effect on the environmental fate and concomitant toxicological effects. CNTs are known to be hydrophobic and favour covalent bonding due to the relatively inert nature of graphene. This means that surface treatments to improve dispersions can include surface coatings or oxidation to effect dispensability, thus influencing the stability of the material in solution.

The levels of graphene and indeed its appropriate characterization in the aquatic environment are largely unknown (Laux et al., 2018). The limited ability to quantify small concentrations, particularly CNTs, means that effective predicted models are very difficult to develop and validate. As such, ecotoxicological testing at environmentally relevant concentrations may be near impossible without addressing the knowledge gap in this area. In a review carried out by Markovic et al. (2018), PECs for CNTs are modelled to be in the range of 0.23–0.28 ng/L, but the study highlights that it may not be possible to discern graphene nanomaterials such as graphene oxide from CNTs in solution as they have similar physico-chemical properties.

8.4.2 Ecotoxicology of Graphene-Related Materials

A substantial number of studies have been carried out to investigate the toxicological effects of graphene-related products such as fullerenes and CNTs on aquatic organisms. Given the low PECs, most traditional ecotoxicological assessments show endpoints well in excess of the PECs as is typical for several nanomaterials. However, this does not necessarily imply that ENMs are non-toxic or harmless. The suite of tests available to ecotoxicologists is limited by those validated by organizations such as the ISO and OECD. Traditionally, these tests have been available for several decades in their validated form and were originally designed for chemical pollutants and effluents, not specifically ENMs. As such, their degree of sensitivity may be insufficient to fully assess chronic and long-term effects of ENM on aquatic organisms. It may be necessary at this point in the development of toxicants to reassess the suitability of these tests, their endpoints, and exposure timeframes. For example, tests in excess of 3 days' exposure are typically considered to be chronic, and anything 3 days and under is considered acute. Indeed, the longest reported test, the chronic *Daphnia magna*, is just 21 days in length.

Algae are the basis of most aquatic ecosystems and are one of the key tests used in a multi-trophic test battery. The species most often used is *Selenestrum capricornutum*, also known as *Raphidocelis subcapitata* or *Pseudokirchneriella subcapitata*, tested in accordance with ISO 8692: 2012 or OECD 201 (International Organisation for Standardisation, 2012a; OECD, 2011). The EC_{50} for these algae was reported to be 20 µg/L when assessed with graphene oxide (Nogueira, Nakabayashi, & Zucolotto, 2015). The EC_{50} reduced by half, to 10 µg/L when assessment was carried out using auto-fluorescence as a biomarker instead of cell number. The authors go on to report that the toxicity is likely to be as a result of reactive oxygen species (ROS) generation and membrane damage. The potential of the shading effect of graphene oxide agglomeration was also noted to cause a reduction in light transmission, which may have influenced the growth rate of these photosynthetic algae.

The next stage up in the food chain is the primary consumers level of the multi-trophic test battery. This trophic level usually includes invertebrate species such as *Daphnia*. This is a validated test system under ISO 6341:2012 and OECD 202 (International Organisation for Standardisation, 2012b; OECD, 2004). A number of studies have evaluated the toxic effects of graphene-related materials on *Daphnia* with varying endpoints. Guo et al. (2013) investigated the uptake by *Daphnia magna* of radiolabelled graphene. No immobilization was observed even after 48 hours at the test concentration of 250 µg/L, which would be a traditional acute endpoint; however, it was found that following just 24 hours exposure to concentrations of 250 µg/L, the *Daphnia* had taken up nearly 1% of their dry mass in graphene. Although much of the graphene body burden was shown to be released in a 2-hour depuration window, some uptake from gravid females to eggs and neonates was reported. This may be a cause for concern because although the test concentration (250 µg/L) was several orders of magnitude higher than the PECs, it does show a potential for bioaccumulation, particularly in consistently polluted habitats where depuration is not possible or indeed likely. Some uptake was also reported at concentrations as low as 25 µg/L, but it was also noted that some clearance occurred in the second half of the 48-hour exposure period. It is hypothesized in the study that this may be because of settlement of the graphene in solution and whilst this may seem to mitigate the likelihood of extensive toxicity, it should be noted that continuous sources of graphene pollution to watercourses may not afford the same clearance opportunity to wild populations subjected to ongoing exposure.

The acute and chronic toxicity of graphene was compared with fullerenes and single and multiwall CNTs by Fan et al. (2016). It was demonstrated that although the test concentrations and relevant median toxicity values were again much higher than the PECs, the toxicity of graphene was higher than the fullerenes and CNTs; however, bioaccumulation of graphene was lower. Additionally, it was noted that graphene promoted growth and reproduction at low concentrations and inhibited it at high concentrations. It was proposed that this was as a result of nutrients adsorbed to the material surfaces essentially 'super feeding' the organisms concomitantly with ingestion of the CNMs.

The third trophic level assessed as part of the traditional multi-trophic test battery is the secondary consumer which is often the apex predator. Typically, Zebrafish *(Danio rerio)* are the model of choice for this trophic level. Several studies using secondary consumers have reported median toxicity concentrations significantly higher than the PECs.

Chen, Hu, Zhang, Huang, Luo, & Huang et al. (2012) compared the toxicity of graphene oxide and multi-walled CNTs using two human cell lines: bone marrow neuroblastoma (SK-N-SH) and epithelial carcinoma (HeLa) and zebrafish embryos. Toxic effects in the form of growth inhibition were observed for the human cell lines in the range of 50 mg/L, but interestingly it was demonstrated that the CNT were more toxic than the graphene oxide. As the two materials are chemically identical, the differential toxicity must be because of their geometry. When tested with zebrafish embryos, a substantial amount of adsorption to the larvae was observed with greater adsorption and indeed penetration seen with the CNT structures than the planar graphene oxide. Both species of carbon yielded increased apoptosis although the graphene oxide effects were more localized to the forehead and ocular areas of the zebrafish. The zebrafish also demonstrated delays to hatching when exposed to 50 mg/L with morphological aberrations observed at the same concentration. Although these concentrations were substantially higher than the PECs, the adsorption findings are of concern, as this suggests a potential to bioaccumulate and biomagnify.

Like many other nanomaterials, CNM are not immune to the effects of water chemistry on their toxicity. In a study by Chen, Ren, Ouyang, Hu, & Zhou (2015), natural organic matter such as humic acid was shown to attenuate the toxic effects of graphene oxide, using the zebrafish model and concentrations of CNMs of up to 100 mg/L. The mitigation of humic acid on graphene oxide toxicity was demonstrated using several techniques and examines several mechanisms including oxidative stress, mitochondrial effects. The concentrations of humic acid used in the study were in the range 0.01–100 mg/L and were reported to be environmentally relevant, but the concentration of graphene used was significantly higher than the PECs, like most ecotoxicological studies.

Although it may seem obvious to search for toxicological or ecotoxicological studies for answers on the effects materials have on organisms, studies on relevant cytological applications can provide insights into the mechanisms involved in the toxicity, often using novel approaches to developing research and therapeutic methods. One such study investigated the use of ultra-small graphene oxide (USGO) as a tool for gene delivery in cells and zebrafish embryos and demonstrated that transfection efficacy rates in zebrafish embryos were improved by 60% when the Polyethyleneimine-functionalized USGO was used as a carrier (Zhou et al., 2012). Whilst this finding has obvious benefits in the approach of synthetic gene delivery, gene therapy, and research, it also provides some insight into the potential mechanism of USGO toxicity as it suggests that these materials can selectively target DNA, the implications being that there is a potential for them to induce mutations and hence genotoxicity.

8.5 Ecotoxicity Silver Nanomaterials

The global increase in the use of nanosilver-containing products suggests that individuals and the environment are increasingly being exposed to silver nanoparticles (AgNP) (Cunningham & Joshi, 2015; EFSA, 2013; Farkas, Christian, Gallego-Urrea, Roos, Hassellöv, Tollefsen, & Thomas, 2011; Reidy, Haase, Luch, Dawson, & Lynch, 2013). This exposure may occur over several phases of the product life cycle from synthesis to use of consumer products and, disposal. The Scientific Committee on Emerging and Newly Identified Health Risks (SCENHIR) Report of 2014 highlights the potential usefulness of silver nanoparticles as a delivery system for silver ions in soil, water, and sediment over time, which suggests that this may be a source of concern in the future as methods and applications are developed and implemented.

8.5.1 Sources of Silver Nanomaterials

Typically, silver is used for its antimicrobial activity. This property results in its widespread use in textiles such as antimicrobial clothing and undergarments, commercial linen such as those used in hospital and hotels, hospital curtains, and medical dressings such as those used for the burns, as well as antimicrobial paints. The use of AgNP adsorbed to textiles has a limited functional lifespan as the nanoparticles eventually will elute out in the wash. This has led to enhancements in the functionalizing technology and the use of nanowires to increase the longevity of the antimicrobial properties in the fabric. Hence, these materials may tend to stay in the textile for longer periods of time and release silver ions instead of AgNPs during the washing process.

Like graphene nanomaterials, the release of ions and speciated derivatives of the ENM can result in the release of Ag^+ (silver ions) from Ag-functionalized products (Kaegi, Sinnet, Zuleeg, Hagendorfer, Mueller, Vonbank, Boller, & Burkhardt, 2010; Lombi, Donner, Scheckel, Sekine, Lorenz, Goetz, Von, & Nowack, 2014; Lorenz, Windler, von Goetz, Lehmann, Schuppler, Hungerbühler, Heuberger, & Nowack, 2012; Nowack, Ranville, Diamond, Gallego-Urrea, Metcalfe, Rose, Horne, Koelmans, & Klaine, 2012). The leaching effect of silver released from socks and its fate in WWTP was analysed by Benn and Westerhoff (2008) using six different sock types. The fabric was found to contain up to 1360 μg Ag/g of textile and leached almost half of that value into distilled water. Microscopic analysis determined that the silver particles ranged in size from 10 to 500 nm in diameter. Physical separation and ion-selective electrode (ISE) analyses suggest that both colloidal and ionic silver leached from the fabric. Variable leaching rates from <1% to almost 100% suggested that the textile manufacturing process may control the longevity of the functionalization and hence the overall severity of the leaching effect and concomitant silver speciation and release. The use of silver nanoparticles in paints and outdoor facades was reviewed by Kaegi et al. (2010), who found that the maximum leaching effect was 145 μg/L/year. This potentially could contribute substantially to surface water as rainwater typically evades wastewater treatment in the modern approach of separating rain water from foul sewerage in drainage systems.

Silver ions that are leached from nano-functionalized fabrics during washing with a hypochlorite/detergent solution rapidly react with chlorides present in the wash water to form AgCl (Impellitteri, Tolaymat, & Scheckel, 2009). The same study suggests that eluted Ag is likely to be in AgCl form and remain in that form in biosolids. However, it has been found that in the basic conditions of the hydroxide rich wash water little dissolution of Ag and instead larger Ag agglomerates (>450 nm) were released (Geranio, Heuberger, & Nowack, 2009).

AgNP-functionalized food packaging has been shown to pose little toxic risk to humans from studies investigating the amount of silver leaching into chicken meat (Cushen, Kerry, Morris, Cruz-Romero, & Cummins, 2013, 2014; Fröhlich & Roblegg, 2012; Hannon, Cummins, Kerry, Cruz-Romero, & Morris, 2015). As the global improvement of recycling rates increases, these packaging materials are cleaned and treated, possibly adding to the load of silver nanoparticles in the aquatic environment.

Models predicting elution of AgNP from particular sources or application types have been developed. From a bench reactor study of AgNP textiles, using a model wastewater treatment plant (WWTP), Benn and Westerhoff (2008) found that from a 5 μg/L spike, only 0.01 μg/L was released in the effluent, representing a 500-fold reduction with the majority of silver eluted as biosolids at 2.8 mg Ag/kg biosolids. This suggests that toxic effects are most likely to be associated with sludge, which is often dried, pelleted, and spread as agricultural fertilizer (European Commission, 2007).

In addition to the efforts made by the scientific community, industry have a significant role to play also. The characteristics of silver nanoparticles commonly used are not widely shared or indeed known. This is most likely due to the commercial need to protect intellectual property and for proprietary reasons; however, this hampers the proper risk assessment of nanomaterials as ecotoxicologists are left guessing as to which forms of silver and indeed other ENMs to test. An EU reference material NM-300 K, a citrate-coated AgNP <20 nm diameter silver nanoparticle, is widely referenced in ecotoxicological studies, although many researchers cite the change in state from nanoparticle to ions, $AgNO_3$ or AgCl as the source of the toxicity (Geranio et al., 2009). As the characteristics of the most widely used AgNPs in consumer products remain the closely guarded secret of industry protecting their intellectual property, the relevance of this reference material is not really known either.

8.5.2 Environmental Fate of Silver Nanomaterials

Predictive models evaluating Ag concentrations in geographical regions have emerged, possibly in response to the specificities pertinent to the water chemistry in different water courses. A mass flow analysis was carried out by Blaser et al. (2008), who used a model of the Rhine River based on a projected population for the EU25 of 469 million and incorporating the aquatic eco-load of anti-microbial plastics and textiles, which they state contributes to 15% of the silver load in rivers. The environmental fate of silver nanoparticles was also studied by Blaser et al. (2008), who reported that the majority of soluble silver in natural freshwater is expected to be in the form of silver sulphide and that free silver ions are only a potential problem if silver concentrations exceed the ambient sulphide concentrations.

In a field study investigating the sewage sludge in a municipal WWTP, Kim, Park, Murayama, & Hochella (2010) found Ag to be in the form of α-Ag_2S nanoparticles. Levard, Hotze, Lowry, & Brown (2012) were in agreement with this assessment and stated that silver nanoparticles are transformed by WWTP to the essentially non-toxic silver sulphate before being eluted to surface waters. This complexing effect of Ag with sulphur is reported by Nowack (2010) to reduce bioavailability due to the insoluble nature of Ag-sulfide complexes.

Another model of environmental concentrations carried out by Gottschalk, Sondere, Schols, & Nowack (2009) on AgNPs reiterated the likelihood of a high proportion of partitioning to WWTP sludge and sludge-treated soil as being the most likely destination of AgNP. This treated soil has the potential to behave as a repository for silver pending future infiltration of ground water and in turn surface water, as the two are intrinsically linked.

Cunningham and Joshi (2015) contend that AgNPs released to the atmosphere, lithosphere, or hydrosphere can be transported or transformed by chemical, physical, or biological processes. They also maintain that the fundamental properties governing the environmental fate of AgNP are not thoroughly understood. Solid waste landfill may present deposition risk, with landfill sewage sludge potentially leading to AgNP release into subsoils and groundwater (Blaser et al., 2008).

Tiede, Boxall, Wang, Gore, Tiede, Baxter, Tear, & Lewis (2010) reported that more than 90% of AgNP is excluded from the WWTP elute by clearing sludge from the plant. However, it should also be noted that as this sludge is frequently dried, pelleted, and used as agricultural fertilizer, it is hypothesized that the silver could infiltrate surface waters from contaminated ground waters.

8.5.3 Silver Nanomaterials in the Aquatic Environment

Having seen the variety of influences water chemistry can have on the nature and chemical speciation of AgNP from the time it is eluted or leached from the functionalized product or material, once in the receiving environment further variables influence the silver. These variables can have a significant effect on bioavailability (Walser, Demou, Lang, & Hellweg, 2011), which is of particular relevance to ecotoxicologists.

Depending on the water chemistry of the receiving environment, AgNPs may remain soluble, stay in suspension, precipitate, aggregate, dissolve, or form complexes with ions and ligands present in the water (Cunningham & Joshi, 2015; Luoma, 2008; Montano et al., 2014). In Section 8.5.1, the effect of high chloride concentration in wash water was discussed. Once in the receiving environment, chloride content or salinity can again have a significant effect on the environmental fate of Ag, which is particularly relevant to estuarine waters. In high-salinity water, such as estuaries or brackish areas, AgNPs and indeed most ENMs tend to aggregate or agglomerate more readily (Luoma, 2008) and thus will tend to settle and be sequestered (possibly temporarily) into the benthos. AgNP were shown to be more stable in waters with lower chloride concentrations, i.e., freshwater (Chinnapongse, MacCuspie, & Hackley, 2011), thus increasing Ag solubility and the tendency for AgCl formations to precipitate. A thorough understanding of the effects in estuaries is critical from a food safety point of view as a substantial amount of aquaculture is carried out in estuarine areas and with the tendency for shellfish to bioaccumulate, the risks imposed by high Ag levels are raised. The complex and dynamic behaviour of Ag/AgNP in the freshwater ecosystem has not fully been elucidated and significant research is required (Zhang et al., 2016).

Dissolution begins with an oxidation reaction influenced by a number of properties, both of the particles themselves and the surrounding water chemistry. They include particle size, particle concentration, temperature, dissolved oxygen levels, pH, presence of ligands, and ionic strength (He, Bligh, & Waite, 2013; Liu & Hurt, 2010; Zhang et al., 2016). It has been noted that particle aggregation reduces the relative surface area, which in addition to affecting particle reactivity and hence toxicity also slows and eventually stops the rate of dissolution in larger (>450 nm) particles (Liu & Hurt, 2010; Zhang et al., 2016).

Seasonal temperature variations and local variations around industrial effluents may also increase AgNP dissolution. Higher solubility of AgNP was noted at higher temperatures by Liu and Hurt (2010). Acidic conditions have also been shown to increase the rate of dissolution (Liu & Hurt, 2010; Zhang et al., 2016). This property suggests that geomorphological features are also an important factor on the fate of silver in the aquatic environment; for example, low-pH wetlands/bogs may be more affected by soluble silver than basic regions such as limestone-rich river basins.

In addition to the effects of the receiving environment on the particles, the physical properties of the particles themselves also contribute to the environmental fate. Coatings for example have been used to cap or stabilize particles and there are numerous coatings available, each exerting different influences on downstream speciation. Citrate-coated AgNPs were shown to facilitate faster dissolution rates than polyvinylpyrrolidine (PVP)-coated particles (Angel et al., 2013).

The presence of ligands can influence the dissolution. Sigg and Lindauer (2015) found that the presence of cysteine resulted in an initial increase of dissolved Ag while chloride and fulvic acids had little effect on the dissolution of AgNPs. Gunsolus et al. (2015) demonstrated the effects of humic acids on AgNP stability and fate. This provides further evidence for temporal variation in AgNP fate as humic acids increase in autumn when leaf litter is at its highest and geographical variation as heavily forested areas are likely to have higher humic acid concentrations. Humic acid is sometimes used in the production of nanoparticles and hence may encourage their formation in solution (Dubas & Pimpan, 2008; Gunsolus et al., 2015).

Blaser et al. (2008) estimated a PEC range of 0.040–0.320 μg/L AgNP depending on loading assumptions, with a PEC of 0.140 μg/L suggested for an intermediate Ag load. This is lower than the predictions of Gottschalk et al. (2009), who reiterated the likelihood of a high degree of partitioning to WWTP sludge and sludge-treated soil and predicted a surface water concentration of 0.588–2.16 ng/L AgNP and in sludge-treated soil of 1.581 μg AgNP/kg/year. Again, this treated soil has the potential to behave as a repository for silver which may influence future infiltration of ground water and, in turn, surface water as the two are intrinsically linked.

In the absence of a definitive answer on which types of AgNPs are potential toxicants of concern, field sampling of rivers downstream of WWTP effluent and characterization of the size and state of silver nanoparticles that occur in the Irish freshwater aquatic ecosystem may provide sufficient direction for a toxicological assessment provided that capture detection/quantification technologies of sufficient sensitivity are developed (McGillicuddy et al., 2018).

8.5.4 Ecotoxicology of Silver Nanomaterials

Given the dynamic nature of silver speciation in the environment, it has become necessary to investigate the toxicity of many different forms of silver in a plethora of test systems taking into account silver speciation, matrix interference, and environmental relevance. As a result, a diverse set of ecotoxicological data has been elucidated covering a substantial number of the available AgNPs across a wide range of test systems (Asghari, Johari, Lee, Kim, Jeon, Choi, Moon, & Yu, 2012; Binaeian, Rashidi, & Attar, 2012; Blinova, Niskanen, Kajankari, Kanarbik, Käkinen, Tenhu, Penttinen, & Kahru, 2013; Bondarenko,

Juganson, Ivask, Kasemets, Mortimer, & Kahru, 2013; Fabrega et al., 2011; Griffitt, Luo, Gao, Bonzongo, & Barber, 2008). These studies employed a wide variety of test systems and attempted to address the paucity of knowledge in the area of aquatic nanotoxicology. However, in the absence of actual measured environmental concentrations, the characteristics of AgNP used and the environment coupled with the likely interferences in traditional test systems and environmental influences, the relevance of this data remains unclear.

In the majority of ecotoxicological risk assessments, a multi-trophic test battery is the most widely used methodology employed, usually covering the three primary trophic levels: primary producer (algae), primary consumer (invertebrates), and secondary consumer or apex predator (fish or *in-vitro* fish models proxy).

The most commonly used test for primary producers in the environment is algae, typically utilizing *Pseudokirchneriella subcapitata* (freshwater green algae) to model algal toxicity in fresh water and *Skeletonema costatum* (marine diatom) in marine waters. Substantial variation in results has been observed, often depending on differing characteristics of the tested particles.

Using the freshwater algae *Pseudokirchneriella subcapitata,* Griffitt et al. (2008) reported an EC_{50} of 190 μg/L for 20–30 nm AgNPs. In contrast, Ribeiro, Gallego-Urrea, Jurkschat, Crossley, Hassellöv, Taylor, & Soares (2014) found an EC_{50} of 32 μg/L AgNP for 3–8 nm AgNPs. This increased toxicity in the latter study can be accounted for in the smaller particle size as it would be expected that smaller particles yield a greater relative surface area and hence higher toxicity. This size dependent toxicity was demonstrated by Ivask et al. (2014) using the same algae who report a range of EC_{50} of 180–1140 μg/L AgNP in a range of citrate-coated nanoparticle sizes of 10–80 nm, behaving in a size-dependent fashion. Size-dependant toxicity was also demonstrated in the primary consumer *Daphnia magna* by Kennedy et al. (2010), who found that $AgNO_3$ (ionic silver which is frequently used as an AgNP proxy) was the most toxic with size-dependant toxicity reported in AgNPs ranging from 10nm - 80nm in diameter. In contrast, the toxicity of AgNPs was assessed with the saltwater (marine) diatom *Skeletonema costatum* by Huang, Cheng, and Yi (2016). This study, which would be of particular relevance in estuarine or transitional waters, found that concentrations of 500 μg/L induced ROS and reduced cell viability by 28%. Down-regulation of photosynthetic genes was reported at concentrations of 5 mg/L and that physical adsorption of particles to diatom cells may result in shading and reduced photosynthetic efficiency, a phenomenon of Ag particles, and not ionic silver. The AgNPs used in that study were 10 nm in diameter and coated in oleylamine. The same study also highlighted the differences in hydrodynamic diameter and other properties of AgNPs in different test matrices (media) over 24 hours and 15 days. The EC_{50} reported by Huang, Cheng, and Yi (2016), the same endpoint as in the freshwater green algae mentioned above, was 25.77 mg/L which is, at least three orders of magnitude higher. The EC_{50} was substantially lower and comparable at 48 μg/L, suggesting that solubility is a key factor in toxicity and hence, the lower solubility in salt water yields lower toxicity in algae. Lee et al. (2005) investigated the effects of chloride concentration in test media on algal toxicity and concluded that a reduced toxic effect was evident in the presence of chloride. Aside from the conclusions that can be drawn about the chloride content of test matrices, this would also suggest that Ag is likely to exert lower toxicity in estuarine or transitional waters than freshwater.

The diversity of results and median effective toxicities is evident and given the many different influences on toxicity described previously, it seems that even very small differences in AgNP characteristics such as coatings, sizes, etc., can have profound effects on the toxicity. Similarly, minor differences in test methodologies, i.e., test matrices/lighting, can also influence toxicity.

In addition to the size-dependent toxicity reported by Kennedy, Hull, Bednar, Goss, Gunter, Bouldin, Vikesland, & Steevens (2010) mentioned above, many other studies investigated the toxicity of AgNP to *Daphnia magna*, alternative Daphnid species, and other invertebrates as models of primary consumers which are, the second trophic level in the multi-trophic test battery. Substantial diversity in the results again demonstrates the variability in different test models and analyte characteristics.

The importance of understanding the dynamic nature of AgNP prior to their interactions with biota as described in Sections 8.5.2 and 8.5.3 become particularly evident in the second trophic level of primary consumers. The differences in median toxicity values for different AgNMs is demonstrated in several studies across a range of well-characterized AgNMs, not just polydisperse colloids and powdered suspensions. Alkane-coated AgNP in the 3–8 nm range yielded an IC_{50} value of 11.02 μg/L in *Daphnia magna* (Ribeiro et al., 2014) whilst AgNP and Ag nanowires yielded IC_{50} values of 7 μg/L and 117 μg/L, respectively (Sohn, Johari, Kim, Kim, Kim, Lee, Chung, & Yu, 2015).

In a comparison of polydisperse silver NP toxicity, between the EC_{50} values of colloidal AgNP and suspended powder carried out by Asghari et al. (2012) using *Daphnia* magna, the Ag colloids were found to be significantly more toxic with a mean EC_{50} of 2–4 μg/L compared to 187 μg/L for suspended powder, i.e., the colloids were about 60 times more toxic. A comparison between the *Daphnia magna* and *Thamnocephalus platyurus*, the freshwater fairy shrimp yielded relatively similar IC_{50} values of 15–17 μg/L and 20–27 μg/L, respectively (Blinova et al., 2013).

Differences in species sensitivity have been reported using PVP coated nanosilver NM-300 <20 nm as a test compound by Völker, Boedicker, Daubenthaler, Oetken, & Oehlmann (2013). In a comparison of toxicity between three daphnid species, *Daphnia pulex, Daphnia galeata* yielded similar results with IC_{50} values of 8.95 μg/L and 13.9 μg/L, respectively; however, *Daphnia magna* was significantly less sensitive with an IC_{50} of 121 μg/L. This is a significant finding as typically, *Daphnia magna* and *Daphnia pulex* have been considered relatively interchangeable, based upon research by Lilius, Tom, & Isomaa (1995) and the fact that *Daphnia pulex* is an alternative daphnid species frequently used in toxicity testing and recommended in OECD Test No. 202 (OECD, 2004).

The next trophic level in the multi-trophic test battery for the aquatic environment is secondary consumers. These are organisms that consume primary consumers. In the freshwater system this usually refers to fish as they are considered the apex predator. Traditionally, piscine studies are carried out *in vivo* using various species including but not limited to silver barbs (*Barbonymus gonionotus*), zebrafish (*Danio rerio*), and goldfish (*Carassius auratus*). The selection of species is usually, though not always, based on fish size, husbandry requirements, and cost, although sometimes environmental relevance overrules cost and convenience and species like common carp (*Cyprinus carpio*) and rainbow trout (*Oncorhynchus mykiss*) at various stages of their life cycles are used (Khan, Jabeen, Qureshi, Asghar, Shakeel, & Noureen, 2015). Zebrafish is probably the most widely studied species at virtually all stages of its life cycle, particularly embryonic stages as this pre-vertebrate stage does not require an animal testing licence in many jurisdictions. It should also be noted at this stage that solubility issues may interfere with test fidelity as many of the species listed are considered 'tropical freshwater' fish in the aquarist industry and require temperature conditions to be about 25°C, which is temperate. However, if the geographical area being considered is Northern Europe or Britain and Ireland where typical river temperatures could be half that value at most, the ecological relevance of these species is questionable. Solubility is likely to be higher in higher temperature test systems whilst precipitation are likely to be more problematic in colder climes.

In an investigation of the toxicity of silver ions and silver 'nanoparticles' carried out by Ribeiro et al. (2014), AgNPs of 127–132 nm diameter were utilized, despite not fitting the EU definition of nanomaterials (European Commission, 2011), although the particles were measured to be 80–100 nm at day 0 before agglomerating and reaching 350 nm at day 3 and hence are included here. Zebrafish embryo 96-hour (hatching rate) LC_{50} was 78.32 μg/L for $AgNO_3$. This median lethal concentration for this test was higher than the highest tested concentration of 128.54 μg/L.

Johari, Kalbassi, & Yu (2013) reported on the toxicity of colloidal AgNPs to rainbow trout at three stages of the life cycle: embryonic, larval, and juvenile. Median lethal concentrations (LC_{50}) ranged from 250–710 μg/L for the two younger stages, and 2160 μg/L for the more advanced juveniles with gill pathologies observed. The particle sizes were measured to be predominately between 1 and 13 nm in diameter.

There has been a move away from *in vivo* testing of vertebrates and attention has now focussed on *in vitro* testing utilizing piscine cell lines. However, this approaches several problems, most notably test interference from cell culture media, which affects bioavailability (Liu, Yu, Yin, & Chao, 2012; von der Kammer, Ferguson, Holden, Masion, Rogers, Klaine, Horne, & Unrine, 2012).

Coupled with the potential for interference from culture media is the need for ratification of continuous cell models with wild type primary cells. To address this, Connolly et al. (2015) compared the toxicity of AgNPs between a selection of continuous rainbow trout cell lines and primary rainbow trout hepatocytes. The tested silver forms were Ag ions, and the EU reference material NM-300 K, a polydisperse combination of two sub populations of 6 and 25 nm diameters. Substantial agglomeration was noted in test media over 24 hours, with AgNPs of an average diameter of 25 nm increasing to more than 467–570 nm. The cells used were RTL-W1, RTH-149, RTG-2 and primary hepatocytes with IC_{50} values 1.1–75.9 mg/L reported for the cell lines and 2–30.6 mg/L for the primary cells over a number of different endpoints. General agreement or only moderate differences between the cell lines and primary cells suggest that they are reasonable analogues of each other. However, the media interference is substantial and the median effect concentrations significantly higher than test systems at lower trophic levels (µg/L range) and significantly higher again than the PECs in the ng/L range.

In addition to the traditional acute ecotoxicity data described above, AgNP risk assessments should also account for the effects of Ag speciation and concomitant effects such as sedimentation or other partitioning within the water column, on the bioavailability to test biota.

Powdered suspensions of polydisperse AgNP and larger AgNP agglomerates will tend to precipitate more than more soluble colloidal silver AgNPs. This was demonstrated by Kalbassi, Johari, Soltani, & Yu (2013) using the effects of this tendency on the toxicity to rainbow trout (*Oncorhynchus mykiss*). As rainbow trout alevins do not actively swim, they are more sensitive to the increased exposure of benthic AgNP toxicants which sediment from the water column.

The optimization of culture media to limit interference and concomitantly improve the sensitivity and environmental relevance of AgNP cytological toxicity studies is critically important in the development of valid, relevant tests capable of informing intelligent risk assessment (Connolly et al., 2015; Loza, Diendorf, Sengstock, Ruiz-Gonzalez, Gonzalez-Calbet, Vallet-Regi, Köller, & Epple, 2014).

Given the variation in test methodologies, the plethora of comparisons and isolated conclusions drawn, it remains impossible to properly risk assess silver nanoparticles. There is a clear need to standardize methods both in terms of analyte characterization and test methods and matrices to facilitate true replication and relevant conclusions on their toxicity to be drawn (Barnard, 2006; Behra & Krug, 2008; Morris, Willis, De Martinis, Hansen, Laursen, Sintes, Kearns, & Gonzalez, 2011; Rauscher, Rasmussen, & Sokull-Klüttgen, 2017).

8.6 Conclusions

It is evident from the plethora of ecotoxicological data available that a full understanding of the concentration, behaviour, fate, and toxicity of ENMs remains elusive. It is almost impossible to assess the risk of any toxicant if the exact nature of the toxicant is unknown. Until such time as a clear picture emerges of the specific ENMs being used in products, all the research community can do is continue to risk assess as many ENMs as possible so that when regulation and mandatory reporting comes, the data will be readily available for regulators. Whilst the current situation very much involves putting the 'cart before the horse' is not ideal, it is the only avenue available to ecotoxicologists as they strive to effectively catch up with the technological developments and ever-increasing applications of nanomaterials and their derivatives.

In addition to the limited knowledge regarding nanomaterials of concern, the standardized test strategies traditionally employed also need to be re-examined in terms of their suitability for nanomaterial risk assessment. There are significant ranges diversity in the median toxicity values observed, for both carbon and silver nanomaterials and their base elements. These variations do not seem to follow any particular trend to allow regulators cut off above or below a certain size. The dynamic nature of the nanomaterials in solution means that the sizes are constantly in flux through various process of agglomeration, dissolution, or ionization.

The problems and uncertainty with the studies completed is addressed in an article by Krug (2018), who apportions some of the blame onto ecotoxicologists for adopting a 'publish or perish' attitude to nanomaterials testing. He states that testing of high doses of ENMs is designed simply to produce publishable, toxic effects. This may be a somewhat unfair appraisal of the situation ecotoxicologists are in. Thorough ecotoxicology studies typically begin with range finding usually at a log scale dose to determine roughly, the concentration range that the chosen test organism is likely to display toxic effects to the chosen chemical. Then, with this knowledge, a more detailed titre of the toxic effects using a geometric dose range is carried out to determine at what concentration the toxic effect occurs. It would be a pointless exercise to carry out a toxicity study with no endpoint achieved. Hence, sensationalizing of results is more likely to be as a result of mainstream media picking up on the toxicity study. Ecotoxicologists have a case to answer, but that case surrounds effective communication of results, their meaning, and context to avoid them being sensationalized irrationally.

Modern approaches to material risk assessment is to validate and prescribe methodologies and test matrices; however, as has been shown, this may not be effective in providing a true picture of the toxicity of the material in a specific location. The changes in water chemistry are significant and have been shown to have an effect on bioavailability and levels of toxicity. This is evident from trends in ecotoxicological assessments of ENMs to now include testing in natural river water in specific geographical regions (Zeng et al., 2018).

In addition to these spatial anomalies in nanomaterial toxicity, there are also temporal variations that must be accounted for. Investigations by Gottschalk, Ort, Scholz, & Nowack (2011) highlight some of these variations based on Swiss models. Giese et al. (2018) further highlight temporal and spatial pressures and the need to allow for them in a risk assessment model. Temporal or seasonal variation can be exemplified by natural organic matter, such as humic acid, which vary in concentration and have been shown to have an effect on the speciation and hence toxicity of nanomaterials.

Given these stark differences in water chemistry and concomitant toxicity, it may be necessary to forgo the traditional 'EU-wide' approach, or global approach to standardize testing and instead allow jurisdictional regulation on ENMs. However, some 'top-down' regulation will still be required to span transnational boundaries and empower enforcement agencies to compel reporting. However, this reporting will be necessary to standardize the quality of ENM characterization to allow relevant ecotox testing to be undertaken. This would allow local regulators to assess the effects in their area based on the toxicity levels experienced in that area at given times of the year allowing for both spatial and temporal considerations.

Currently, ecotoxicologists are trying to elucidate the 'big picture' of the ecotoxicity of nanoparticles. A typical 'big picture' approach is summarized in Figure 8.4. It essentially examines what is happening at a given stage of a nanomaterials lifecycle and how those influences can alter the toxicological profile.

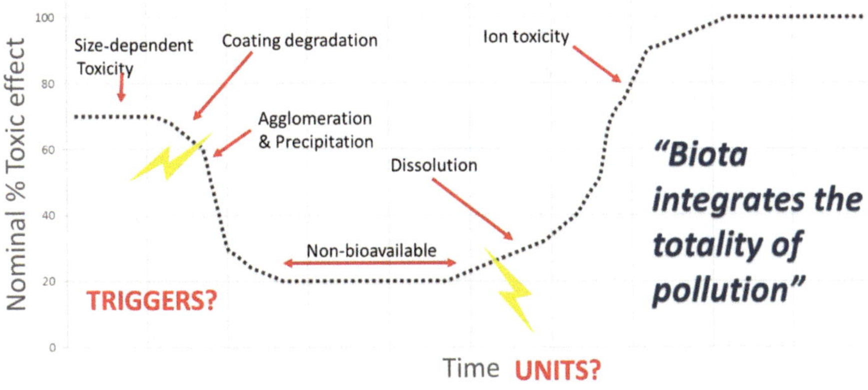

FIGURE 8.4 The 'big picture' in nanomaterial ecotoxicology.

The lightning bolts indicate extrinsic influences such as humic acid concentration or other natural organic materials concentrations or water chemistry. To realistically mitigate from potential ecological and health damage from ENMs, it is imperative that local regulators with the help of scientists determine the nature of the major contributing factors influencing the ecotoxicity of nanomaterials in their own jurisdiction. If necessary, EU-wide or globally standardized 'broad stroke' strategies may be utilized to determine a worst-case-scenario and regulate accordingly to avoid downstream effects not currently measureable.

Nanomaterials are an exciting and innovative development on the world stages of electronics, science, technology, and biomedicine, to name but a few. However, it is essential that careful oversight and reporting is employed to prevent them from becoming the asbestos or Minimata disease of the twenty-first century.

References

Akaighe, N., Maccuspie, R. I., Navarro, D. A., Aga, D. S., Banerjee, S., Sohn, M., & Sharma, V. K. (2011). Humic acid-induced silver nanoparticle formation under environmentally relevant conditions. *Environmental Science & Technology, 9*(45), 3895–3901.

Angel, B., Batley, G., Jarolimek, C., & Rogers, N. (2013). The impact of size on the fate and toxicity of nanoparticulate silver in aquatic systems. *Chemosphere*, 93(2), 359–365. doi: 10.1016/j.chemosphere.2013.04.096.

Arvidsson, R., Molander, S., & Sandén, B. A. (2013). Review of potential environmental and health risks of the nanomaterial graphene. *Human and Ecological Risk Assessment: An International Journal, 19*(4), 873–887. doi:10.1080/10807039.2012.702039.

Asghari, S., Johari, S. A., Lee, J. H., Kim, Y. S., Jeon, Y. B., Choi, H. J., Moon, M. C., & Yu, I. J. (2012). Toxicity of various silver nanoparticles compared to silver ions in Daphnia magna. *Journal of Nanobiotechnology, 10*(14), 1–14. doi:10.1186/1477-3155-10-14.

Barnard, A. S. (2006). Nanohazards: Knowledge is our first defence. *Nature Materials, 5*(4), 245–248. doi:10.1038/nmat1615.

Behra, R., & Krug, H. (2008). Nanoparticles at large. *Nature Nanotechnology, 3*, 253–254. doi:10.1021/es.7032718.

Benn, T. M., & Westerhoff, P. (2008). Nanoparticle silver released into water from commercially available sock fabrics. *Environmental Science and Technology, 42*(11), 4133–4139. doi:10.1021/es7032718.

Binaeian, E., Rashidi, A. M., & Attar, H. (2012). Toxicity study of two different synthesized silver nanoparticles on bacteria vibrio fischeri. *World Academy of Science, 67*(29), 1012–1018. doi:10.5897/AJB11.4050.

Blaser, S. A., Scheringer, M., MacLeod, M., & Hungerbühler, K. (2008). Estimation of cumulative aquatic exposure and risk due to silver: Contribution of nano-functionalized plastics and textiles. *Science of the Total Environment, 390*(2–3), 396–409. doi:10.1016/j.scitotenv.2007.10.010.

Blinova, I., Niskanen, J., Kajankari, P., Kanarbik, L., Käkinen, A., Tenhu, H., Penttinen, O. P., & Kahru, A. (2013). Toxicity of two types of silver nanoparticles to aquatic crustaceans Daphnia magna and Thamnocephalus platyurus. *Environmental Science and Pollution Research, 20*(5), 3456–3463. doi:10.1007/s11356-012-1290-5.

Bondarenko, O., Juganson, K., Ivask, A., Kasemets, K., Mortimer, M., & Kahru, A. (2013). Toxicity of Ag, CuO and ZnO nanoparticles to selected environmentally relevant test organisms and mammalian cells *in vitro*: A critical review. *Archives of Toxicology, 87*(7), 1181–1200. doi:10.1007/s00204-013-1079-4.

Boxall, A., Chaudhry, Q., Sinclair, C., Jones, A., Aitken, R., Jefferson, B., & Watts, C. (2007). *Current and Future Predicted Environmental Exposure to Engineered Nanoparticles.* Central Science Laboratory, New York.

Bury, N. R., Shaw, J., Glover, C., & Hogstrand, C. (2002). Derivation of a toxicity-based model to predict how water chemistry influences silver toxicity to invertebrates. *Comparative Biochemistry and Physiology Part C: Toxicology & Pharmacology*, *133*(1–2), 259–270.

Chen, L. Q., Hu, P. P., Zhang, L., Huang, S. Z., Luo, L. F., & Huang, C. Z. (2012). Toxicity of graphene oxide and multi-walled carbon nanotubes against human cells and zebrafish. *Science China Chemistry*, *55*(10), 2209–2216. doi:10.1007/s11426-012-4620-z.

Chen, Y., Ren, C., Ouyang, S., Hu, X., & Zhou, Q. (2015). Mitigation in multiple effects of graphene oxide toxicity in zebrafish embryogenesis driven by humic acid. *Environmental Science and Technology*, *49*(16), 10147–10154. doi:10.1021/acs.est.5b02220.

Chinnapongse, S. L., MacCuspie, R. I., & Hackley, V. A. (2011). Persistence of singly dispersed silver nanoparticles in natural freshwaters, synthetic seawater, and simulated estuarine waters. *Science of the Total Environment*, *409*(12), 2443–2450. doi:10.1016/j.scitotenv.2011.03.020.

Connolly, M., Fernandez-Cruz, M.-L., Quesada-Garcia, A., Alte, L., Segner, H., & Navas, J. (2015). Comparative cytotoxicity study of silver nanoparticles (AgNPs) in a variety of rainbow trout cell lines (RTL-W1, RTH-149, RTG-2) and primary hepatocytes. *International Journal of Environmental Research and Public Health*, *12*(5), 5386–5405. doi:10.3390/ijerph120505386.

Cunningham, S., & Joshi, L. (2015). *Assessment and Exposure of Marine and Freshwater Model Organisms to Metallic Nanoparticles*. Galway. Retrieved from https://www.epa.ie/pubs/reports/research/water/Research Report 150.pdf.

Cushen, M., Kerry, J., Morris, M., Cruz-Romero, M., & Cummins, E. (2013). Migration and exposure assessment of silver from a PVC nanocomposite. *Food Chemistry*, *139*(1–4), 389–397. doi:10.1016/j.foodchem.2013.01.045.

Cushen, M., Kerry, J., Morris, M., Cruz-Romero, M., & Cummins, E. (2014). Evaluation and simulation of silver and copper nanoparticle migration from polyethylene nanocomposites to food and an associated exposure assessment. *Journal of Agricultural and Food Chemistry*, *62*(6), 1403–1411. doi:10.1021/jf404038y.

Dubas, S. T., & Pimpan, V. (2008). Humic acid assisted synthesis of silver nanoparticles and its application to herbicide detection. *Materials Letters* (Vol. 62). doi.:10.1016/j.matlet.2008.01.033.

EFSA. (2013). Annual report of the EFSA scientific network of risk assessment of nanotechnologies in food and feed for 2013. *The EFSA Journal*, *EN-531*, 1–58.

Ellis, L.-J. A., Baalousha, M., Valsami-Jones, E., & Lead, J. R. (2018). Seasonal variability of natural water chemistry affects the fate and behaviour of silver nanoparticles. *Chemosphere*, *191*, 616–625. doi:10.1016/J.CHEMOSPHERE.2017.10.006.

European Chemicals Agency (ECHA). (2014). *Guidance on information requirements and chemical safety assessment Chapter R.7b: Endpoint specific guidance*. Retrieved from https://echa.europa.eu/documents/10162/13632/information_requirements_r7b_en.pdf.

European Commission. (2007) Follow-up study on the implementation of Directive 1999/31/EC on the landfill of waste in EU-25, Pub. L. No. 1999/31, European Comission.

European Commission. (2011). Commission recommendation on the definition of nanomaterial. *Official Journal of European Union*, *2011/696/E*. Retrieved from http://eur-lex.europa.eu/LexUriServ/LexUriServ.do?uri=OJ:L:2011:275:0038:0040:en:PDF.

European Observatory for Nanomaterials. (2018). EU nanomaterials observatory updated with two searchable databases. Retrieved June 18, 2018, from https://euon.echa.europa.eu/view-article/-/journal_content/title/eu-nanomaterials-observatory-updated-with-two-searchable-databases.

Fabrega, J., Luoma, S. N., Tyler, C. R., Galloway, T. S., & Lead, J. R. (2011). Silver nanoparticles: Behaviour and effects in the aquatic environment. *Environment International*, *37*(2), 517–531. doi:10.1016/j.envint.2010.10.012.

Fan, W., Liu, Y., Xu, Z., Wang, X., Li, X., & Luo, S. (2016). The mechanism of chronic toxicity to Daphnia magna induced by graphene suspended in a water column. *Environmental Science: Nano*, *3*(6), 1405–1415. doi:10.1039/C6EN00361C.

Farkas, J., Christian, P., Gallego-Urrea, J. A., Roos, N., Hassellöv, M., Tollefsen, K. E., & Thomas, K. V. (2011). Uptake and effects of manufactured silver nanoparticles in rainbow trout (Oncorhynchus mykiss) gill cells. *Aquatic Toxicology, 101*(1), 117–125. doi:10.1016/j.aquatox.2010.09.010.

Fröhlich, E., & Roblegg, E. (2012). Models for oral uptake of nanoparticles in consumer products. *Toxicology, 291*(1–3), 10–17. doi:10.1016/j.tox.2011.11.004.

Geranio, L., Heuberger, M., & Nowack, B. (2009). The behavior of silver nanotextiles during washing. *Environmental Science and Technology, 43*(21), 8113–8118. doi:10.1021/es9018332.

Giese, B., Klaessig, F., Park, B., Kaegi, R., Steinfeldt, M., Wigger, H., Gleich, A., & Gottschalk, F. (2018). Risks, release and concentrations of engineered nanomaterial in the environment. *Scientific Reports, 8*(1), 1565. doi:10.1038/s41598-018-19275-4.

Gottschalk, F., Ort, C., Scholz, R. W., & Nowack, B. (2011). Engineered nanomaterials in rivers – Exposure scenarios for Switzerland at high spatial and temporal resolution. *Environmental Pollution, 159*(12), 3439–3445. doi:10.1016/j.envpol.2011.08.023.

Gottschalk, F., Sondere, T., Schols, R., & Nowack, B. (2009). Modeled environmental concentrations of engineered nanomaterials for different regions. *Environmental Science & Technology, 43*(24), 9216–9222. doi:10.1021/es9015553.

Gottschalk, F., Sun, T., & Nowack, B. (2013a). Environmental concentrations of engineered nanomaterials: Review of modeling and analytical studies. *Environmental Pollution, 181*, 287–300. doi:10.1016/j.envpol.2013.06.003.

Gottschalk, F., Sun, T., & Nowack, B. (2013b). Environmental concentrations of engineered nanomaterials: Review of modeling and analytical studies. *Environmental Pollution, 181*, 287–300. doi:10.1016/j.envpol.2013.06.003.

Griffitt, R. J., Luo, J., Gao, J., Bonzongo, J.-C., & Barber, D. S. (2008). Effects of particle composition and species on toxicity of metallic nanomaterials in aquatic organisms. *Environmental Toxicology and Chemistry/SETAC, 27*(9), 1972–1978. doi.org/10.1897/08-002.1.

Gunsolus, I. L., Mousavi, M. P. S., Hussein, K., Bühlmann, P., & Haynes, C. L. (2015). Effects of humic and fulvic acids on silver nanoparticle stability, dissolution, and toxicity. *Environmental Science & Technology, 49*(13), 8078–8086. doi:10.1021/acs.est.5b01496.

Guo, X., Dong, S., Petersen, E. J., Gao, S., Huang, Q., & Mao, L. (2013). Biological uptake and depuration of radio-labeled graphene by Daphnia magna. *Environmental Science & Technology*, 47(21), 12524–12531. doi:10.1021/es403230u.

Handy, R. D., Von Der Kammer, F., Lead, J. R., Hassellöv, M., Owen, R., & Crane, M. (2008). The ecotoxicology and chemistry of manufactured nanoparticles. *Ecotoxicology, 17*(4), 287–314. doi:10.1007/s10646-008-0199-8.

Hannon, J. C., Cummins, E., Kerry, J., Cruz-Romero, M., & Morris, M. (2015). Advances and challenges for the use of engineered nanoparticles in food contact materials. *Trends in Food Science & Technology, 43*(1), 43–62. doi:10.1016/j.tifs.2015.01.008.

He, D., Bligh, M. W., & Waite, T. D. (2013). Effects of aggregate structure on the dissolution kinetics of citrate-stabilized silver nanoparticles. *Environment Science & Technology*, (47), 9148–9156. doi:10.1021/es400391a.

He, K., Chen, G., Zeng, G., Peng, M., Huang, Z., Shi, J., & Huang, T. (2017). Stability, transport and ecosystem effects of graphene in water and soil environments. *Nanoscale, 9*(17), 5370–5388. doi:10.1039/C6NR09931A.

Huang, J., Cheng, J., & Yi, J. (2016). Impact of silver nanoparticles on marine diatom *Skeletonema Costatum. Journal of Applied Toxicology, 36*(10), 1343–1354. doi:10.1002/jat.3325.

Hussain, S. M., Warheit, D. B., Ng, S. P., Comfort, K. K., Grabinski, C. M., & Braydich-Stolle, L. K. (2015). At the crossroads of nanotoxicology in vitro: Past achievements and current challenges. *Toxicological Sciences: An Official Journal of the Society of Toxicology, 147*(1), 5–16. doi:10.1093/toxsci/kfv106.

Impellitteri, C. A., Tolaymat, T. M., & Scheckel, K. G. (2009). The speciation of silver nanoparticles in antimicrobial fabric before and after exposure to a hypochlorite/detergent solution. *Journal of Environmental Quality, 38*(4), 1528–1530. doi:10.2134/jeq2008.0390.

International Organisation for Standardisation. (2012a). *ISO 8692:2012 Water Quality: Fresh Water Algal Growth Inhibition Test with Unicellular Green Algae*. Retrieved from https://www.iso.org/standard/54150.html.

International Organisation for Standardisation. (2012b). *ISO 6341:2012 Water Quality: Determination of the Inhibition of the Mobility of Daphnia Magna Straus (Cladocera, Crustacea)—Acute Toxicity Test*. Retrieved from https://www.iso.org/standard/54614.html.

Ivask, A., Kurvet, I., Kasemets, K., Blinova, I., Aruoja, V., Suppi, S., Vija, H. et al. (2014). Size-dependent toxicity of silver nanoparticles to bacteria, yeast, algae, crustaceans and mammalian cells *in vitro*. *PLoS ONE, 9*(7), 2–14. doi:10.1371/journal.pone.0102108.

Jarvie, H. P., Al-Obaidi, H., King, S. M., Bowes, M. J., Lawrence, M. J., Drake, A. F., Green, M. A., & Dobson, P. J. (2009). Fate of silica nanoparticles in simulated primary wastewater treatment. *Environmental Science & Technology, 43*(22), 8622–8628. doi:10.1021/es901399q.

Johari, A., Kalbassi, R., & Yu I, R. (2013). Toxicity comparison of colloidal silver nanoparticles in various life stages of rainbow trout (Oncorhynchus mykiss). *Iranian Journal of Fisheries Sciences* (Vol. 12), pp. 76–95. Retrieved from http://aquaticcommons.org/22580/1/IFRO-v12n1p76-en.pdf.

Kaegi, R., Sinnet, B., Zuleeg, S., Hagendorfer, H., Mueller, E., Vonbank, R., & Boller, M., Burkhardt, M. (2010). Release of silver nanoparticles from outdoor facades. *Environmental Pollution, 158*(9), 2900–2905. doi:10.1016/j.envpol.2010.06.009.

Kaegi, R., Voegelin, A., Sinnet, B., Zuleeg, S., Hagendorfer, H., Burkhardt, M., & Siegrist, H. (2011). Behavior of metallic silver nanoparticles in a pilot wastewater treatment plant. *Environmental Science and Technology, 45*(9), 3902–3908. doi:10.1021/es1041892.

Kalbassi, M. R., Johari, S. A., Soltani, M., & Yu, I. J. (2013). Particle size and agglomeration affect the toxicity levels of silver nanoparticle types in aquatic environment, *Ecopersia, 1*(3), 247–264.

Kennedy, A. J., Hull, M. S., Bednar, A. J., Goss, J. D., Gunter, J. C., Bouldin, J. L., Vikesland, P. J., & Steevens, J. A. (2010). Fractionating nanosilver: Importance for determining toxicity to aquatic test organisms. *Environmental Science and Technology, 44*(24), 9571–9577. doi:10.1021/es1025382.

Khan, M. S., Jabeen, F., Qureshi, N. A., Asghar, S., Shakeel, M., & Noureen, A. (2015). Toxicity of silver nanoparticles in fish: a critical review. *Journal of Biodiversity and Environmental Sciences, 6*(5), 211–227.

Kim, B., Park, C.-S., Murayama, M., & Hochella, M. F. (2010). Discovery and characterization of silver sulfide nanoparticles in final sewage sludge products. *Environmental Science & Technology, 44*(19), 7509–7514. doi:10.1021/es101565j.

Klaine, S. J., Alvarez, P. J. J., Batley, G. E., Fernandes, T. F., Handy, R. D., Lyon, D. Y., McLaughlin, M. J., & Lead, J. R. (2008). Nanomaterials in the environment: Behavior, fate, bioavailability, and effects. *Environmental Toxicology and Chemistry/SETAC, 27*(9), 1825–1851. doi:10.1897/08-090.1.

Krug, H. F. (2018). The uncertainty with nanosafety: Validity and reliability of published data. *Colloids and Surfaces B: Biointerfaces, 172*, 113–117. doi:10.1016/j.colsurfb.2018.08.036.

Kwak, J. I., & An, Y. J. (2015). A review of the ecotoxicological effects of nanowires. *International Journal of Environmental Science and Technology, 12*(3), 1163–1172. doi:10.1007/s13762-014-0727-4.

Laux, P., Riebeling, C., Booth, A. M., Brain, J. D., Brunner, J., Cerrillo, C., Creutzenberg, O. et al. (2018). Challenges in characterizing the environmental fate and effects of carbon nanotubes and inorganic nanomaterials in aquatic systems. *Environmental Science: Nano, 5*(1), 48–63. doi:10.1039/C7EN00594F.

Lee, D.-Y., Fortin, C., & Campbell, P. G. C. (2005). Contrasting effects of chloride on the toxicity of silver to two green algae, Pseudokirchneriella subcapitata and Chlamydomonas reinhardtii. *Aquatic Toxicology (Amsterdam, Netherlands), 75*(2), 127–135. doi:10.1016/j.aquatox.2005.06.011.

Levard, C., Hotze, E. M., Lowry, G. V., & Brown, G. E. (2012). Environmental transformations of silver nanoparticles: Impact on stability and toxicity. *Environmental Science and Technology, 46*(13), 6900–6914. doi:10.1021/es2037405.

Lilius, H., Tom, H., & Isomaa, B. (1995). A comparison of the toxicity of 30 reference chemicals to Daphnia magna and Daphnia pulex. *Environmental Toxicology and Chemistry, 14*(12), 2085–2088. doi:10.1002/etc.5620141211.

Liu, J., & Hurt, R. H. (2010). Ion release kinetics and particle persistence in aqueous nano-silver colloids. *Environmental Science and Technology, 44*(6), 2169–2175. doi:10.1021/es9035557.

Liu, J., Yu, S., Yin, Y., & Chao, J. (2012). Methods for separation, identification, characterization and quantification of silver nanoparticles. *TrAC Trends in Analytical Chemistry, 33*, 95–106. doi:10.1016/j.trac.2011.10.010.

Lombi, E., Donner, E., Scheckel, K. G., Sekine, R., Lorenz, C., Goetz, N. Von, & Nowack, B. (2014). Silver speciation and release in commercial antimicrobial textiles as influenced by washing. *Chemosphere, 111*, 352–358. doi:10.1016/j.chemosphere.2014.03.116.

Lorenz, C., Windler, L., von Goetz, N., Lehmann, R. P., Schuppler, M., Hungerbühler, K., Heuberger, M., & Nowack, B. (2012). Characterization of silver release from commercially available functional (nano)textiles. *Chemosphere, 89*(7), 817–824. doi:10.1016/j.chemosphere.2012.04.063.

Loza, K., Diendorf, J., Sengstock, C., Ruiz-Gonzalez, L., Gonzalez-Calbet, J. M., Vallet-Regi, M., Köller, M., & Epple, M. (2014). The dissolution and biological effects of silver nanoparticles in biological media. *Journal of Materials Chemistry B, 2*(12), 1634. doi:10.1039/c3tb21569e.

Luoma, S. N. (2008). *Silver Nanotechnologies and the Environment: Old Problems or New Challenges?* Washington, DC: Project on Emerging Nanotechnologies of the Woodrow Wilson International Center for Scholars.

Maillard, J.-Y., & Hartemann, P. (2012). Silver as an antimicrobial: Facts and gaps in knowledge. *Critical Reviews in Microbiology, 39*, 1–11. doi:10.3109/1040841X.2012.713323.

Markovic, M., Kumar, A., Andjelkovic, I., Lath, S., Kirby, J. K., Losic, D., Batley, G. E., & McLaughlin, M. J. (2018). Ecotoxicology of manufactured graphene oxide nanomaterials and derivation of preliminary guideline values for freshwater environments. *Environmental Toxicology and Chemistry, 37*(5), 1340–1348. doi:10.1002/etc.4074.

McGillicuddy, E., Morrison, L., Cormican, M., Dockery, P., & Morris, D. (2018). Activated charcoal as a capture material for silver nanoparticles in environmental water samples. *Science of the Total Environment, 645*, 356–362. doi:10.1016/j.scitotenv.2018.07.145.

McGillicuddy, E., Murray, I., Kavanagh, S., Morrison, L., Fogarty, A., Cormican, M., Rowan, N., & Morris, D. (2016). Silver nanoparticles in the environment: Sources, detection and ecotoxicology. *Science of the Total Environment, 575*, 231–246. doi:10.1016/j.scitotenv.2016.10.041.

Mcgillicuddy, E., Murray, I., Shevlin, D., Morrison, L., Cormican, M., Fogarty, A., Rowan, N., & Morris, D. (2018). *Report No. 259 Detection, Toxicology, Environmental Fate and Risk Assessment of Nanoparticles in the Aquatic Environment (DeTER).* (EPA, Ed.). Government of Ireland.

Minghetti, M., & Schirmer, K. (2016). Effect of media composition on bioavailability and toxicity of silver and silver nanoparticles in fish intestinal cells (RTgutGC). *Nanotoxicology, 10*(10), 1526–1534. doi:10.1080/17435390.2016.1241908.

Montano, M. D., Ranville, J., Lowry, G. V, Blue, J., Hiremath, N., Koenig, S., & Tuccillo, M. (2014). *Detection and Characterization of Engineered Nanomaterials in the Environment: Current State-of-the-art and Future Directions Report, Annotated Bibliography, and Image Library.* U.S. Environmental Protection Agency, Washington, DC, EPA/600/R-14/244. Retrieved from https://cfpub.epa.gov/si/si_public_record_report.cfm?Lab=NERL&dirEntryId=286070.

Morris, J., Willis, J., De Martinis, D., Hansen, B., Laursen, H., Sintes, J. R., Kearns, P., & Gonzalez, M. (2011). Science policy considerations for responsible nanotechnology decisions. *Nature Nanotechnology, 6*(2), 73–77. doi:10.1038/nnano.2010.191.

Nanowerk. (2016). *Nanotechnologies and Emerging Technologies*. Retrieved 10 March 2016, from http://www.nanowerk.com/.

Nogueira, P. F. M., Nakabayashi, D., & Zucolotto, V. (2015). The effects of graphene oxide on green algae Raphidocelis subcapitata. *Aquatic Toxicology, 166*, 29–35. doi:10.1016/j.aquatox.2015.07.001.

Nowack, B., & Bucheli, T. D. (2007). Occurrence, behavior and effects of nanoparticles in the environment. *Environmental Pollution, 150*(1), 5–22. doi:10.1016/j.envpol.2007.06.006.

Nowack, B., & Mueller, N. C. (2008). Exposure modeling of engineered nanoparticles in the environment. *EMPA Activities, 3*(2008–2009), 63. doi:10.1021/es7029637.

Nowack, B., Ranville, J. F., Diamond, S., Gallego-Urrea, J. a., Metcalfe, C., Rose, J., Horne, N., Koelmans, A. A., & Klaine, S. J. (2012). Potential scenarios for nanomaterial release and subsequent alteration in the environment. *Environmental Toxicology and Chemistry, 31*(1), 50–59. doi:10.1002/etc.726.

OECD. (2011). Test No. 201: *Freshwater Alga and Cyanobacteria, Growth Inhibition Test.* Paris: OECD Guidelines for the Testing of Chemicals, Section 2, OECD Publishing. Retrieved from https://doi.org/10.1787/9789264069923-en.

OECD. Test No. 202: Daphnia sp. Acute Immobilisation Test, Guidelines 202 § (2004). doi:10.1787/9789264069947-en.

Park, S., Woodhall, J., Ma, G., Veinot, J. G. C., Cresser, M. S., & Boxall, A. B. A. (2014). Regulatory ecotoxicity testing of engineered nanoparticles: Are the results relevant to the natural environment? *Nanotoxicology, 8*(5), 583–592. doi:10.3109/17435390.2013.818173.

Rauscher, H., Rasmussen, K., & Sokull-Klüttgen, B. (2017). Regulatory aspects of nanomaterials in the EU. *Chemie-Ingenieur-Technik, 89*(3), 224–231. doi:10.1002/cite.201600076.

Reidy, B., Haase, A., Luch, A., Dawson, K. A., & Lynch, I. (2013). Mechanisms of silver nanoparticle release, transformation and toxicity: A critical review of current knowledge and recommendations for future studies and applications. *Materials, 6*(6), 2295–2350. doi:10.3390/ma6062295.

Ribeiro, F., Gallego-Urrea, J. A., Jurkschat, K., Crossley, A., Hassellöv, M., Taylor, C., & Soares, A.M. & Loureiro, S. (2014). Silver nanoparticles and silver nitrate induce high toxicity to Pseudokirchneriella subcapitata, Daphnia magna and Danio rerio. *Science of the Total Environment*, 466–467, 232–241. doi:10.1016/j.scitotenv.2013.06.101.

SCENHIR. (2014). Nanosilver: Safety, health and environmental effects and role in antimicrobial resistance. doi:10.2772/76851.

Sigg, L., & Lindauer, U. (2015). Silver nanoparticle dissolution in the presence of ligands and of hydrogen peroxide. *Environmental Pollution, 206*, 582–587. doi:10.1016/j.envpol.2015.08.017.

Sohn, E. K., Johari, S. A., Kim, T. G., Kim, J. K., Kim, E., Lee, J. H., Chung, Y.S., & Yu, I. J. (2015). Aquatic toxicity comparison of silver nanoparticles and silver nanowires, *BioMed Research International 2015*, 1–12. doi:10.1155/2015/893049.

Taniguchi, N. (1974). On the basic concept of nanotechnology. In *Proceedings of the International Conference on Production Engineering* (pp. 18–23). Tokyo, Japan: Japan Society of Precision Engineering.

The European Commission. (2011). Commission recommendation of 18 October 2011 on the definition of nanomaterial. *Official Journal of the European Union*, 696, 38–40.

Tiede, K., Boxall, A. B. A., Wang, X., Gore, D., Tiede, D., Baxter, M., Tear, S. P., & Lewis, J. (2010). Application of hydrodynamic chromatography-ICP-MS to investigate the fate of silver nanoparticles in activated sludge. *The Journal of Analytical Atomic Spectrometry, 25*, 1149–1154. doi:10.1039/b926029c.

Tomaszewska, E., Soliwoda, K., Kadziola, K., Tkacz-Szczesna, B., Celichowski, G., Cichomski, M., Szmaja, J., & Grobelny, J. (2013). Detection limits of DLS and UV-Vis spectroscopy in characterization of polydisperse nanoparticles colloids. *Journal of Nanomaterials, 2013*, 1–10. doi:10.1155/2013/313081.

Völker, C., Boedicker, C., Daubenthaler, J., Oetken, M., & Oehlmann, J. (2013). Comparative toxicity assessment of nanosilver on three Daphnia species in acute, chronic and multi-generation experiments. *PLoS One*, *8*(10). doi:10.1371/journal.pone.0075026.

von der Kammer, F., Ferguson, P. L., Holden, P. a., Masion, A., Rogers, K. R., Klaine, S. J., Horne, N., & Unrine, J. M. (2012). Analysis of engineered nanomaterials in complex matrices (environment and biota): General considerations and conceptual case studies. *Environmental Toxicology and Chemistry*, *31*(1), 32–49. doi:10.1002/etc.723.

Walser, T., Demou, E., Lang, D. J., & Hellweg, S. (2011). Prospective environmental life cycle assessment of nanosilver. *Environmental Science & Technology*, *45*(10), 4570–4578. Retrieved from http://search.ebscohost.com/login.aspx?direct=true&db=bth&AN=61438130&site=ehost-live.

Weinberg, H., Galyean, A., & Leopold, M. (2011). Evaluating engineered nanoparticles in natural waters. *TrAC – Trends in Analytical Chemistry*, *30*(1), 72–83. doi:10.1016/j.trac.2010.09.006.

Wijnhoven, S. W. P., Peijnenburg, W. J. G. M., Herberts, C. a, Hagens, W. I., Oomen, A. G., Heugens, E. H. W., Roszek, B. et al. (2009). Nano-silver: A review of available data and knowledge gaps in human and environmental risk assessment. *Nanotoxicology*, *3*(2), 109–138. doi:10.1080/17435390902725914.

Yin, Y., Yang, X., Zhou, X., Wang, W., Yu, S., Liu, J., & Jiang, G. (2015). Water chemistry controlled aggregation and photo-transformation of silver nanoparticles in environmental waters. *Journal of Environmental Sciences*, *34*, 116–125. doi:10.1016/j.jes.2015.04.005.

Zeng, J., Xu, P., Chen, G., Zeng, G., Chen, A., Hu, L., Huang, Z. et al. (2018). Effects of silver nanoparticles with different dosing regimens and exposure media on artificial ecosystem. *Journal of Environmental Sciences*. doi:10.1016/J.JES.2018.03.019.

Zhang, C., Hu, Z., & Deng, B. (2016). Silver nanoparticles in aquatic environments: Physiochemical behavior and antimicrobial mechanisms. *Water Research*, *88*, 403–427. doi:10.1016/j.watres.2015.10.025.

Zhang, J., Gong, J. L., Zenga, G. M., Ou, X. M., Jiang, Y., Chang, Y. N., Guo, M., Zhang, C., & Liu, H. Y. (2016). Simultaneous removal of humic acid/fulvic acid and lead from landfill leachate using magnetic graphene oxide. *Applied Surface Science*, *370*, 335–350. doi:10.1016/j.apsusc.2016.02.181.

Zhang, Z., Yang, X., Shen, M., Yin, Y., & Liu, J. (2015). Sunlight-driven reduction of silver ion to silver nanoparticle by organic matter mitigates the acute toxicity of silver to Daphnia magna. *Journal of Environmental Sciences*, 35, 1–7, 62–68. doi:10.1016/j.jes.2015.03.007.

Zhou, X., Laroche, F., Lamers, G. E. M., Torraca, V., Voskamp, P., Lu, T., Chu, F. et al. (2012). Ultra-small graphene oxide functionalized with polyethylenimine (PEI) for very efficient gene delivery in cell and zebrafish embryos. *Nano Research*, *5*(10), 703–709. doi:10.1007/s12274-012-0254-x.

9

Lab-on-a-Chip-Based Devices for Rapid and Accurate Measurement of Nanomaterial Toxicity

Mehenur Sarwar,
Amirali Nilchian,
and Chen-zhong Li

9.1 Nanomaterials and Toxicity

9.1.1 Nanoparticles

Nanomaterials are widely applied in diagnosis, drug delivery, imaging, cosmetic, sports electronics, and various other research fields. Unfortunately, these nanomaterials are often not widely explored for their adverse effects, creating a significant need for a methodology to detect and monitor accumulation and toxicity of nanoparticles (NPs). Various physicochemical properties of nanomaterials determine how they will interact with the ecosystem. However, in the case of absorption and elimination within the biological system, the best predictors of toxicity include the NP composition, shape, size, surface charge, surface functional group, and reactivity. To study the effect of shape on cellular uptake and cell function, Huang et al. (2010) used three different shapes of monodisperse mesoporous silica NPs which have similar particle diameter, chemical composition, and surface charge but with different aspect ratios

(Huang et al. 2010). They found that particles with larger aspect ratios had the fastest internalization rates. In another study, researchers observed the biodistribution of NPs with different sizes, shapes, and surface charges (Blanco et al. 2015). Particles larger than 2,000 nm aggregate promptly inside the spleen and liver, and in addition in the linings of the lungs. NPs between the sizes of 100–200 nm have been appeared to impact the enhanced permeability and retention (EPR) and break filtration by liver and spleen. As size increments past 150 nm, more NPs are captured inside the liver and spleen. On the other hand, kidneys can filter out NPs smaller than 5 nm. In another study, circular gold NPs of various sizes were found to be non-poisonous to human skin cells whereas gold nanorods were reported to be extremely toxic for the same origin of cells. The nanorods crossed the cell membrane, aggregated inside the cells, and caused cell death. It was explained the toxicity might be due to the use of cetyltrimethylammonium bromide (CTAB) as covering material. The fabrication of nanorods requires CTAB, which functions as aqueous growth solution. This can be overcome by using poly(styrenesulfonate) (PSS) on the gold NPs and later coating with CTAB. This approach delivered non-toxic nanorods, which can be utilized to study living cells (Wang et al. 2008).

A unique pattern of bioaccumulation could be attributed to individual changes in size, shape, and surface chemistry. Therefore, the fate of NPs hugely depends on their physical characteristics as well as their chemical composition.

9.1.2 Importance of Nanomaterials in the Modern Era

Engineered nanomaterials with distinctive size, shape, surface structure, aggregation, and solubility parameters can be dedicated to a wide variety of important industrial processes. While these materials bring many advantages in current technology and research, the lack of a full understanding of their risks is a prevalent danger. Despite this there is evidence to support the fact that the benefits may outweigh the risks for possible applications in therapeutics. For example, Doxorubicin, a chemotherapeutic drug which exerts a high level of toxicity, can be delivered to the target cell types precisely using integrated nanomaterials without harming the neighbouring healthy cells (Prados et al. 2012). The cytotoxicity induced by nanomaterial exposure is believed to originate from the interaction of an electron donor or acceptor with an oxygen group on the NM, which can create a superoxide radical. Once formed, superoxide radicals are able to create more reactive oxygen species (ROS). ROS induces protein degradation, denaturation, DNA damage, mutagenesis, carcinogenesis, and altered cell cycle regulation (Manke et al. 2013). In addition, they can be uptaken by neuronal tissue, leading to brain and/or nervous system injury (Buzea et al. 2007) Once inside the mitochondria, NM can also cause energy failure (Nel 2007). For these diverse reasons, it is evident a thorough understanding of nanotoxicity must be developed in parallel with biomedical sensors that can measure it.

9.1.3 Nanotoxicity

Nanotoxicity can be initiated by exposure to nanomaterials in various ways. For example, after ingestion of a metal oxide, polymeric and carbon-based NPs, the undifferentiated human colon adenocarcinoma cell line known as Caco-2 was studied to observe the effect of ingestion to the animal models (Gerloff et al. 2009). The need for this research is especially evident, considering that silicon dioxide (SiO_2) and titanium oxide (TiO_2) nanomaterials are found in many foods as additives. Zinc oxide (ZnO) and magnesium oxide (MgO) are similarly used within food packaging. All of these NP sources were found to be cytotoxic to the Caco-2 cell line, except for MgO. Other researchers have also studied nanotoxicity by microinjection of NPs directly into the cells. For example, quantum dots (QDs) have been microinjected into primary human umbilical vein epithelial cells (Qin et al. 2016).

Even the modern cosmetic industry frequently relies on NPs. TiO_2 and ZnO are the most common NPs used in cosmetics – especially in sunscreen products. To assess the impact of these NPs, a human-derived keratinocyte HaCa T-cell line was used to investigate the effect of TiO_2 NPs. The necrotic cell death was observed only when cells were exposed for a long time (~7 days) at very high concentrations of TiO_2 (Crosera et al. 2015).

Carbon-based nanomaterials, which are often found to be in fine dust, were also studied to assess their toxic effect. Carbon black NPs have also been applied to macrophage cell lines (RAW 264.7) and primary human alveolar macrophages, and later the researchers studied caspase-1 activity and cell death in those cell lines (Hiraku et al. 2017). Their study revealed that carbon black induces nitrative DNA damage and clathrin-mediated endocytosis in cultured cells.

9.1.4 Nanotoxicity Assessment

Various equipment and techniques have been utilized to measure nanotoxicity. For example, NP uptake can be observed by TEM and fluorescence imaging. These are used to pinpoint the exact location of the NPs within the cells. The specific characteristics of the NP (i.e., surface charge, size and shape, etc.) determine their accumulation in particular organs. For example, TEM has already been used to study the uptake of Au, SiO_2 and TiO_2 in mouse peritoneal mast cells (MPMCs) (Marquis et al. 2009; Maurer-Jones et al. 2010).

A common trait of nanomaterials toxicity is chromosomal damage. Scientists often use DNA synthesis assays to investigate the proliferative state and general health of dividing cells. The genotoxicity from Ag NPs and polyethylene glycol (PEG)-coated cadmium selenide/zinc sulphide (CdSe/ZnS) QDs was investigated on A549 cells and skin epithelial cells (HSF-42), respectively, and no genotoxicity from the solvent or NPs apparently was noticeable (Zhang et al. 2006; Oostingh et al. 2011). However, the assay result reported to be flawed due to several challenges, ranging from particle agglomeration in biological media and optical interference with assay systems.

The most commonly used method to investigate DNA damage is the Comet Assay. The assay is based on gel electrophoresis and can determine the damage within individual eukaryotic cells. The limit of detection of this assay is approximately 50 strand breaks per diploid mammalian cell (Olive and Banáth 2006). Another method of measuring cytotoxicity of cells following nanomaterial exposure is through a Trypan Blue Exclusion assay. When cells are treated with Trypan Blue, dead cells uptake trypan blue whereas the viable cells remain unstained. This has been validated through observations of the toxicity of multiwalled carbon nanotubes (MWCNT), $CuZnFe_2O_4$, ZnO, and CuO NPs on human lung epithelial tumour cell line A549. Using the Trypan Blue Exclusion assay, they found a significant decrease of cell viability compared to the control group when the cell was exposed to 40 $\mu g/cm^2$ of each MWCNT, $CuZnFe_2O_4$, ZnO, and 20 $\mu g/cm^2$ CuO NPs. We have reported the fabrication of lab-on-chip devices with independently addressable microwell electrodes for high-throughput single-cell nanotoxicity studies. Control over individual microwells was achieved through the use of positive-dielectrophoresis (DEP) and verified by examining impedance properties. The device presents a novel model for single-cell assays of nanotoxicity and drug testing (Karlsson et al. 2008; Shah et al. 2014a, 2014b). This approach has also been taken by a wide variety of other researchers. Shukla et al. 2009 found that exposure to crocidolite asbestos reduced cell viability by 50%. Additionally, Bejjani et al. (2005) reported that poly(lactic) acid NPs (PLAs) at 4 mg/mL do not affect cell viability, utilizing the Trypan Blue Assay. However, with trypan blue assay it is not possible to assess any level of cell damage; this type of assay only can differentiate between dead or live cells.

Cytotoxicity assessment in cells also can be evaluated using the Microculture Tetrazolium Assay (MTA). After treating with MTA, live cells metabolize it into a purple-coloured product which can be dissolved in dimethyl sulfoxide (DMSO) and detergent for spectrometer measurement. However, there are fundamental limitations on this assay: First, not all cell lines metabolize MTA efficiently. Second, the protocol is time-consuming (2–4 hours incubation period). Third, the usage of DMSO to solubilize the end product is extremely toxic. Thus, this assay can subject lab personnel to health risks. To solve this problem, XTT (2,3-bis(2-methoxy-4-nitro-5-sulfophenyl)-5-[(phenylamino)carbonyl]-2H-tetrazolium hydroxide) is used instead of MTA. XTT, when metabolized, can produce a water-soluble purple product. Therefore, it eliminates the need of extra step to solubilize the end product with DMSO.

These various nanotoxicity assessment strategies all provide unique perspectives as to the health of cells. Due to the diverse ways that NMs can damage cell lines, it is essential to have a variety of techniques available that can measure cellular-level toxicity. NPs being smaller than microparticles holds the potential to move freely within the human body.

9.1.5 Nano-particle vs Micro-particle Toxicity

The need for NP-exposure testing schemes is clear when comparing their toxicity to their larger counterparts, microparticles. A recent study examining relative toxicity and bio-distribution identified stark differences between copper microparticles and copper NPs after oral exposure in rats (Lee et al. 2016). Copper microparticles were found to be primarily excreted in faeces and were only found in the liver and kidney in low, harmless concentrations. However, copper NPs were excreted in urine but bio-accumulated significantly in nearly all organs. Additionally, the copper NPs caused specific damage to a variety of regions, such as red blood cells, the thymus, the kidney, the liver, and the spleen. Comparatively, the microparticles did not create any similar damage – even at a higher dosage. Hence, the smaller sizes of NPs evidently had significantly more negative impacts, due to their enhanced permeability.

9.1.6 Governmental Regulation Strategy

The rapid development of NPs in recent years has created a significant struggle for governmental agencies, who must learn the best way to regulate their manufacture and distribution despite the still incomplete knowledge of their potential dangers. Nanomaterials have already pervaded a variety of industries, including catalysts for manufacturing processes (Xia et al. 2013), optical components (Kaczmarek et al. 2009), groundwater mediation strategies (Karn et al. 2009), semiconductors (Suresh 2013), sporting goods (Kessler et al. 2011), personal care products (Lohani et al. 2014), and clothing (Marijaa et al. 2009). Much of the challenge is also due to the precise definition of a NP since it encompasses a wide variety of materials, sizes, and shapes. Additionally, the use of typical thresholds based on weight/concentration is inadequate when used to describe NP dangers, since their specific size and shape is a far better predictor of their toxicity then solely examining their concentration (Clark 2011).

Despite these challenges, governmental regulation strategies have typically proceeded through a three-pronged approach, composed of information-gathering rules, pre-manufacture notifications, and international cooperation (EPA 2017; Bell et al. 2006). First, the information-gathering rule requires manufacturers or importers of products containing nanoscale materials to notify the EPA of specific details, including the specific chemical structure, quantity produced, method of manufacturing, exposure potential, and available safety data. While this is useful for generating a broad database of NP production, it does not generate new information regarding potential toxicity. The second prongs, pre-manufacture notifications, are provided by industries to the EPA to give them the opportunity to take action prior to NP production. Since 2005, actions the EPA have taken based on these notices have included requiring the usage of specialty personal protective equipment, incorporating additional safeguards to reduce the chance of environmental release, and forcing post-production testing to monitor worker health. Finally, international cooperation forms the third prong and is composed of the work of various committees to collectively develop broad regulation strategies for NP usage. For example, the International Organization for Standardization (ISO) has developed a technical committee ISO/TC 229 to identify the gold standard NP exposure testing methodologies, modelling and simulation techniques, and safety practices. Additionally, the Canada-U.S. Regulatory Cooperation Council (RCC) Nanotechnology Initiative led to a collective set of policy principles which satisfied both the Canadian Environmental Protection Act and the corresponding EPA of the U.S. This use of international policy alignment ensures that NP manufacturing and use is as safe as possible – especially in emerging economies where manufacturing may outpace regulatory oversight.

NPs are likely to be regulated in the same way as other pollutants under the Clean Water Act, Clean Air Act, and Resources Conservation and Recovery Act. Additionally, environmental remediation will be carried out through the Comprehensive Environmental Response, Compensation & Liability Act (CERCLA). Hence, it can be seen that the framework already exists to fund investigatory and remediation efforts. However, there exist underlying technical issues, as governmental agencies may have significant difficulty purifying and identifying NPs already embedded in the environment. These agencies must additionally contend with the presence of naturally-occurring NPs as well. Due to this uncertainty, a wide variety of agencies and legal frameworks have independently adopted protective frameworks.

The most recent effort from the EPA is composed of a final rule regarding the Toxic Substances Control Act, where in addition to the normal reporting processes described earlier organizations will need to carry out additional testing on their product to definitively characterize its properties. Additionally, NPs containing pesticides will be regulated under the Federal Insecticide, Fungicide & Rodenticide Act (FIFR) (U.S. Forest Service 2014). The FDA has also developed specific principles in order to determine whether a product requires their regulation, focusing on the presence of physical or biochemical properties induced by the presence of NPs. Finally, the Occupational Safety and Health Administration (OSHA) and the National Institute for Occupational Safety and Health (NIOSH) have developed worker safety standards, beginning with the 2016 regulation of multi-walled carbon nanotubes in Minnesota due to concerns of carcinogenicity (Taylor et al. 2017). Ultimately, this complex framework of policies composes existing governmental regulation standards.

9.2 Fundamental Technologies and Commercialization

The past three decades can be recognized as the 'sensitivity' era, during which most of the industry investments and scientific community efforts were devoted towards the improvement of biodetector sensitivity. However, accuracy always comes with a price. The more sensitive sensor typically requires larger and more sophisticated instrumentation. Thus, miniaturization of biosensors without compromising their sensitivity and reliability has been a persistent challenge for years. The problem is even further compounded due to the difficulty of accurately identifying a wide range of organic and inorganic compounds.

9.2.1 Lab-on-a-Chip (LoC) Technology

A lab-on-chip is a class of device, measuring a maximum a few square centimetres in area, which has the capability to automate and integrate several laboratory techniques on a chip. Due to its miniature size, minimal resources are required, which generates a low amount of waste. These devices also often contain the ability of rapid heating and mixing (Sackmann et al. 2014) and thus provide a platform for chemical reactions. Often, microfluidic (MF) technologies are incorporated to manage the various compartments and reagents. In addition, these chips often include various types of sensors such as gas sensors, humidity sensors, temperature sensors, flow meters, and viscometers.

The lab-on-a-chip concept was first introduced by Terry (1979) at Stanford University. Terry developed an integrated gas chromatography system capable of separation and detection of the gaseous mixture of organic solvents. Although he published his report in 1983, LoC technology did not receive much attention from the scientific community until the mid-1990s (Terry et al. 1979; Angell et al. 1983). The beginning of the twenty-first century was the flourishing era for LoC technology, during which implementation of LoC technology in biosensing systems extended from point-of-care (POC) diagnostic and cancer research to the pharmaceutical industry and even military applications. Some examples of employing biosensors on a chip technology in recent years are for rapid identification of bacterial pathogen using DEP (Cheng et al. 2013), label-free DNA detection using electric impedance variation (Berdat et al. 2008), amperometric high-throughput cholesterol detection using carbon nanotube electrode (Wisitsoraat et al. 2010) and high-sensitive electrochemical-based detection of H_2O_2 (Chikkaveeraiah et al. 2009).

The LoC concept evolved because of the convergence of three major fields: electronics technology and integrated circuits (IC), micro-electro-mechanical systems (MEMS), and biological sciences. LoC is a general term which encompasses various branches of biosensing technology. Sometimes LoC is referred to as one of the subsets of MEMS systems, which is not an accurate categorization. In fact, MEMS is one of the major technologies along with electronics, software programming, bio-compound immobilization, and computational analysis, which are incorporated in design and fabrication of LoC devices (Bhalinge et al. 2016; Chen and Shamsi 2017) The unique advantages of LoC devices are as follows:

1. LoC devices often require a very small volume of reagents for sample analysis. This approach tremendously reduces the cost of the experiment. At the end the waste produced is minimal.
2. LoC devices allow low to zero exposure to dangerous chemicals.
3. Fast responses can be achieved by the integration of multiple sensors within the LoC devices. High-throughput analysis is enabled with intricate parallelization.
4. Disposable low-cost LoC devices can be fabricated in automated mass production.

Due to the above-mentioned attributes of LoC systems, they are best suited for nano-materials toxicity measurement.

Utilizing a MFs platform can also be useful in the effort to detect the adverse effect of nanotoxicity in cells. In a MF device, it is possible to mimic a natural (bio)-chemical processes inside a single chip, due to the ability to precisely control small volumes in chemical reactions. This will be advantageous to assess toxicity in a miniaturized copy of the biological system. Among the commercially available MF devices, their applications are diverse. These products can be used for genotyping microarray analysis and cell analysis and benefit from the ability to precisely control and mix very small volumes. As manufacturing costs have fallen, many firms offer to make custom MF devices, which can be extremely helpful for researchers trying to develop a flow system with unique characteristics (Volpatti and Yetisen 2014).

Besides accuracy, durability, and reliability, competency of a sensing technology to be utilized for portable applications is mainly determined by its miniaturization capability. Among numerous methods of nanomaterials toxicity assessment, the electrochemical and optical-based techniques have the most promising potential for transformation into LoC systems.

It is additionally important to mention that the label-free characteristic of electrochemical methods make them an appropriate option for a variety of applications – particularly environmental. In contrast, not being label-free is the major drawback of optical measurement techniques, which limits their capability for toxicity assessment applications. Thus, the next two sections of this chapter are dedicated to electrochemical and optical-based technologies (Marquis, et al. 2009; Sadik et al. 2009).

9.2.2 Impedance-Based Electrochemical Sensing Systems

Electric impedance is representative of resistance in an alternating current (AC) circuit. The term impedance refers to the frequency-dependent resistance to the flow of current through circuit elements such as a capacitor, resistor, or inductor. Calculating electrical resistance for a simplified DC circuit (assuming all circuit elements are an ideal resistor) is as easy as dividing voltage by current ($R = V/I$). An ideal resistor has a frequency-independent resistance value, which follows Ohm's Law at all voltage and current levels. However, most real applications comprise intricate circuit design including several electronic compartments with non-ideal and complex behaviour. Hence, impedance replaces resistance as a more general circuit parameter.

In biosensing applications, the impedance measurement is a practical measure of electrons or charged ions transporting through the interfaces of the sample and electrodes. Thus, an equivalent circuit is often employed to simulate the impedance spectrum of the system.

On a mathematical note, the impedance (Z) is determined based on the application of a low-amplitude electrical voltage to the system and reading the current response of the measured sample. The excitation signal is expressed as a function of time, as shown in equation 9.1 below, where E_t is the potential at time t, E_0 is the amplitude of the signal, and ω is the radial frequency. The response signal, I_t, is shifted in phase (ϕ) and has a different amplitude than I_0, as demonstrated in equation 9.2.

$$E_t = E_0 \sin(\omega t) \tag{9.1}$$

$$I_t = I_0 \sin(\omega t + \phi) \tag{9.2}$$

By applying principles of Ohm's Law, the impedance of a defined system is calculated by dividing the voltage-time function $\omega(t)$ to the resulting current-time function $i(t)$. This is provided below in equation 9.3. As the impedance is a complex value, it can be transformed into the complex domain format, demonstrated in equation 9.4.

$$Z = \frac{v(t)}{i(t)} = \frac{V_m \sin(\omega t)}{I_m \sin(\omega t + \theta)} \tag{9.3}$$

$$Z(\omega) = \frac{v(j\omega)}{i(j\omega)} = Z' + jZ'' = |Z|\cos(\theta) + j|Z|\sin(\theta) \tag{9.4}$$

Impedance-based electrochemical measurement techniques are extensively utilized for numerous biosensing applications, particularly in cell-based systems such as cell proliferation and death, drug effect, immune-cell signalling, cell adhesion and migration, and cytotoxicity (Hondroulis et al. 2010; Radhakrishnan et al. 2014).

9.2.3 Electric Cell-Substrate Impedance Sensing System

Giaever and Keese (1993) reported the first quantitative analytical measurement using cells' electrical impedance to track morphological changes of cells in real-time. Soon after, the Electric Cell-Substrate Impedance Sensing System (ECIS) method became popular and widely approved by the scientific community, which triggered an unprecedented leap forward in the biosensing field.

The ECIS principle is based on small changes in the electrical properties of cells' body, membrane, and intra/extra-cellular matrix. Multiple complex interactions occurring at the cellular level fundamentally alter the electrical impedance of cells. Features of different biochemical reactions are characterized and distinguished by correlating small variations in impedance values with changes in cellular morphology, signalling pathway, and movement functions. Generally, an ECIS biochip consists of a biocompatible substrate with an array of electrodes on which a monolayer of cells is cultured (Wang and Liu 2010; Shah 2014).

9.2.3.1 ECIS Sensing Mechanism

Cultured cells on an ECIS biochip spread out and cover the electrode area. Therefore, the applied electric current constrained by cell layers has no path except by passing around or through the cells' membrane. Primarily, the static impedance value is associated with the number of cells and the extent of coverage over the electrode. The impedance change is mainly correlated with electro-mechanical and biochemical alteration of cells. Understanding the basics of cellular interactions with their surrounding environment is essential for better understanding the mechanism behind ECIS measurements.

The cell membrane has an extremely complex structure known as the bilayer lipid membrane (BLM). The BLM encapsulates cells and acts as an ultimate barrier that defends cell compartments by establishing stringent selective permeability. Membrane permeability is determined by cell function and activity. BLM has a thickness of about 7 nm and electrical properties which can be modelled as a high-volume capacitor with a capacity of about 1 $\mu F/cm^2$ and very low conductivity of 6–10 S/m (Grimnes and Martinsen 2008).

Cells alter their morphology through their consistent interaction with the substrate and neighbouring cells. Consequently, the mechanical and electrical properties of cell membranes change according to different factors such as substrate stiffness and adjacent cells junctions. Exposing cells to compounds such as nanotoxins leads to biochemical and electromechanical alteration of the cell membrane. Another major pathway for impedance change due to cytotoxicity is through the introduction of the nanotoxins into cellular subcompartments, which ultimately kills the cell. Once the cell dies, it detaches itself from the substrate floating in the media, which decreases the impedance readout from the surface. Changes in the impedance can also correlate with different size, concentration, or morphology of the exposed nanomaterial which is compared under a controlled system. Accordingly, the impedance value changes are recorded continuously throughout the test period. The final step in ECIS detection is to interpret impedance variations to reveal important information about cellular behaviours and ongoing events (Halliwell et al. 2014).

9.2.3.2 ECIS Applications and Toxicity Measurement

Conventional toxicity assessment techniques only provide the endpoint result, which is their major disadvantage for toxicity assessment. The ability of the ECIS system for real-time measurement and data collection makes it a favourable technique for toxicity assessment, particularly for an environmental sample containing a complex mixture of various compounds. Moreover, ECIS technology has shown great capacity for integration with MFs platforms to form a self-contained all-in-one biosensor device capable of rapid detection, accurate identification, and analysis of the desired sample. There are several ECIS-based devices commercialized for several applications and designs of experiments (Giaever and Keese 1993; Brennan et al. 2012).

Real-time monitoring and analysis of biological activity at the cellular level has been carried out by other researchers using ECIS-based assays. The effect of different environmental water samples on ACHN and HepG2 cell lines were studied and quantitatively analysed (Pan et al. 2015). It should be pointed out that a major disadvantage of the proposed method is the lack of ability to identify the individual effect of different substances or potential toxicants on cells.

To resolve this problem, a multi-sample biochip was designed and fabricated using gold microelectrodes capable of rapid and accurate detection. Cytotoxicity of nanomaterials with different morphology and size was measured on CCL-153 and RTgill-W1 cell lines; single-walled carbon nano-tube (SWCNT), cadmium oxide (CdO), gold (Au NPs), and silver nano-particles (Ag NPs) were chosen as the nanotoxic compounds. The ECIS sensor design was reportedly sensitive enough to be able to measure the kinetic effects of each nanotoxic compounds on the targeted cells. The conventional table-top electrochemical station was used for data analysis (Zhu et al. 2013).

Toxicity assessment of phenol and ammonia on bovine aortic endothelial (BAEC) and rat fat pad endothelial (RFPEC) cells line have also been investigated using commercialized ECIS bio-chips. Cell culture was carried out in common open culture wells as well as closed culture chamber. Using a MF platform provided control over fluid flow rate and shear stress distribution by fabricating micro-array of perfusion barriers within PDMS culture chamber, which resulted in the uniform distribution of toxicant over targeted cells (Zhang et al. 2015).

In another study, ECIS MF biochips seeded with BLMVEC and RTgill-W1 cell lines were designed for toxicity assessment of several chemical compounds in drinking water. A comparative study of both cell lines revealed higher viability and longevity of RTgill-W1 cells within the desired microenvironment. After culture, RTgill-W1 cells were successfully stored in the sealed fluidic-biochip for over one year without maintenance and feeding (Brennan et al. 2012).

ECIS technology clearly has a broad range of applicable usage scenarios, ranging from nanotoxicity assessment to LoC applications. However, converging the various beneficial features of ECIS to fulfil all qualifications of a reliable LoC device for nanotoxicity assessment requires a more detailed investigation on different design and fabrication parameters.

9.2.3.3 Design and Fabrication Considerations

Prior to design and fabrication of a LoC device, several important considerations must take place. The goal of fabricating a LoC device is to bring a semi-automated, cost-effective, and portable platform to market, compared to its laborious competitors. Learning from previous studies, having a MF device as a LoC is advantageous compared to having an open system, due to fluid control for uniform distribution and uptake of substance. Integrated sensor material selection should be chosen based on the needs of selectivity, sensitivity, and reproducibility. It is also important to consider that possible sensor material consumption, especially in electrochemical-based methods, causes unintended background noise and false signal production. Also, the geometrical shape of the sensor, its biocompatibility, dimensions, and electrical activity puts an extra consideration on the design parameters necessary to be taken into account. In addition, the longevity of the target cell for analysis on such a platform, its compatibility with the sensor substrate and shear stress applied to cell layers due to fluid flow should be of utmost importance, whether one wants to choose LoC device as their design or not.

Generally, in order to make an ECIS biosensor, there is a number of criteria necessary to take into consideration such as biocompatibility, sensing materials, electrode design, cell immobilization, and MFs fabrication. Moreover, to match the design to the investigation's specific needs, other challenges must be addressed. These include maintaining physiochemical micro-environments, developing software that could accompany the design for continuous sensor sampling, and device integration/packaging (Shah et al. 2014a, 2014b).

9.2.3.4 Biocompatibility

Undoubtedly, biocompatibility is the prerequisite attribute for all biosensor components. This is especially true in an ECIS device due to the delicate nature of the measurement mechanism in which even a slight change in cell electro-mechanical or biochemical properties can falsify the test result. In ECIS, since living cells are in an intimate contact with the electrode surface any unwanted interaction between cells and substrate materials causes signal interference, which eventually affects the test result. Biocompatibility of sensor compartments becomes even more important in the case of nanotoxicity assessment due to the low concentration of existing nanotoxic compounds in some complex environmental samples. In such circumstances, false signals from other sources can easily outweigh the generated signal resulting from slight changes in targeted cells' properties caused by toxic nanomaterials (Hondroulis et al. 2010; Shah et al. 2014a, 2014b).

9.2.3.5 Materials Selection

Various insulating materials such as ceramics (e.g., silicon wafer), sapphire glass, and polymers like silicone-based elastomers are good candidates for a substrate material in an ECIS sensor. Newer approaches are mostly focused on the synthesis and application of soft and flexible materials because of their advantages for portable applications. Unsurprisingly, the appropriate choice for electrode material is thus narrowed down to a few options. Conventionally, gold (Au) and platinum (Pt) along with indium tin oxide (ITO) are utilized as electrode material since extensive studies support their promising biocompatibility, durability, and reliability (Sadik et al. 2009; Hondroulis et al. 2010). Furthermore, in the past few years, carbon-base materials with diverse nano-morphologies (nano-tube, nano-sphere, nano-star, etc.) have been at the centre of attention as new candidates for electrodes material (Sadik et al. 2009; Hondroulis et al. 2010).

9.2.3.6 Electrode Design and Fabrication

At first glance, it simply seems that as electrodes become smaller, their performance increases. However, the correlation between electrodes dimensions and sensor efficiency is much more complicated than a linear relation. Therefore, utilizing software assisted designs along with mathematical methods is required for producing an efficient, high-performance system. Fundamentally, smaller and higher-throughput electrodes are more convenient for integration into the MF platform and eventually integrated LoC devices.

There are two major electrode systems: *monopolar* and *inter-digitated (IDE).* Monopolar systems comprise a pair of working and counter electrodes (CEs) with a circular configuration. The size ratio of working to CE is normally smaller than 1–100. Due to its small size, the working electrode (WE) has a current density much higher than the CE. Consequently, a higher voltage drop occurs on the surface of WE which ultimately results in dominating the total impedance variation (Shah 2014; Shah et al. 2014a, 2014b).

IDE includes two electrodes with identical branch-like layout. A multiplicity of independently operating strips connects to one terminal to form one branch of an electrode whereas the other identical counterpart is placed parallel within the space between strips of the first electrode.

The resolution and sensitivity of IDEs are mainly determined by the number of its fingers (strips). However, there are other significant parameters such as electrode width, length of fingers, and the distance between neighbouring electrodes which must be taken into consideration (Radhakrishnan et al. 2014).

Optimization of the electrode has a significant influence on the overall performance of the ECIS sensor. Investigation of the effect of electrode dimensions on detection sensitivity of ECIS using mathematical modelling is among recent approaches to this matter (Zhang et al. 2017). Current flow through an insulating layer reduces electrode performance and sensitivity. To achieve the optimal sensor design, noise generated by coating-layer capacitance component must be eliminated (Price et al. 2009). As has been emphasized previously, the design of the electrode is possibly considered the most vital component of an LoC platform. With the advancement of today's MEMS technology, complicated geometrical designs, combined electrode material and fully optimized current flow and detection are completely possible. This makes sensor-integrated LoC devices the leading option in multiplexing and potent analysis for several applications.

9.2.3.7 Device Preparation and Cell Immobilization

The first step to utilize ECIS for monitoring cell toxicity is chip design and cell immobilization. After design and fabrication of the chip integrated MF, the device is sterilized using ethanol (depending on the thickness of the electrode on the surface) and UV exposure. Cells of interest are cultured on to the device, either using direct culture or cell introduction using a MF inlet and outlet seeding device. Surface modifications such as fibrin, collagen, or attachment proteins such as cystamine might take place prior to cell culture, depending on the substrate material. This will help the efficient attachment of cells to the chip's surface and a more reliable data output. In some cases, the surroundings of the culture vessel, except for the surface of the sensor, will be coated with blocking agents such as BSA for decreasing the cell's crosstalk and increasing specificity of attachment to the sensor surface. Conventional quantification methods (microscopy) could be used for evaluating the yield of attached cells. Once satisfactory, the process moves on to electrical and impedance monitoring of cells behaviour, with and without introducing chemical agents or other stimuli (Radhakrishnan et al. 2014).

9.2.4 Microfluidics Platforms and Bio-chip Integration

Recent improvements in nano-electro-mechanical systems (NEMS) and soft lithography combining with nano-technology has provided great opportunities for miniaturization of MF platforms. Therefore, more complicated patterns on planar and 3D platforms can be designed and fabricated in much smaller feature size and higher resolution. There are several parameters involving the design and fabrication of MFs for biosensors. Arguably, the most essential attribute of an MF platform in LoC devices is the capability of integration with other compartments of a biosensor such as transducers to form a self-contained portable instrument. Low-cost, fast, and efficient fabrication procedure is among other considerable factors of MF technology.

MF systems are divided into several formats, such as channelled, paper-based, digital, and continuous flow. However, all formats can be categorized under two major platform types: open-channel and closed-channel MFs. Open-channel platforms are commonly recognized as paper-based MFs, which are mostly known for their increasing usage in medical and POC diagnostics. On one hand, low-cost production, easy fabrication, and disposal ability of paper-based platforms make them advantageous for inexpensive and portable biosensors. On the other hand, unprotected open-channel design and poor integration capability hinder utilization of paper-based MFs for sophisticated systems. Thus, so far, only closed-channel MFs platforms have been employed for most LoC devices. Current ECIS-based MF devices require improvements in their capability of biochip conjugation with electrical components, biocompatibility of substrate material for better cell adhesion and viability, development of hybrid platforms, and high-throughput fabrication methods (Sadik et al. 2009; Chen and Shamsi 2017; Rackus et al. 2015).

9.2.5 Optical Sensing System

Various optical detection techniques have shown reliable functionality for biotoxicity assessment. Some of these methods could be integrated within a lab-on-a-chip device, though it is a less common approach, due to the arduous integration of various optical and electrical components as well as several interference factors. Accordingly, the following sections will cover the basics of how these technologies work, along with their integration into LoC devices and their possible applications for nanotoxicity screening.

The majority of optical detection methods are labelled, meaning that in order to perform quantitative analysis specific recognition agents need to be attached to the target or probe. As a result of this, these target-specified systems have limited applications. Nonetheless, in recent years, development of new label-free optical transducers, advancement in miniaturization of optoelectronic sensors such as CCDs along with digital signal processing for signal amplification, laid the foundations for new applications of optical biosensors (Mirasoli et al. 2014; Chen and Shamsi 2017).

A label-free detection mechanism resolves the shortcoming of labelled methods by providing the opportunity to identify a wide-range of non-specified toxic materials. However, emerging miniaturized sensing apparatuses with ever-increasing computational processing power improve the integration feasibility of different components into a self-contained sensing platform that can collect, detect, identify, and analyze. Moreover, utilization of sensitive smartphone cameras along with sophisticated image analysis software is becoming a new trend to improve detection and analysis ability of both colorimetric and luminescence LoC systems (Zhang et al. 2018).

Toxic compounds are consistently interacting with different biomolecules throughout highly dynamic micro-environments of cells. Some of the resulting compounds (residue substances or byproducts) produced by specific biochemical reactions can emit light. Analysis of the emitted light provides crucial information about the biochemical reactions at the cellular-molecular level. Light as an electromagnetic wave can be characterized by its two key attributes: wavelength and amplitude.

While wavelength of the emitted light is specific for each compound, the amplitude of emission can vary depending on the intensity of interactions. Thus, while the wavelength of emitted light helps to indicate the nature of biochemical interactions due to the spectral properties of the reaction product, amplitude represents the strength of the emission source, which is correlated to a number of reacting substances (Shah 2014).

9.2.5.1 Colorimetry and Fluorescence

Two optical detection methods which are commonly used include colorimetry and fluorescence. In colorimetry, a biochemical interaction results in a reaction product which absorbs light in a specific wavelength. This leads to it appearing to the naked eye as a unique colour. On the other hand, in a fluorescence event the reaction compound absorbs light in the non-visible range and then emits light at a higher wavelength due to radiative relaxation.

Colour density in the colorimetric system and the number of emitted photons in a fluorescence assay are proportional to the amount of reaction product. An example of a technology incorporating these techniques is the enzyme-linked fluorescent immunoassays assay (ELFIA), where an enzyme converts a substrate to a reaction product that fluoresces when excited by light of a specific wavelength. The fluorescent unit or intensity that is detected is proportional to the amount of measured analyte (Dennis Driscoll 2013).

9.2.5.2 Bio-luminescence and Chemi-luminescence

The word luminescence is a general term which indicates the emission of light from an electronically excited substance as it returns to its ground state. The different forms of luminescence are determined and specified by the mechanism that initiated excitation. For instance, fluorescence is also called photo-luminescence, since fluorescence is a form of luminescence excited by photons.

Bioluminescence and chemiluminescence represent similar phenomena. Because of a series of chemical reactions, reactive compounds give rise to a molecule which is in an electronically excited state. Ultimately, the excited molecule returns to its ground states (lower energy level) and emits characteristic photons of light. In fact, bioluminescence is a subcategory of chemiluminescence, where the source of emission is a bio-molecule. For example, a class of different bio-compounds called Luciferin are the source of bioluminescence in several living organisms. Light-emitting Luciferin molecules reach their excited state as a result of oxidative-chemical reactions. Typically, an enzyme from a class of oxidative-enzymes called Luciferase couples with the corresponding Luciferin molecule to catalyse the reaction. In summary, chemiluminescence is the emission of light due to chemical reactions while bioluminescence is the emission of light from living organisms (Mirasoli et al. 2014; Roggo and van der Meer 2017).

9.2.5.3 Optical Biosensor Application and Toxicity Measurement

As an alternative to conventional electrochemical methods, efforts in the past decade have led to the development of optical-based biosensors for various environmental applications for toxicity assessment.

In a common approach, different types of bacterial strains were utilized as bioreporters for water toxicity assessment. The bacteria type was chosen based on the targeted toxic substances. A complementary metal–oxide–semiconductor (CMOS) photodetector connected to the computer station along with custom-designed software was employed to collect and analyse emitted light from stimulated cells. Usually, the whole sensing setup is placed into a light-tight box, preventing signal interferences generated by other light sources (Elad et al. 2011; Axelrod et al. 2016).

Sometimes a similar approach is used to address different issues. For example, a luminescence bacterial biosensor primary developed for measurement of water toxicity was modified and utilized for air toxicity determination. The sensing device comprises multiple components integrated into two major parts: disposable and non-disposable. Disposable parts composed of calcium alginate pads encompass immobilized target-specified bioluminescent bacteria. The non-disposable part includes the photomultiplier tube (PMT), liquid light guide, commercialized photosensor module, and corresponding electrical circuitries. Two types of sensors eventually integrate and are placed into a light-tight cylindrical chamber. Photons generated from a biochemical interaction of toxicant and cells pass through the liquid light guide interface and reach the PMT for signal amplification. Real-time monitoring and data analysis are made possible using a custom-built driver (Eltzov et al. 2015a, 2015b).

In the past decade, several LoC optical-based biosensors have been successfully designed and applied for single-use measurement or disposable applications. However, new approaches are more focused on utilizing MFs platforms to develop more efficient systems for multi-sample measurements in a parallel or consecutive manner. Although multiplex sensing systems are more difficult to design and fabricate, their advantages make them preferential for several applications. For instance, target-specificity as one of the major limitations of luminescence-based biosensors for toxicity assessment can be easily overcome by using multi-sample sensing module (Mirasoli et al. 2014; Roggo and van der Meer 2017).

As is a vital matter with all LoC devices, their ease of use, novel packaging, automatic recording, and portability are necessary to remain competitive. On the other hand, most of the bacterial bioreceptor assays need a level of cellular manipulation, leading to limited field applicability. A research group headed by Dr. Truffer fabricated a standalone portable *E. coli* bioreceptor-based biosensor to measure arsenic (As (III)) concentrations in water. The device consisted of two parallel PDMS-based MF channels, mounted on a glass slide and isolated within a cage. Genetically modified *E. coli* bacterial cells that produce a green fluorescent protein (GFP) in response to As(III) were filled into the MF chip. Light was excited to cell lines through a laser diode and another detector was used to collect the emitted fluorescent light. Embedded signal amplification electronics, microprocessors, and microchips for a USB connection to a computer make this device a strong point-of-care testing (POCT) toxicity assessor (Truffer et al. 2014).

As a more advanced application of optical-integrated LoC devices, laser-based platforms have been used as a state of the art technology for monitoring chemical interactions binding kinetics, vesicular release, exocytosis events, and cellular-level nanotoxicity. A common example, surface plasmon resonance (SPR) technology, is a label-free technique used to measure the binding kinetics between biomolecules. This is carried out by monitoring real-time optical signals arising from a prism while a solution with ligand molecules flows over a sensor substrate functionalized with purified receptor molecules. A laser light is shined on a specified point on the prism and reflected in a known angle, which will be detected by a detector camera. The laser causes the sensor molecules (mostly gold thin films or gold NPs) to resonate in their close vicinity. Their resonance is highly sensitive to small changes on the medium that flows above the sensor (i.e., gas, fluid), such as absorption of small molecules on the surface. Such incidents will cause the laser refraction angle to shift higher or lower, which would be detected. Further data analysis could give the kinetics of the events occurring at the surface of the chip.

The technology was used by Li et al. (2011) to integrate live SKOV-3 cells and monitor the release of VEGF from them due to a chemical stimulus for fast exocytosis. The same platform could be used as a new method for analysing the toxicity of nanomaterials on cells on the SPR gold films. Detachment of cells due to nanotoxicity could cause the angle to shift from its original reflecting point. This could then be analysed to determine the number of cells detached from the surface due to the nanotoxicity of a known concentration of material (Liu et al. 2014).

9.2.5.4 Design and Fabrication Considerations

Apart from biocompatibility and other discussed considerations regarding integration of biochips into a MFs platform to achieve optimal LoC design, the optical property of substrate materials should also be considered for optical biosensors. Hence, a substrate of such property should encompass low background luminescence in the ground state, the ability to produce intense light in its active state, and ability to produce stable light emission over a long period of time. Signal transmission from luminescence bioreceptors to detectors should be amplified to increase the sensor's sensitivity and cells should be maintained in the appropriate microenvironment to minimize the background noise from cell/tissue auto-luminescence.

On the other hand, a tradeoff for using bioreporter-based LoCs is their low response time. In cases of portable systems (in-field measurements), where a response time of minutes is required, the requirement of 30 minutes to a few hours for most engineered bioreceptors to produce sufficient reporter proteins for a readable signal becomes especially problematic. It is also worth mentioning that with these bacterial-based platforms, achieving detection limits of submicromolar levels, which are required for environmental standards, is difficult to attain.

9.3 Current Progress: Detecting Nanomaterial Toxicity

9.3.1 Stage 1: Lateral Flow Immunoassay

NPs in contact with cells can generate ROS, including peroxides, superoxides, hydroxyl radicals, singlet oxygen, and alpha oxygen. Once inside a cell, NPs can damage DNA, a common pathway of nanotoxicity. Oxidation of guanine bases of DNA can produce 8-hydroxyguanine and its nucleoside 8-hydroxy-2′-deoxyguanosine (8-OHdG). Measurement of these oxidized products gives an insight of the nanotoxicity level within the body. Conventional methods of measuring 8-OHdG include high-performance liquid chromatography (HPLC), gas chromatography-mass spectrometry (GCMS), HPLC tandem mass spectrometry, and enzyme-linked immunosorbent assay (ELISA). Unfortunately, the equipment handling requires skilled personnel, and the overall procedure is time-consuming and often extremely expensive.

Our lab in the past has developed a novel lateral flow immunoassay (LFIA) to determine the genomic level of nanotoxicity (Zhu et al. 2013, 2014; Leichner et al. 2017). The test strip is based on five compartments: (1) sample loading zone, (2) conjugate pad, (3) test line, (4) control line, and (5) absorption pad. Each of these compartments plays an important role in terms of genotoxicity determination. To test a sample, the sample loading zone is immersed in the sample. Due to capillary force, the sample moves along the strip and reaches the conjugate pad. This movement of liquid relies on the physical properties of the material (wettability/pore size). The gold NP-tagged monoclonal antibodies (AuNPs-Abs) reside in the conjugate pad and form a complex with the biomarkers. Unbound AuNPs-Abs accumulate in the test line and forms a complex with BSA-8-hydroxyguanosine. It can be identified by a red colour change, even with the naked eye. The colour intensity is inversely proportional to the concentration of the analyte. In the control line, polyclonal goat anti-mouse IgG can bind with the unbound AuNP-Abs and tri-complex producing red line in the control zone. This phenomenon is utilized to depict that the test was performed successfully, and the sample was sufficient to flow from the starting point (sample zone) towards the test line and control line. The limit of detection of 8-OHdG was found to be 2.07 ng mL^{-1} in PBS on the paper fluidic device by the colorimetric method.

LFIA has been an interesting platform for numerous applications (Prabhulkar and Li 2010; Li et al. 2011). This device produces results that can easily be read unaided by the human eye or with an electrochemical analyser. Although these devices have numerous advantages, the colorimetric platform can show false positive or false negative results in real samples (urine and blood). Additionally, the antibodies can become physically adsorbed on the surface of the gold NP with hydrophobic and ionic interactions. Finally, samples with low pH can disrupt the binding of AuNPs to Abs, leading to a false result (Figure 9.1).

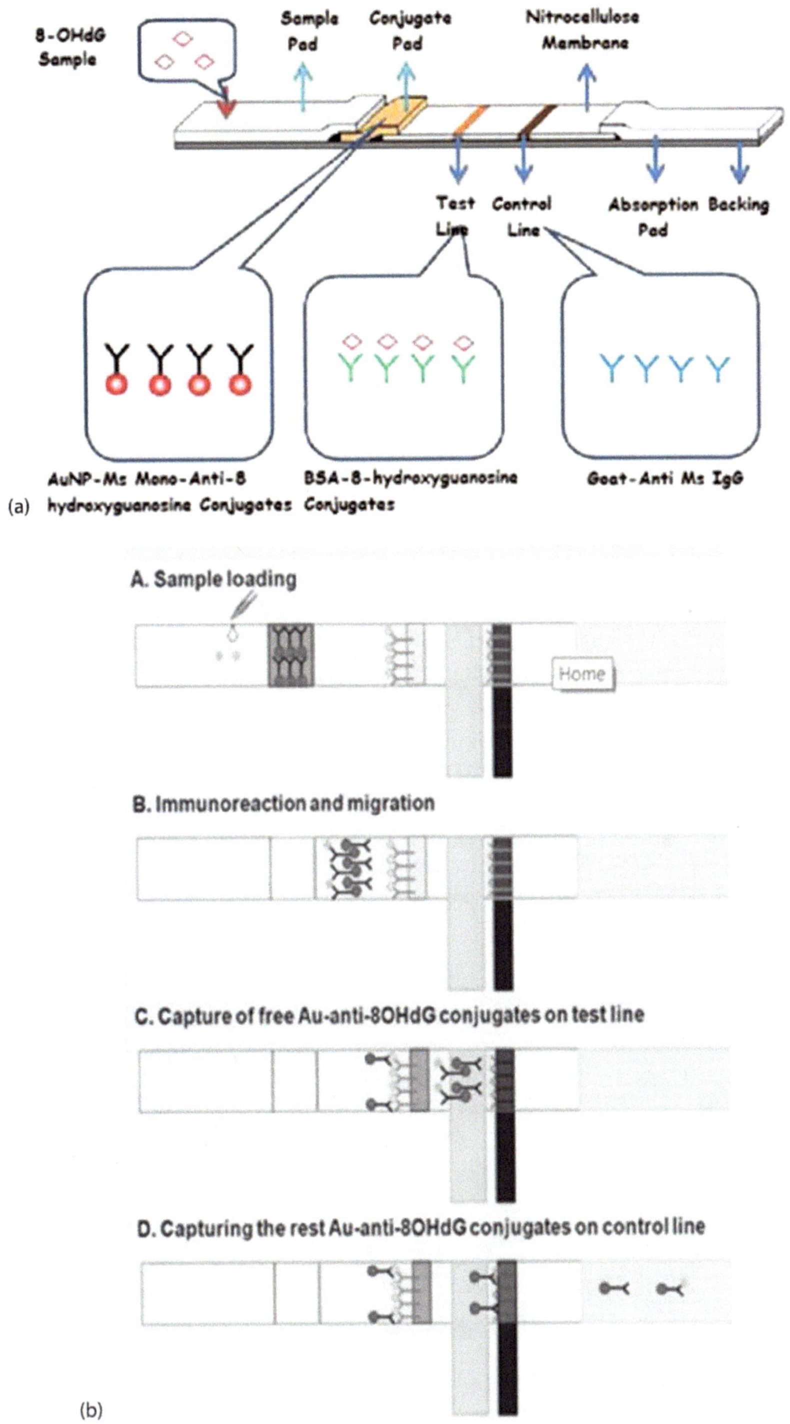

FIGURE 9.1 (a) Schematic representation of the paper fluidic device to detect nanotoxicity in a urine sample (b) Each zone participates in important immunoreaction to provide a colorimetric response on the test line. The intensity of colour developed is inversely proportional to the concentration of the analyte. (A) Sample loading, (B) Immunoreaction and migration, (C) Capture of free AU-anti-8OHdG conjugates on test line, (D) Capturing the rest AU-anti-8OHdG conjugates on control line. *(Continued)*

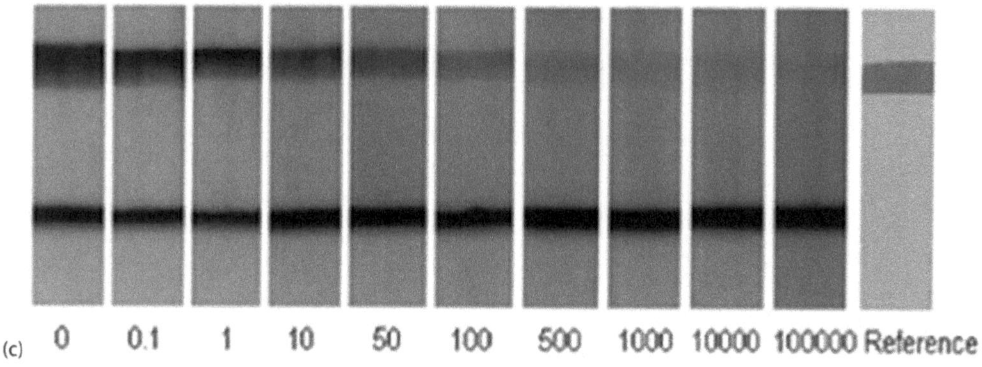

FIGURE 9.1 (Continued) (c) Different concentration of 8-OH DG develops lighter to a darker colour on the paper strip. (From Zhu, X. et al., *The Analyst* 139, 2850, 2014.)

9.3.2 Stage 2: Electrochemical Integration

Later, these paper strips were modified to generate an electrochemical response by using activated carbon (Zhu et al. 2014). The fabrication of CNT conductive paper integrated immunostrip (ECIS) is similar to that of the traditional flow test strip discussed previously. A CNT-conductive paper was used as the WE. The sensing side was placed facing down on the control line. An electrochemical analyser was used to laminate WE with a silver-plated copper wire. Ag/AgCl ink-painted copper paper was integrated 1 mm away from the WE to be used as a reference/CE. The limit of detection was determined to be 3.11 ng mL^{-1} with an electrochemical analyser (Figure 9.2).

The capability of this sensor relies on the property of 8-OHdG, which when oxidized produces two protons and two electrons. These entities can easily be detected with an electrochemical analyser.

Additionally, the structure of CNT paper under SEM shows a forest-like pattern, which can provide a larger surface area with higher sensitivity (Figures 9.3 and 9.4).

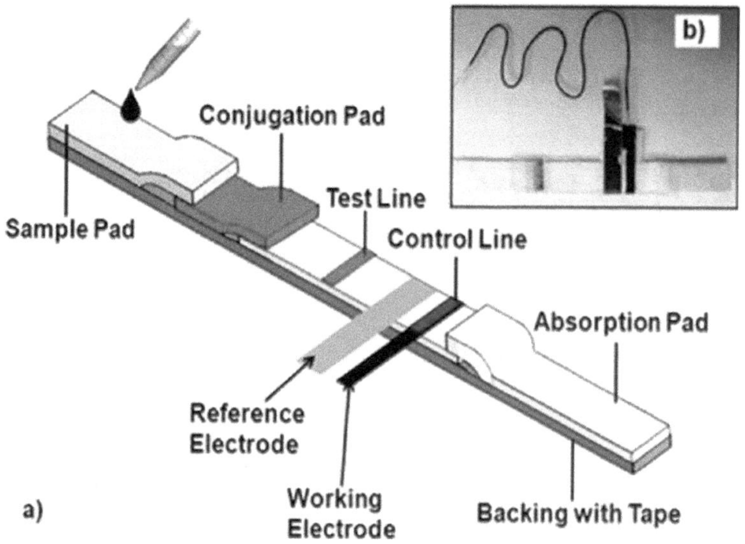

FIGURE 9.2 Paper-based ECIS. (a) Schematic representation of different zones. (b) Picture of the paper fluidic device with ECIS system. *(Continued)*

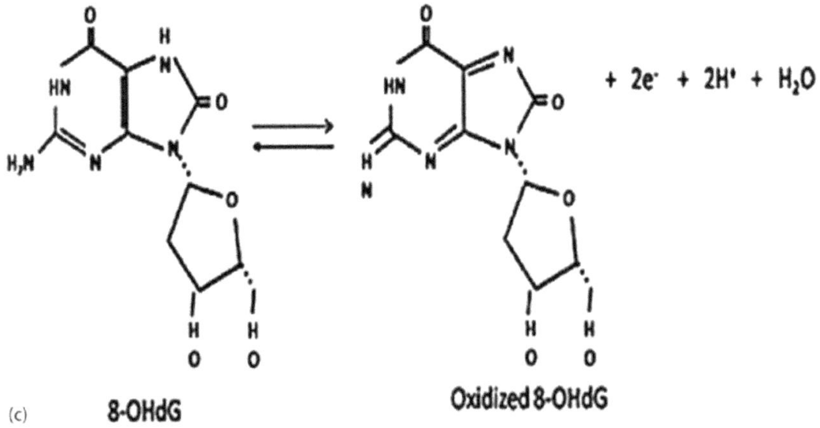

FIGURE 9.2 (Continued) Paper-based ECIS. (c) 7 8-OHdG when oxidized produces two protons and two electrons. (From Zhu, X. et al., *The Analyst* 139, 2850, 2014.)

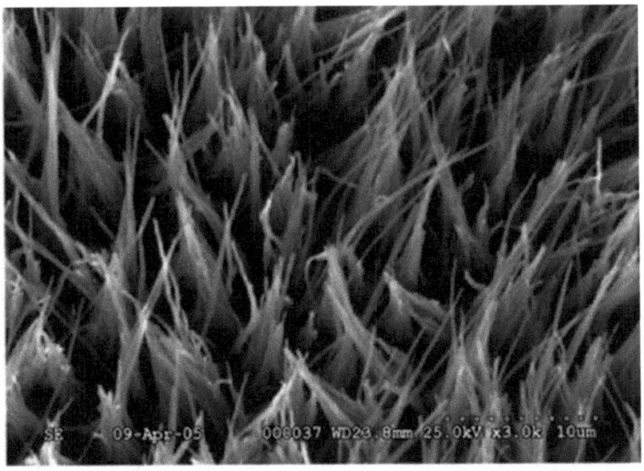

FIGURE 9.3 CNT under scanning electron microscopy (SEM) shows a forest-like pattern, instead of the sheet or net structure which is often seen in CNT paper. (From Zhu, X. et al., *The Analyst* 139, 2850, 2014.)

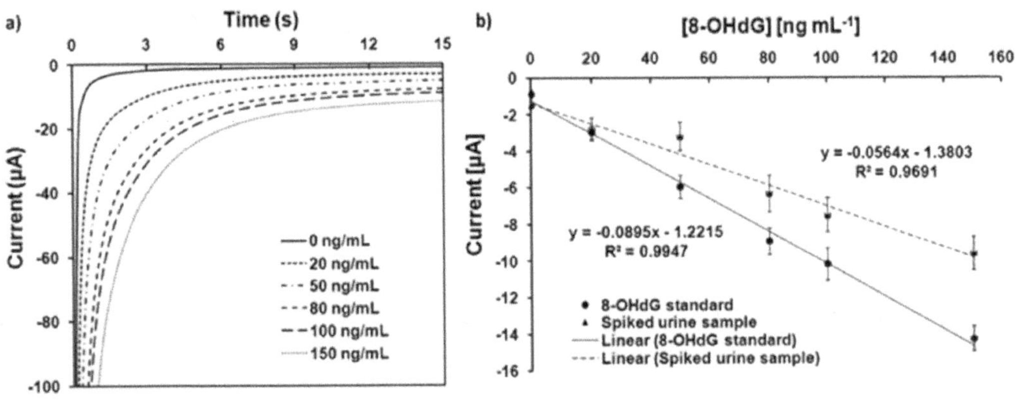

FIGURE 9.4 Detection of 8-OHdG in the ECIS system with paperfluidic device. (a) The curve shows the response of current to five different concentrations (ngmL^{-1}) of 8-OHdG (0, 20, 50, 80, 100, and 150). (b) The constructed calibration plot both in standard samples and urine samples. (From Zhu, X. et al., *The Analyst* 139, 2850, 2014.)

Through this integration of electrical measurements, this miniature platform gained significant capabilities. However, it became necessary to also develop a parallel platform which could assess single-cell nanotoxicity.

9.3.3 Stage 3: Single-Cell Nanotoxicity Platform Fabrication

To manipulate cells with NP exposure and analyze their behaviour, micro-well electrodes are incorporated into the chip (Shah et al. 2014a, 2014b). This enables the isolation of single cells in each well by applying positive-DEP. Finally, the impedance was recorded to test cell activity in presence of NPs as well as to various drugs. The MF chip was composed of two rows of four electrodes, placed in 200-μm gaps. This approach allows the whole system to isolate individual cells, thus preventing cross-talk between them. A reference electrode of 400 μm was made part of the device to avoid the need for an external electrode. Additionally, a WE of 20 μm was also employed in the system. Figure 9.5 depicts the fabrication of the microchips with lithography and also shows the system when wired to the PLCC adapter.

The fabrication process started with using a double-sided, polished, 4-inch quartz glass wafer (University Wafers, USA). This wafer was then immersed in piranha solution for 30 minutes. After that, the wafer was rinsed thoroughly with de-ionised water and dried with nitrogen gas. To remove any residual liquid from the wafer, it was then placed on a hot plate for 5 minutes at 115°C. An ion beam evaporator (JEOL, Japan) was used to deposit 25 nm of Cr adhesion layer and 250 nm of Au thin films. AZ1518 (MicroChem Co.) positive photoresist is spin-coated on the wafer and exposed to a pattern glass-chrome mask using a mask aligner (OAI 800). The developer, AZ400 used with DI water in a ratio of 1:4. Etching was performed by an Au etchant and Cr etchant to remove undesired Au and Cr, respectively. It was further cleaned under sonication for 5 minutes each time in solvents like Acetone, Methanol, and DI water. Finally, a plasma cleaning step was done in a reactive oxygen chamber.

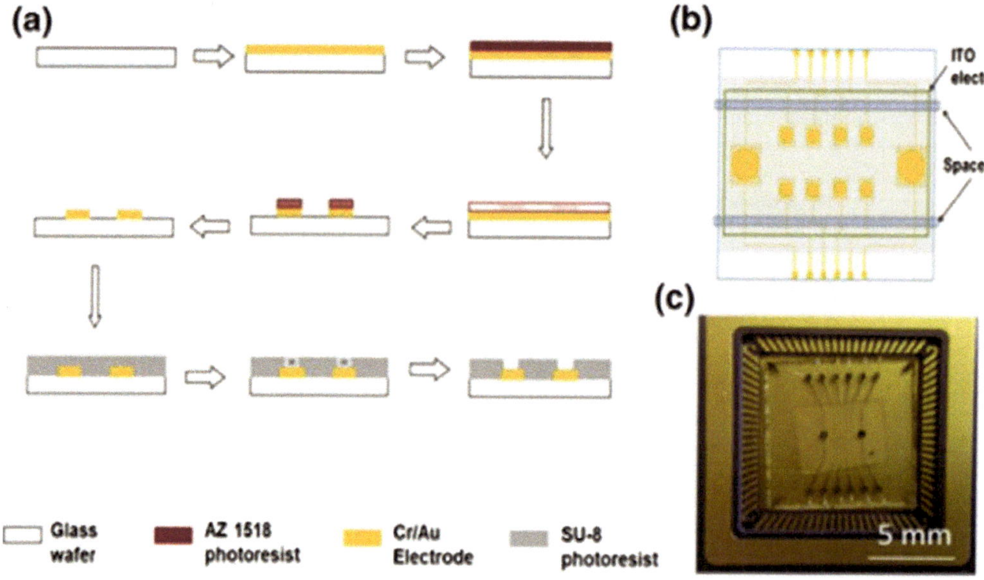

FIGURE 9.5 Fabrication process of microchip. (a) Stepwise method of lithography technique to create a pattern on microchip. (b) Assembly of microchip. (c) Set up of the microchip wired with PLCC adapter. (From Shah, P. et al., *Biomed. Microdevices*, 16, 35–41, 2014.)

A negative photoresist, SU-8 2025 (MicroChem Co.) was spin coated for 30°s at 3,500 rpm. This step achieved 25 μm thickness of the photoresist layer on the wafer. The wafer then exposed to the UV light, hot baked, and developed using SU-8 developer. The plasma chamber was used to remove uncross-linked SU-8 and organic contamination. The thickness of SU-8 was measured by a profilometer (Alpha step). SU-8 pattern covers most of the wafers except the 20 μm long sensing zone and the connection (bond pad). The reason behind this design of the chip was to block cross-talk between cells. The gold microelectrodes were found to have 5 μΩ-cm of resistance.

Functionalization of the gold micro-electrode was done by surface modification with L-cysteine. Thiol (–SH) groups were introduced by immersing the electrodes under 10 mM L-cysteine for an hour. The thiol group attaches on the gold covalently and forms a self-assembled monolayer (SAM). The cell membrane was slightly negative in nature; therefore, the attachment of cells was favoured by the presence of positive charges like free amine. Microelectrodes were characterized via optical, electrochemical, single-cell trapping and lastly for electrical impedance sensing of a single cell.

The MF channel of cross-section 0.4 mm^2 (4 mm wide and 0.1 mm height) and 1 cm length was fabricated using double-sided adhesive paper. An ITO electrode (Delta Technologies Inc., CO, USA) was placed in the chip to generate a non-uniform AC field. The cell solution contained 2×10^5 cell/mL in 0.2 M sucrose buffer. The flow rate was controlled to remain within 5–20 uL/min. The DEP force was applied to trap the single cell. Within 1 minute, the cell becomes trapped in the microchannel and unattached cells were flushed out with sucrose buffer. The microchip was wired to a PLCC adapter to simplify the system for analysis.

A particle with a neutral charge tends to remain stable in dipole moment when placed in a uniform electric field whereas, in a non-uniform electric field, different force magnitude is experienced by the neutral particle. This phenomenon gives rise to a different net force. Therefore, the DEP force can be used to manipulate a polarized entity (Jones 2003) (Figure 9.6).

The time-averaged dielectric field, F_{DEP} when applied to cell suspension gives rise to the following mathematical equation:

$$\mathbf{F}_{DEP} = 2\,\pi\, R^3 \mathcal{E}_m\, Re\,[f_{CM}\,(\omega)]\, \nabla E^2$$

where R = cell radius; $\mathcal{E}_m$ = permittivity of cell suspension medium; $\mathcal{E}_c$ = permittivity of cell; ∇E = rms value of applied A.C. field; ω = angular velocity of the applied field, $Re\ [f_{CM}\ (\omega)]$ which is part of 'Clausius–Mossotti' factor (polarization factor) can be derived by the equation below:

$$f_{CM}\,(\omega) = (\mathcal{E}_c^* - \mathcal{E}_m^*)/(\mathcal{E}_c + 2\mathcal{E}_m)$$

where, $\mathcal{E}_c^*$ *and* $\mathcal{E}_m^*$ are the complex electrical permittivity of cell and cell suspended medium, respectively. $* = [\varepsilon - j\sigma/\omega]$, where σ is electrical conductivity and $j = \sqrt{-1}$.

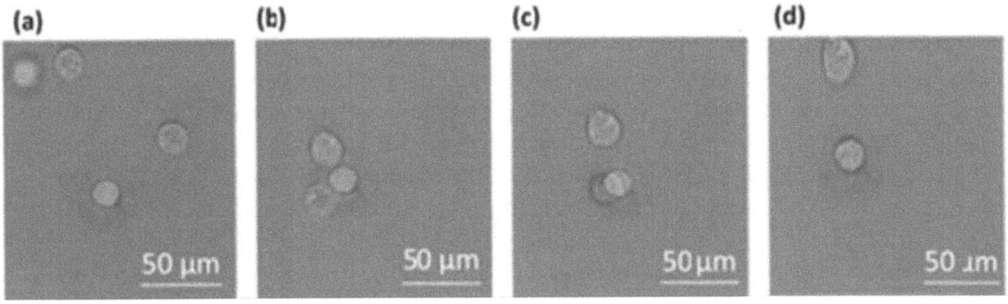

FIGURE 9.6 P_{DEP} technique allows single-cell trapping in various time point shown in the images taken from cell capturing video above (a) 10s-time interval (b) 15s time interval (c) 19s time interval and (d) 20s. (From Shah, P. et al., *Biomed. Microdevices*, 16, 35–41, 2014.)

Cell medium buffer inherently possesses a high resistance and, therefore, favours a positive DEP where $\mathcal{E}_c$ is greater than $\mathcal{E}_m$.

Following this process, a cell viability test was performed with trypan blue assay, where cells were treated with 1: 1 ratio of phosphate buffered saline and trypan blue solution after removing medium and incubated for 5 minutes. Next, the cells were washed with PBS buffer several times. The cells were checked for viability under the microscope. Dead cells accumulated trypan blue in their cytoplasm whereas the viable cells remained unstained. The impedance from the microchip electrodes was measured against 0.015 V reference electrode. The data were collected at frequencies ranging from 1 to 10^5 Hz for each measurement. PBS buffer was used in every condition to measure the impedance at room temperature.

The change in impedance arises when cells proliferates or experiences a change in the environment, which causes damage to the cell. The impedance recorded in the absence of cell is 1.51 MΩ and with a single cell attached on the microelectrode the impedance increased to 17 MΩ. The single-cell analysis enables researchers to reveal the complex processes, in cell behaviour like neurotransmitter kinetics, ion channel functions, and cell communications.

9.3.4 Stage 4: Testing of Single-Cell Nanotoxicity Platform with PC12 Cells

To investigate nanotoxicity further, our lab investigated PC12 cells upon TiO_2 and CuO NPs exposure. The CuO used were 208 nm in size with a zeta potential of −30.4 mV and TiO_2 were 190 nm in size and had −26.9 mV of zeta potential (Mathie et al. 2006). Cu from CuO has been reported to exert an effect on voltage-gated ion channels (K^+, Na^+, and Ca^{2+}) and on ligand-gated ion channels (GABA receptors, glutamate receptors and P2X receptors). CuO NPs have been reported to effect PC12 cells viability and functionality by producing ROS inside the cells (Xu et al. 2012). Much research has been done in a large volume of cells and fails to observe the individual cells behaviour upon exposure to NPs. Our custom microchips, whose construction was discussed previously, are able to measure the effects of NPs on the individual PC12 cell. These cells were exposed to 100 µg/mL CuO-NPs. Similarly, TiO_2 NPs were found to alter the exocytosis function of mast cells (Maurer-Jones et al. 2012). In the case of PC12 cells, dose-dependent decrement in viability was found (Liu et al. 2010).

After exposure to 100 µg/mL of CuO-NPs, a faster exocytosis release kinetics and slower quantal release per spike were detected. Exposure to TiO_2-NPs led to increased catecholamine release from the cells. On the other hand, the release of granules decreased by 23% when compared to control. The changes appeared to alter the amperometric signal. NPs produced irreversible changes in the PC12 cells and washing of the NPs after 3 minutes of exposure had no effect. Our experiments on single cells after exposure to NPs correlated to the other studies. NPs cause DNA damage, ROS generation, and cell death in PC12 cells when exposed for a few hours to days (Figure 9.7).

The cell impedance data were collected for 5 hours and then exported for analysis.

In our lab, we also tested silver NPs (Ag-NPs) toxicity on PC12 cells (Shah et al. 2014a, 2014b). Ag-NPs are commonly found in cosmetics, dental care products, and many more consumer products for its antimicrobial effect (Sondi and Salopek-Sondi 2004). The Ag-NPs were measured in a zetasizer instrument for their size and surface charge. The NPs were on average 25 nm in diameter and had a surface charge of −37 mV. The PC12 cells were trapped in a microchip fabricated in our lab as described previously. With DEP technique the cells were trapped individually on a gold electrode (Shah et al. 2015). Once cells were trapped, they were incubated for 15 minutes in a cellular incubator to enhance the cell attachment on the electrode. The ITO electrode was removed carefully, and the medium was changed to isotonic buffer. The cell was then treated with 100 mmol/L KCl solution to increase the porosity and induce the release of catacaloamines. At a constant voltage of 0.7 V, released catacalomines were oxidized under gold microelectrode (Figure 9.8).

Average half-maximum time $t_{1/2}$ (ms) in the control cell was 4.48 ± 0.19 and in the Ag-NPs treated cell was 4.30 ± 0.10. Average catacalomine release Q (*fC*) in the control cell was 80.42 ± 6.19 and in the

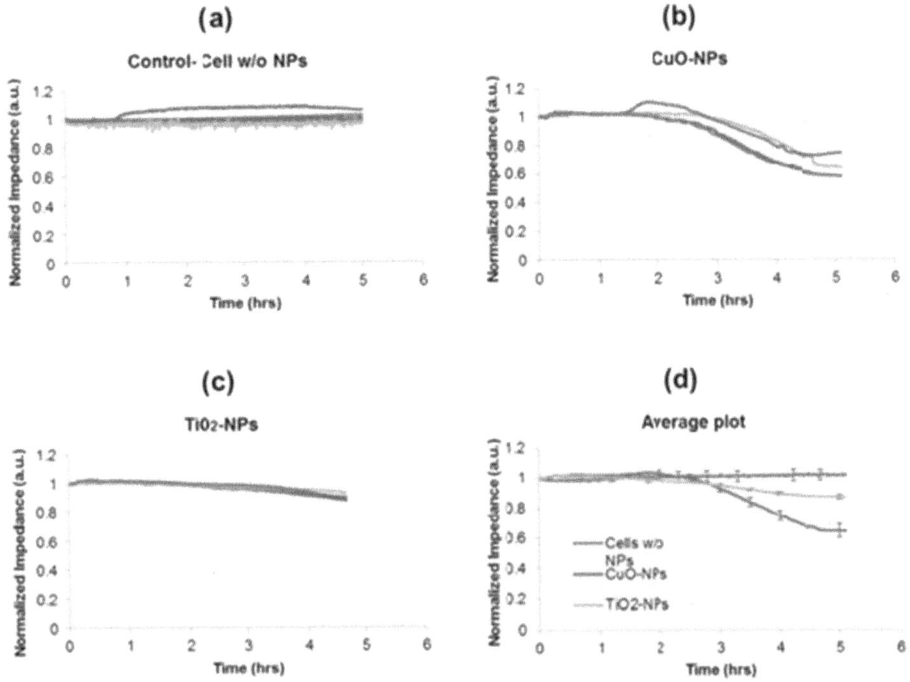

FIGURE 9.7 Impedance change was recorded when PC12 cells are exposed to NPs. (a) Cells with no NPs. (b) Cells treated with 100 μg/mL CuO. (c) Cells treated with 100μg/ml TiO_2. (d) Merged average plot of three conditions reported in (a)–(c). (From Shah, P. et al., *Biomed. Microdevices*, 16, 35–41, 2014.)

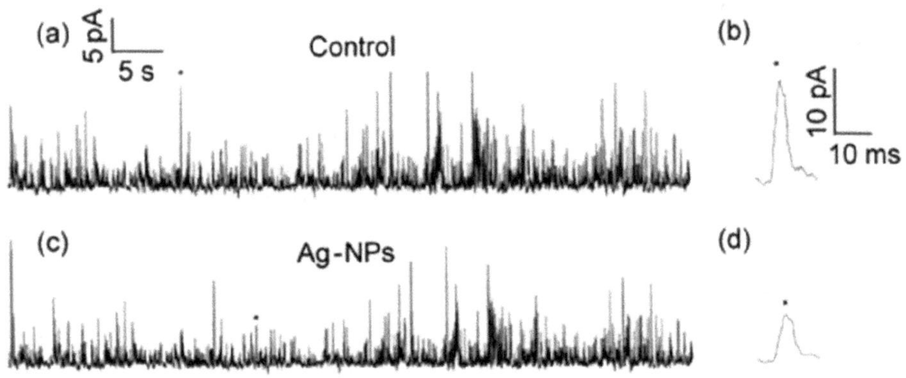

FIGURE 9.8 Amperometric measurement of catacalomines with microelectrodes. (a) 100 mmol/L KCL treated cells. (c) 100 μg/mL Ag-NPs and 100 mmol/L KCl (b, d) shows peak from the recorded signal. (From Shah, P. et al., *Sci. China Chem.*, 58, 1600–1604, 2015.)

Ag-NPs treated cell was 24.07 ± 1.12. The average exocytosis spike frequency in the control cell detected was 0.47 ± 0.03 Hz and in the Ag-NPs treated cell was 0.43 ± 0.029 Hz.

These novel devices generated within our laboratory for the first time make it possible to evaluate nanomaterial toxicity on the chip-based cell-integrated device. It was found that at 10 μg/mL, citrate coated Ag-NPs were not toxic to PC12 cells. However, higher concentrations of NPs reduced neurotransmitter release by 70%. Without any animal testing LoC devices were able to reveal important cell characteristic upon exposure to NPs.

9.4 Conclusions

This book chapter aims to provide a comprehensive perspective on the use of Lab on Chip technology for nanotoxicity assessment. A detailed discussion of nanotoxicity and a description of the fundamental technological advances in the underlying platforms are accompanied by a thorough review of the work performed in our own laboratory. Ultimately, LoC devices are excellent tools for screening nanomaterial toxicity in a rapid, simple, and cost-effective way. Additionally, with their limitless customizability, the applications of these devices are endless. Previously, our laboratory fabricated lab-on-chip devices with independently addressable microwell electrodes for high-throughput single-cell nanotoxicity studies. Control over individual microwells was achieved through the use of positive-DEP and verified by examining impedance properties. The device presents a novel model for single-cell assays of nanotoxicity.

References

Angell, James B., Stephen C. Terry, and Phillip W. Barth. 1983. "Silicon Micromechanical Devices." *Scientific American* 248 (4): 44–55. doi:10.1038/scientificamerican0483-44.

Axelrod, Tim, Evgeni Eltzov, and Robert S. Marks. 2016. "Bioluminescent Bioreporter Pad Biosensor for Monitoring Water Toxicity." *Talanta* 149: 290–297. doi:10.1016/j.talanta.2015.11.067.

Bell, C.L., M.N. Duvall, J.C. Chen, J. Votaw. 2006. "Regulation of Nanoscale Materials under the Toxic Substances Control Act." *Reviewing New Chemicals under the Toxic Substances Control Act (TSCA)*. http://www.abanet.org/environ/nanotech/pdf/TSCA.pdf.

Bejjani, Riad, David BenEzra, Hagit Cohen, Jutta Rieger, Charlotte Andrieu, Jean-Claude Jeanny, Gershon Gollomb, and Francine Behar-Cohen. 2005. "Nanoparticles for Gene Delivery to Retinal Pigment Epithelial Cells." *Molecular Vision* 11 (14): 124–132.

Berdat, Daniel, Ana C. Martin Rodríguez, Fernando Herrera, and Martin A. M. Gijs. 2008. "Label-Free Detection of DNA with Interdigitated Micro-Electrodes in a Fluidic Cell." *Lab Chip* 8 (2): 302–308. doi:10.1039/B712609C.

Bhalinge, Payoshnee, Sourab Kumar, Abhishek Jadhav, Shilpi Suman, Pavan Gujjar, and Nitesh Perla. 2016. "Biosensors: Nanotools of Detection: A Review." *International Journal of Healthcare and Biomedical Research* 4: 26–39.

Blanco, Elvin, Haifa Shen, and Mauro Ferrari. 2015. "Perspective Principles of Nanoparticle Design for Overcoming Biological Barriers to Drug Delivery" *Nature Biotechnology* 33 (9): 941–951. doi:10.1038/nbt.3330.

Brennan, Linda M., Mark W. Widder, Lucy E.J. Lee, and William H. van der Schalie. 2012. "Long-Term Storage and Impedance-Based Water Toxicity Testing Capabilities of Fluidic Biochips Seeded with RTgill-W1 Cells." *Toxicology in Vitro* 26 (5): 736–745. doi:10.1016/j.tiv.2012.03.010.

Buzea, Cristina, Ivan I Pacheco, and Kevin Robbie. 2007. "Nanomaterials and Nanoparticles: Sources and Toxicity." *Biointerphases* 2 (4): MR17–71. doi:10.1116/1.2815690.

Chen, Sensen, and Mohtashim H Shamsi. 2017. "Biosensors-on-Chip: A Topical Review." *Journal of Micromechanics and Microengineering* 27 (8): 083001. doi:10.1088/1361-6439/aa7117.

Cheng, I. Fang, Hsien Chang Chang, Tzu Ying Chen, Chenming Hu, and Fu Liang Yang. 2013. "Rapid (<5 Min) Identification of Pathogen in Human Blood by Electrokinetic Concentration and Surface-Enhanced Raman Spectroscopy." *Scientific Reports* 3: 1–8. doi:10.1038/srep02365.

Chikkaveeraiah, Bhaskara V., Hongyun Liu, Vigneshwaran Mani, Fotios Papadimitrakopoulos, and James F. Rusling. 2009. "A Microfluidic Electrochemical Device for High Sensitivity Biosensing: Detection of Nanomolar Hydrogen Peroxide." *Electrochemistry Communications* 11 (4): 819–822. doi:10.1016/j.elecom.2009.02.002.

Clark, Joseph A. 2011. "Nanotechnology and the State of Government Regulation—Past, Present, and Future; International Risk Management".

Crosera, Matteo, Andrea Prodi, Marcella Mauro, Marco Pelin, Chiara Florio, Francesca Bellomo, Gianpiero Adami et al. 2015. "Titanium Dioxide Nanoparticle Penetration into the Skin and Effects on HaCaT Cells." *International Journal of Environmental Research and Public Health* 12 (8): 9282–9297. doi:10.3390/ijerph120809282.

Dennis Driscoll, Fengqiang Wang. 2013. "The Comparison of Chemiluminescent- and Colorimetric-Detection Based ELISA for Chinese Hamster Ovary Host Cell Proteins Quantification in Biotherapeutics." *Journal of Bioprocessing & Biotechniques* 3 (3). doi:10.4172/2155-9821.1000136.

Elad, Tal, Ronen Almog, Sharon Yagur-Kroll, Klimentiy Levkov, Sahar Melamed, Yosi Shacham-Diamand, and Shimshon Belkin. 2011. "Online Monitoring of Water Toxicity by Use of Bioluminescent Reporter Bacterial Biochips." *Environmental Science & Technology* 45 (19): 8536–8544. doi:10.1021/es202465c.

Eltzov, Evgeni, Adiel Yehuda, and Robert S. Marks. 2015b. "Creation of a New Portable Biosensor for Water Toxicity Determination." *Sensors and Actuators B: Chemical* 221 (December): 1044–1054. doi:10.1016/j.snb.2015.06.153.

Eltzov, Evgeni, Avital Cohen, and Robert S. Marks. 2015a. "Bioluminescent Liquid Light Guide Pad Biosensor for Indoor Air Toxicity Monitoring." *Analytical Chemistry* 87 (7): 3655–3662.

EPA. 2017. "Control of Nanoscale Materials under the Toxic Substances Control Act." *United States Environmental Protection Agency.* https://www.epa.gov/reviewing-new-chemicals-under-toxic-substances-control-act-tsca/control-nanoscale-materials-under.

Gerloff, Kirsten, Catrin Albrecht, Agnes W. Boots, Irmgard Förster, and Roel P. F. Schins. 2009. "Cytotoxicity and Oxidative DNA Damage by Nanoparticles in Human Intestinal Caco-2 Cells." *Nanotoxicology* 3 (4): 355–364. doi:10.3109/17435390903276933.

Giaever, Ivar, and Charles R. Keese. 1993. "A Morphological Biosensor for Mammalian Cells." *Nature* 366: 591–592. doi:10.1038/366591a0.

Grimnes, Sverre, and Ørjan Grøttem Martinsen. 2008. "Geometrical Analysis." In *Bioimpedance and Bioelectricity Basics*, 161–204. Elsevier. doi:10.1016/B978-0-12-374004-5.00006-4.

Halliwell, Jennifer, Alison C. Savage, Nicholas Buckley, and Christopher Gwenin. 2014. "Electrochemical Impedance Spectroscopy Biosensor for Detection of Active Botulinum Neurotoxin." *Sensing and Bio-Sensing Research* 2: 12–15. doi:10.1016/j.sbsr.2014.08.002.

Hiraku, Yusuke, Yoshihiro Nishikawa, Ning Ma, Tahmina Afroz, Kosuke Mizobuchi, Ryo Ishiyama, Yuta Matsunaga, Takamichi Ichinose, Shosuke Kawanishi, and Mariko Murata. 2017. "Nitrative DNA Damage Induced by Carbon-Black Nanoparticles in Macrophages and Lung Epithelial Cells." *Mutation Research/Genetic Toxicology and Environmental Mutagenesis* 818: 7–16. doi:10.1016/j.mrgentox.2017.04.002.

Hondroulis, Evangelia, Chang Liu, and Chen-Zhong Li. 2010. "Whole Cell Based Electrical Impedance Sensing Approach for a Rapid Nanotoxicity Assay." *Nanotechnology* 21 (31): 315103. doi:10.1088/0957-4484/21/31/315103.

Huang, Xinglu, Xu Teng, Dong Chen, Fangqiong Tang, and Junqi He. 2010. "The Effect of the Shape of Mesoporous Silica Nanoparticles on Cellular Uptake and Cell Function." *Biomaterials* 31 (3): 438–448. doi:10.1016/j.biomaterials.2009.09.060.

Jones, Thomas B. 2003. "Basic Theory of Dielectrophoresis and Electrorotation." *IEEE Engineering in Medicine and Biology Magazine : The Quarterly Magazine of the Engineering in Medicine & Biology Society* 22 (6): 33–42. http://www.ncbi.nlm.nih.gov/pubmed/15007989.

Kaczmarek, M., and Tomita, Y. 2009. "Optics of Nanocomposite Materials." Journal of Optics A Pure and Applied Optics 11 (2): 020201.

Karlsson, Hanna L, Pontus Cronholm, Johanna Gustafsson, and Lennart Mo. 2008. "Copper Oxide Nanoparticles are Highly Toxic: A Comparison between Metal Oxide Nanoparticles and Carbon Nanotubes." *Chemical Research in Toxicology* 21 (9): 1726–1732.

Karn, B., Kuiken, T., and Otto, M. 2009. "Nanotechnology and in situ Remediation: A Review of the Benefits and Potential Risks." *Environmental Health Perspectives* 1823–1831. doi:10.1289/ehp.0900793.

Kessler, R. 2011. "Engineered Nanoparticles in Consumer Products: Understanding a New Ingredient." *Environmental Health Perspectives* 119 (3). doi:10.1289/ehp.119-a120.

Lee, In-chul, Je-won Ko, Sung-hyeuk Park, Na-rae Shin, In-sik Shin, Changjong Moon, Je-hein Kim, Hyoung-chin Kim, and Jong-choon Kim. 2016. "Comparative Toxicity and Biodistribution Assessments in Rats Following Subchronic Oral Exposure to Copper Nanoparticles and Microparticles." *Particle and Fibre Toxicology*1–16. doi:10.1186/s12989-016-0169-x.

Leichner, Jared, Mehenur Sarwar, Amirali Nilchian, Xuena Zhu, Hongyun Liu, Shaomin Shuang, and Chen-zhong Li. 2017. "Electrochemical Lateral Flow Paper Strip for Oxidative-Stress Induced DNA Damage Assessment." In *Biosensors and Biodetection: Methods and Protocols, Volume 2: Electrochemical, Bioelectronic, Piezoelectric, Cellular and Molecular Biosensors*, (Ed.), Ben Prickril and Avraham Rasooly, pp. 23–39. New York: Springer. doi:10.1007/978-1-4939-6911-1_3.

Li, Chen-zhong, Katherine Vandenberg, Shradha Prabhulkar, Xuena Zhu, Lisa Schneper, Kalai Methee, Charles J. Rosser, and Eugenio Almeide. 2011. "Paper Based Point-of-Care Testing Disc for Multiplex Whole Cell Bacteria Analysis." *Biosensors and Bioelectronics* 26 (11): 4342–4348. doi:10.1016/j.bios.2011.04.035.

Liu, Chang, Subbiah Alwarappan, Haitham A. Badr, Rui Zhang, Hongyun Liu, Jun-Jie Zhu, and Chen-Zhong Li. 2014. "Live Cell Integrated Surface Plasmon Resonance Biosensing Approach to Mimic the Regulation of Angiogenic Switch upon Anti-Cancer Drug Exposure." *Analytical Chemistry* 86 (15): 7305–7310. doi:10.1021/ac402659j.

Liu, Shichang, Lanju Xu, Tao Zhang, Guogang Ren, and Zhuo Yang. 2010. "Oxidative Stress and Apoptosis Induced by Nanosized Titanium Dioxide in PC12 Cells." *Toxicology* 267 (1–3): 172–177. doi:10.1016/j.tox.2009.11.012.

Lohani, Alka, Anurag Verma, Himanshi Joshi, Niti Yadav, and Neha Karki. 2018. "*Nanotechnology-Based Cosmeceuticals*" 2014. Hindawi Publishing Corporation. doi:10.1155/2014/843687.

Manke, Amruta, Liying Wang, and Yon Rojanasakul. 2013. "Mechanisms of Nanoparticle-Induced Oxidative Stress and Toxicity." *BioMed Research International* 2013: 942916. doi:10.1155/2013/942916.

Marijaa, Grancarić Ana, Tarbuk Anitaa, and Kovaček Ivančicab. 2009. "Nanoparticles of Activated Natural Zeolite on Textiles for Protection and Therapy." *Chemical Industry and Chemical Engineering Quarterly* 15: 203–210.

Marquis, Bryce J., Melissa A. Maurer-Jones, Katherine L. Braun, and Christy L. Haynes. 2009. "Amperometric Assessment of Functional Changes in Nanoparticle-Exposed Immune Cells: Varying Au Nanoparticle Exposure Time and Concentration." *The Analyst* 134 (11): 2293. doi:10.1039/b913967b.

Marquis, Bryce J., Sara A. Love, Katherine L. Braun, and Christy L. Haynes. 2009. "Analytical Methods to Assess Nanoparticle Toxicity." *Analyst* 134 (3): 425–439. doi:10.1039/b818082b.

Mathie, Alistair, Gemma L. Sutton, Catherine E. Clarke, and Emma L. Veale. 2006. "Zinc and Copper: Pharmacological Probes and Endogenous Modulators of Neuronal Excitability." *Pharmacology and Therapeutics* 111 (3): 567–583. doi:10.1016/j.pharmthera.2005.11.004.

Maurer-Jones, Melissa A., Jenna R. Christenson, and Christy L. Haynes. 2012. "TiO2 Nanoparticle-Induced ROS Correlates with Modulated Immune Cell Function." *Journal of Nanoparticle Research* 14 (12). doi:10.1007/s11051-012-1291-9.

Maurer-Jones, Melissa A., Yu-Shen Lin, and Christy L. Haynes. 2010. "Functional Assessment of Metal Oxide Nanoparticle Toxicity in Immune Cells." *ACS Nano* 4 (6): 3363–3373. doi:10.1021/nn9018834.

Mirasoli, Mara, Massimo Guardigli, Elisa Michelini, and Aldo Roda. 2014. "Recent Advancements in Chemical Luminescence-Based Lab-on-Chip and Microfluidic Platforms for Bioanalysis." *Journal of Pharmaceutical and Biomedical Analysis* 87: 36–52. doi:10.1016/j.jpba.2013.07.008.

Nanoparticles-glass, Semiconductor, and Optical Properties. 2009. "Optics of Nanocomposite Materials," 8–9. doi:10.1088/1464-4258/11/2/020201.

Nel, Andre. 2007. "Toxic Potential of Materials." *Science* 311 (5726): 622–627. doi:10.1126/science.1114397.

Olive, Peggy L, and Judit P Banáth. 2006. "The Comet Assay: A Method to Measure DNA Damage in Individual Cells." *Nature Protocols* 1 (1): 23–29.

Oostingh, Gertie J., Eudald Casals, Paola Italiani, Renato Colognato, Rene Stritzinger, Jessica Ponti, Tobias Pfaller et al. 2011. "Problems and Challenges in the Development and Validation of Human Cell-Based Assays to Determine Nanoparticle-Induced Immunomodulatory Effects." *Particle and Fibre Toxicology* 8 (1): 8. doi:10.1186/1743-8977-8-8.

Pan, Tianhong, Haoran Li, Swanand Khare, Biao Huang, Dorothy Yu Huang, Weiping Zhang, and Stephan Gabos. 2015. "High-Throughput Screening Assay for the Environmental Water Samples Using Cellular Response Profiles." *Ecotoxicology and Environmental Safety* 114: 134–142. doi:10.1016/j.ecoenv.2015.01.020.

Kessler, R. 2011. "Engineered Nanoparticles in Consumer Products: Understanding a New Ingredient." *Environmental Health Perspectives* 119 (3). doi:10.1289/ehp.119-a120.

Prabhulkar, Shradha, and Chen-Zhong Li. 2010. "Assessment of Oxidative DNA Damage and Repair at Single Cellular Level via Real-Time Monitoring of 8-OHdG Biomarker." *Biosensors and Bioelectronics* 26 (4): 1743–1749. doi:10.1016/j.bios.2010.08.029.

Prados, Jose, Consolación Melguizo, Raul Ortiz, Celia Vélez, Pablo J Alvarez, Jose L Arias, Maria A Ruíz, Visitacion Gallardo, and Antonia Aranega. 2012. "Doxorubicin-Loaded Nanoparticles: New Advances in Breast Cancer Therapy." *Anti-Cancer Agents in Medicinal Chemistry* 12 (9): 1058–1070.

Press, Dove. 2016. "Cytotoxicity of CdTe Quantum Dots in Human Umbilical Vein Endothelial Cells: The Involvement of Cellular Uptake and Induction of pro-Apoptotic Endoplasmic Reticulum Stress," 529–42.

Price, Dorielle T., Abdur Rub Abdur Rahman, and Shekhar Bhansali. 2009. "Design Rule for Optimization of Microelectrodes Used in Electric Cell-Substrate Impedance Sensing (ECIS)." *Biosensors and Bioelectronics* 24 (7): 2071–2076. doi:10.1016/j.bios.2008.10.026.

Qin, H., Liu, K., Zhang, Y., Guo, M., Ge, Y., Yan, M., Xu, M., Sun, Y., and Zheng, X. 2016. "Cytotoxicity of CdTe Quantum Dots in Human Umbilical Vein Endothelial Cells: The Involvement of Cellular Uptake and Induction of Pro-apoptotic Endoplasmic Reticulum Stress." *International Journal Nanomedicine* 529. doi:10.2147/ijn.s93591.

Rackus, Darius G., Mohtashim H. Shamsi, and Aaron R. Wheeler. 2015. "Electrochemistry, Biosensors and Microfluidics: A Convergence of Fields." *Chemical Society Reviews* 44 (15): 5320–5340. doi:10.1039/C4CS00369A.

Radhakrishnan, Rajeswaran, Ian I. Suni, Candace S. Bever, and Bruce D. Hammock. 2014. "Impedance Biosensors: Applications to Sustainability and Remaining Technical Challenges." *ACS Sustainable Chemistry & Engineering* 2 (7): 1649–1655. doi:10.1021/sc500106y.

Roggo, Clémence, and Jan Roelof van der Meer. 2017. "Miniaturized and Integrated Whole Cell Living Bacterial Sensors in Field Applicable Autonomous Devices." *Current Opinion in Biotechnology* 45: 24–33. doi:10.1016/j.copbio.2016.11.023.

Sackmann, Eric K., Anna L. Fulton, and David J. Beebe. 2014. "The Present and Future Role of Microfluidics in Biomedical Research." *Nature* 507 (7491): 181–189. doi:10.1038/nature13118.

Sadik, O. A., A. L. Zhou, S. Kikandi, N. Du, Q. Wang, and K. Varner. 2009. "Sensors as Tools for Quantitation, Nanotoxicity and Nanomonitoring Assessment of Engineered Nanomaterials." *Journal of Environmental Monitoring* 11 (10): 1782. doi:10.1039/b912860c.

Shah, Pratikkumar, Qiaoli Yue, Xuena Zhu, Fangcheng Xu, Hui-Sheng Wang, and Chen-Zhong Li. 2015. "PC12 Cell Integrated Biosensing Neuron Devices for Evaluating Neuronal Exocytosis Function upon Silver Nanoparticles Exposure." *Science China Chemistry* 58 (10): 1600–1604. doi:10.1007/s11426-015-5383-0.

Shah, Pratikkumar, Ajeet Kaushik, Xuena Zhu, Chengxiao Zhang, and Chen-Zhong Li. 2014a. "Chip Based Single Cell Analysis for Nanotoxicity Assessment." *The Analyst* 139 (9): 2088–2098. doi:10.1039/c3an02280c.

Shah, Pratikkumar, Xuena Zhu, Chunying Chen, Ye Hu, and Chen Zhong Li. 2014b. "Lab-on-Chip Device for Single Cell Trapping and Analysis." *Biomedical Microdevices* 16 (1): 35–41. doi:10.1007/s10544-013-9803-7.

Shah, Pratikkumar. 2014. *Development of a Lab-on-a-Chip Device for Rapid Nanotoxicity Assessment In Vitro.* Miami, FL: Florida International University. doi:10.25148/etd. FI15032160.

Shukla, Arti, Maximilian B. MacPherson, Jedd Hillegass, Maria E. Ramos-Nino, Vlada Alexeeva, Brooke T. Mossman, Pamela M. Vacek, Jeffrey P. Bond, Harvey I. Pass, and Chad Steele. 2009. "Alterations in Gene Expression in Human Mesothelial Cells Correlate with Mineral Pathogenicity." *American Journal of Respiratory Cell and Molecular Biology* 41 (1): 114–123.

Sondi, Ivan, and Branka Salopek-Sondi. 2004. "Silver Nanoparticles as Antimicrobial Agent: A Case Study on E. Coli as a Model for Gram-Negative Bacteria." *Journal of Colloid and Interface Science* 275 (1): 177–182. doi:10.1016/j.jcis.2004.02.012.

Suresh, Sagadevan. 2013. "Semiconductor Nanomaterials, Methods and Applications: A Review" *Nanoscience and Nanotechnology* 3 (3): 62–74. doi:10.5923/j.nn.20130303.06.

Taylor, Alicia A., and Elaine L. Freeman. 2017. "New Nanomaterial Regulations Require Detailed Information from Industry." *Exponent.*

Taylor, A.A., and Freeman, E. L. 2019. "New Nanomaterial Regulations Require Detailed Information from Industry." https://www.exponent.com/knowledge/alerts/2017/06/new-nanomaterial-regulations/?pageSize=NaN&pageNum=0&loadAllByPageSize=true.

Terry, S.C., J.H. Jerman, and J.B. Angell. 1979. "A Gas Chromatographic Air Analyzer Fabricated on a Silicon Wafer." *IEEE Transactions on Electron Devices* 26 (12): 1880–1886. doi:10.1109/T-ED.1979.19791.

Truffer, Frederic, Nina Buffi, Davide Merulla, Siham Beggah, Harald van Lintel, Philippe Renaud, Jan Roelof van der Meer, and Martial Geiser. 2014. "Compact Portable Biosensor for Arsenic Detection in Aqueous Samples with Escherichia Coli Bioreporter Cells." *Review of Scientific Instruments* 85 (1): 015120. doi:10.1063/1.4863333.

U.S. Forest Service. 2014. "Nanomaterials in Government Regulations & Programs." *United States Department of Agriculture.*

U.S. Forest Service. 2019. "Nanomaterials in Government Regulations & Programs." https://www.fs.fed.us/research/nanotechnology/nanomaterials.php.

Volpatti, Lisa R., and Ali K. Yetisen. 2014. "Commercialization of Microfluidic Devices." *Trends in Biotechnology* 32 (7): 347–350. doi:10.1016/j.tibtech.2014.04.010.

Wang, Ping, and Qingjun Liu. 2010. *Cell-Based Biosensors: Principles and Applications.* Boston, MA: Artech House.

Wang, Shuguang, Wentong Lu, Oleg Tovmachenko, Uma Shanker Rai, Hongtao Yu, and Paresh Chandra Ray. 2008. "Challenge in Understanding Size and Shape Dependent Toxicity of Gold Nanomaterials in Human Skin Keratinocytes." *Chemical Physics Letters* 463 (1–3): 145–149. doi:10.1016/j.cplett.2008.08.039.

Wisitsoraat, A., P. Sritongkham, C. Karuwan, D. Phokharatkul, T. Maturos, and A. Tuantranont. 2010. "Fast Cholesterol Detection Using Flow Injection Microfluidic Device with Functionalized Carbon Nanotubes Based Electrochemical Sensor." *Biosensors and Bioelectronics* 26 (4): 1514–1520. doi:10.1016/j.bios.2010.07.101.

Xia, Younan, Hong Yang, and Charles T. Campbell. 2013. "Nanoparticles for Catalysis." *Accounts of Chemical Research* 46 (8): 1671–1672.

Xu, Pengjuan, Jing Xu, Shichang Liu, Guogang Ren, and Zhuo Yang. 2012. "In Vitro Toxicity of Nanosized Copper Particles in PC12 Cells Induced by Oxidative Stress." *Journal of Nanoparticle Research* 14 (6): 906. doi:10.1007/s11051-012-0906-5.

Zhang, Di, Li Huang, Bing Liu, Haibin Ni, Liangdong Sun, Enben Su, Hongyuan Chen, Zhongze Gu, and Xiangwei Zhao. 2018. "Quantitative and Ultrasensitive Detection of Multiplex Cardiac Biomarkers in Lateral Flow Assay with Core-Shell SERS Nanotags." *Biosensors and Bioelectronics* 106: 204–211. doi:10.1016/j.bios.2018.01.062.

Zhang, Tingting, Jackie L. Stilwell, Daniele Gerion, Lianghao Ding, Omeed Elboudwarej, Patrick A. Cooke, Joe W. Gray, A. Paul Alivisatos, and Fanqing Frank Chen. 2006. "Cellular Effect of High Doses of Silica-Coated Quantum Dot Profiled with High Throughput Gene Expression Analysis and High Content Cellomics Measurements." *Nano Letters* 6 (4): 800–808. doi:10.1021/nl0603350.

Zhang, Xudong, Fang Li, Anis Nurashikin Nordin, John Tarbell, and Ioana Voiculescu. 2015. "Toxicity Studies Using Mammalian Cells and Impedance Spectroscopy Method." *Sensing and Bio-Sensing Research* 3: 112–121. doi:10.1016/j.sbsr.2015.01.002.

Zhang, Xudong, William Wang, Anis Nurashikin Nordin, Fang Li, Sunghoon Jang, and Ioana Voiculescu. 2017. "The Influence of the Electrode Dimension on the Detection Sensitivity of Electric Cell–substrate Impedance Sensing (ECIS) and Its Mathematical Modeling." *Sensors and Actuators B: Chemical* 247: 780–790. doi:10.1016/j.snb.2017.03.047.

Zhu, Xuena, Evangelia Hondroulis, Wenjun Liu, and Chen-zhong Li. 2013. "Biosensing Approaches for Rapid Genotoxicity and Cytotoxicity Assays upon Nanomaterial Exposure." *Small* 9 9–101821–30. doi:10.1002/smll.201201593.

Zhu, Xuena, Pratikkumar Shah, Susan Stoff, Hongyun Liu, and Chen-zhong Li. 2014. "A Paper Electrode Integrated Lateral Flow Immunosensor for Quantitative Analysis of Oxidative Stress Induced DNA Damage." *The Analyst* 139 (11): 2850. doi:10.1039/c4an00313f.

Index